Communications in Computer and Information Science 3023

Series Editors

Rationale

The CCIS series is devoted to the publication of proceedings of computer science conferences. Its aim is to efficiently disseminate original research results in informatics in printed and electronic form. While the focus is on publication of peer-reviewed full papers presenting mature work, inclusion of reviewed short papers reporting on work in progress is welcome, too. Besides globally relevant meetings with internationally representative program committees guaranteeing a strict peer-reviewing and paper selection process, conferences run by societies or of high regional or national relevance are also considered for publication.

Topics

The topical scope of CCIS spans the entire spectrum of informatics ranging from foundational topics in the theory of computing to information and communications science and technology and a broad variety of interdisciplinary application fields.

Information for Volume Editors and Authors

Publication in CCIS is free of charge. No royalties are paid, however, we offer registered conference participants temporary free access to the online version of the conference proceedings on SpringerLink (http://link.springer.com) by means of an http referrer from the conference website and/or a number of complimentary printed copies, as specified in the official acceptance email of the event.

CCIS proceedings can be published in time for distribution at conferences or as post-proceedings, and delivered in the form of printed books and/or electronically as USBs and/or e-content licenses for accessing proceedings at SpringerLink. Furthermore, CCIS proceedings are included in the CCIS electronic book series hosted in the SpringerLink digital library at http://link.springer.com/bookseries/7899. Conferences publishing in CCIS are allowed to use our online conference service (Meteor) for managing the whole proceedings lifecycle (from submission and reviewing to preparing for publication) free of charge.

Publication process

The language of publication is exclusively English. Authors publishing in CCIS have to sign the Springer CCIS copyright transfer form, however, they are free to use their material published in CCIS for substantially changed, more elaborate subsequent publications elsewhere. For the preparation of the camera-ready papers/files, authors have to strictly adhere to the Springer CCIS Authors' Instructions and are strongly encouraged to use the CCIS LaTeX style files or templates.

Abstracting/Indexing

CCIS is abstracted/indexed in DBLP, Google Scholar, EI-Compendex, Mathematical Reviews, SCImago, Scopus. CCIS volumes are also submitted for the inclusion in ISI Proceedings.

How to start

To start the evaluation of your proposal for inclusion in the CCIS series, please send an e-mail to ccis@springer.com

Cecilia Zanni-Merk · Elviz Ismayilov ·
Nigar Ismayilova · Stéphane Genaud ·
Ulviya Abdulkarimova
Editors

Artificial Intelligence for Digital Transformations

First International Conference, AIDT 2026
Baku, Azerbaijan, June 22–24, 2026
Proceedings

Springer

Editors
Cecilia Zanni-Merk
INSA Rouen Normandie
Rouen, France

French-Azerbaijani University
Baku, Azerbaijan

Nigar Ismayilova
Azerbaijan State Oil and Industry University
Baku, Azerbaijan

Ulviya Abdulkarimova
French-Azerbaijani University/Azerbaijan
State Oil and Industry University
Baku, Azerbaijan

Elviz Ismayilov
Azerbaijan State Oil and Industry University
Baku, Azerbaijan

Stéphane Genaud
University of Strasbourg
Strasbourg, France

ISSN 1865-0929 ISSN 1865-0937 (electronic)
Communications in Computer and Information Science
ISBN 978-3-032-31318-8 ISBN 978-3-032-31319-5 (eBook)
https://doi.org/10.1007/978-3-032-31319-5

This Springer imprint is published by the registered company Springer Nature Switzerland AG
The registered company address is: Gewerbestrasse 11, 6330 Cham, Switzerland

Preface

The *International Conference on Artificial Intelligence for Digital Transformations (AIDT 2026)* was held in Baku, Azerbaijan, from 22nd to 24th June 2026. The conference brought together researchers, practitioners, and industry experts to present and discuss recent advances in artificial intelligence and its integration into modern digital systems.

AIDT 2026 aimed to promote high-quality research in core and emerging areas of artificial intelligence, with a particular focus on novel methods, theoretical contributions, and impactful applications. The conference scope reflected the increasing role of AI in domains such as intelligent systems, Internet of Things (IoT), data-driven decision-making, and cyber-physical systems, which are key enablers of digital transformation across industries.

The conference attracted 167 submissions from the international research community. Each paper was carefully reviewed in a double-blind process by at least three members of the International Program Committee. Following a rigorous evaluation process, 28 papers were accepted for publication, resulting in a highly selective acceptance rate of approximately 17%. This reflects both the strong interest in the conference and the high standards maintained by the Program Committee. In line with the conference call for papers, particular attention was given to submissions demonstrating clear and original contributions in artificial intelligence.

The accepted papers cover a broad spectrum of topics, including machine learning, deep learning, explainable AI, AI for IoT, intelligent data processing, and interdisciplinary applications of AI technologies.

We would like to sincerely thank all authors for their submissions, the members of the International Program Committee for their thorough and constructive evaluations, and the organizing committee for their dedication and support in ensuring the success of AIDT 2026.

We hope that this volume will serve as a valuable reference for researchers and practitioners and will contribute to ongoing advances in artificial intelligence and digital technologies.

June 2026

Cecilia Zanni-Merk
Elviz Ismayilov
Nigar Ismayilova
Stéphane Genaud
Ulviya Abdulkarimova

Preface

Organization

Conference Host

Rufat Azizov — Azerbaijan State Oil and Industry University, Azerbaijan

General Chairs

Cecilia Zanni-Merk — INSA Rouen Normandie, France/French-Azerbaijani University, Azerbaijan

Elviz Ismayilov — Azerbaijan State Oil and Industry University, Azerbaijan

Program Chairs

Nigar Ismayilova — Azerbaijan State Oil and Industry University, Azerbaijan

Stéphane Genaud — University of Strasbourg, France

Organization Chair

Martin Godon — French-Azerbaijani University, Azerbaijan

Publication Chair

Ulviya Abdulkarimova — French-Azerbaijani University / Azerbaijan State Oil and Industry University, Azerbaijan

International Program Committee

Ulviya Abdulkarimova — French-Azerbaijani University / Azerbaijan State Oil and Industry University, Azerbaijan

Gabriel Alencar Medeiros	INSA Rouen Normandie, France
Araz Aliev	Azerbaijan State Oil and Industry University, Azerbaijan
Ramiz Aliguliyev	Institute of Information Technology of the Ministry of Science and Education, Azerbaijan
Ali Ayadi	University of Strasbourg, France
Muhammed Ali Aydin	Istanbul University-Cerrahpaşa, Türkiye
Rashid Bakirov	Bournemouth University, UK
Ladjel Bellatreche	ISAE-ENSMA, France
Mathieu Bourgais	INSA Rouen Normandie, France
Raphaël Couturier	Université Marie et Louis Pasteur, France
Evren Daglarli	Istanbul Technical University, Türkiye
Yusuf Yaslan	Istanbul Technical University, Türkiye
Asma Dhaouadi	INSA Rouen Normandie, France
Hassen Drira	University of Strasbourg, France
Valentina Dragos	ONERA (The French Aerospace Lab), France
Wassim El Falou	Lebanese University, Lebanon
Yusuf Yaslan	Istanbul Technical University, Türkiye
Nahla El Kadhi	French-Azerbaijani University, Azerbaijan
Claudia Frydman	Aix-Marseille University, France
Stéphane Genaud	University of Strasbourg, France
Mahmoud Golabi	Université de Haute-Alsace, France
Franco Giustozzi	INSA Strasbourg, France
Maxime Gueriau	INSA Rouen Normandie, France
Jamaladdin Hasanov	ADA University, Azerbaijan
Oleg Huseynov	Azerbaijan State Oil and Industry University, Azerbaijan
Yusif Ibrahimov	French-Azerbaijani University, Azerbaijan
Nigar Ismayilova	Azerbaijan State Oil and Industry University, Azerbaijan
Hind Laghmara	INSA Rouen Normandie, France
Thomas Lampert	University of Strasbourg, France
Alessandro Leite	INSA Rouen Normandie, France
Morgan Madec	University of Strasbourg, France
Abdenacer Makhlouf	Université de Haute-Alsace, France
Mohamed Karray	ESME, France
Rza Mustafayev	French-Azerbaijani University, Azerbaijan
Aziz Nuriyev	Azerbaijan State Oil and Industry University, Azerbaijan
Igor Santos Pereira	Federal University of Uberlândia, Brazil
Fernando Roda	CONICET, Argentina
Rim Saddem	Aix-Marseille University, France

Aya Saad	SINTEF Ocean, Norway
Cesar Sanin	University of New England, Australia
Marion Schaeffer	Independent Consultant, France
Ovidiu Serban	Imperial College London, UK
Lyudmila Suxostat	Institute of Information Technology of the Ministry of Science and Education, Azerbaijan
Rüya Şamlı	Istanbul University-Cerrahpaşa, Türkiye
Tommy Yuan	University of York, UK
Cecilia Zanni-Merk	INSA Rouen Normandie, France/French-Azerbaijani University, Azerbaijan

Contents

AI in Emerging Digital Infrastructures

From Detection to Explanation: Combining Stream Reasoning and Concept Induction in Smart Farming 3
Léa Charbonnier, Franco Giustozzi, and Julien Saunier

Improving Taboo-Word Generation in Turkic Language with Morphology-Aware Filtering and Diversity Selection 19
Anar Mammadli and Elviz Ismayilov

A Domain Ontology for Cooperative Perception in the Internet of Mobile Things 32
Kamran Ismayilzada and Mathieu Bourgais

AI-Driven Automation and Optimization

Cognitive Load Estimation Through Eye-Tracking in Industrial Tasks: A Pilot Study 51
Kanan Gurbanov, Latafat Qardashova, Gregoire Chabrol, Samy Rima, and Rabih Amhaz

A Financial-Risk Aware Machine Learning Framework for Industrial Convergence 65
Hikmat Allahverdiyev

Hybrid Wavelet–LSTM and Weibull Risk Modeling for Early Fault Prediction Using Accelerometer-Based Vibration Data in Rotating Machinery 79
Nail Mammadov, Yunis Guliyev, Laman Guliyeva, and Turker Susmus

AI-Driven Digital Twin Framework for Risk-Aware Predictive Maintenance in Electrical Control Systems 94
Kamal Ahmadov and Aliya Dostalizade

AI-Driven Decision Support for Wind-BESS Dispatch Under Export Caps: A Behavioral Cloning Approach 109
Ahmed M. Jendeya

Core AI Models and Learning Paradigms

Relational Verification, Feature Sensitivity, and Cross-City Transfer in Heterogeneous Graph Learning for Urban Functional Zone Classification ... 127
Etibar Vazirov and Daniele Apiletti

Bridging Prompting and Semantic Grounding: A Unified Evaluation Framework for Document-Level Relation Extraction ... 142
Gabriel Henrique Alencar Medeiros and Franco Giustozzi

A Divide-and-Conquer Framework for Query-Efficient Causal Discovery with Large Language Models ... 157
Louis Hernandez, Matthieu Boussard, and Alessandro Leite

Towards a Non-monotonic Neuro-Symbolic Framework: Application to Outlier Mitigation During Training ... 173
Nathan Olejniczak and Anne Jeannin-Girardon

Lexical Ambiguity in French: Linguistic Issues and Interpretation by Artificial Intelligence Models ... 189
Vafa Quliyeva and Vazeh Asgarov

Improving Named Entity Recognition for the Azerbaijani Language Using Neural and Conditional Random Fields-Based Models ... 205
Tofig Aliyev and Samir Rustamov

An Approach for Evaluating Recommender Systems: Assessing Usefulness and Coherence Through Item Relationships ... 220
Marie Griffon, Nicolas Delestre, and Maxime Guériau

Deep Features, Classical Models: Binary vs. Multiclass Deepfake Detection ... 236
Abbas Aliyev and Samir Rustamov

Can We Trust LLMs for Mental Health-Based Decisions? A Causality Aware Reliability Analysis ... 251
Yusif Ibrahimov, Turan Mutallimov, Seymour Mirzabayov, and Elgun Hasanov

Real-Time Azerbaijani Sign Language Sentence Recognition: Comparative Evaluation of Cosine Retrieval and Seq2Seq Approaches ... 264
Gulshan Sariyeva, Rahima Karimova, Vasila Aliyeva, Ilaha Jamalli, and Jamaladdin Hasanov

Adapting Large-Scale Neural TTS Models to Low-Resource Languages: A Comparative Study 279
Alakbar Valizada

Federated Learning Approach to Autoencoder-Based Feature Extraction for Pneumonia Detection 294
Elshan Naghizade and Samir Aliyev

Benchmarking Algorithmic Fairness in AI Recruitment: A Comparative Study of Mitigation Strategies and the FIS Metric 310
Nigar E. Adilova, Omer Mutlu Turk Kaya, Yunis Allahverdi, Fatima Naghizada, and Nurlan Orujov

Handoff Hallucinations: Taxonomy, Benchmark, and Mitigation for Multi-agent Pipeline Failures 325
Hamida Jafarova and Fatima Alakbarli

AzText: Curating Web-Scale Pretraining Data for a Low-Resource Language 341
Eljan Mahammadli and Samir Rustamov

Governance, Strategy, and Human–Organization Interaction

An Ontology-Driven Recommender System for Enhancing Workplace Practices 359
Asma Dhaouadi and Maria Santilli

Statutory AI: Aligning Large Language Models With Legal Norms: Content Warning: This Paper Includes Examples that May Contain Harmful or Offensive Language. 375
Cindy Delage, Stéphane Canu, Marc Decombas, and Jonathan Foureur

Explainability of Tourism Risk Assessment by Leveraging Fuzzy Ontology Properties 389
Pavel Kosov, Latafat Gardashova, and Aziz Nuriyev

Sustainability and Responsible Innovation

Efficiency Driven Digital Transformation for Sustainable Real Time Monitoring of High Pressure South Caspian Wells 407
Fariz Karimov, Yelena Shmoncheva, Gullu Jabbarova, and Rafail Manafov

A Community-Driven Dataset and Protocol for Azerbaijani Sign Language Recognition: Advancing Digital Inclusion Through Human-Centered AI 423
Nigar Alishzade, Lala Ibadullayeva, Rajab Iskandarli, Jabrayil Babayev, and Humbat Jamalov

Author Index 439

AI in Emerging Digital Infrastructures

From Detection to Explanation: Combining Stream Reasoning and Concept Induction in Smart Farming

Léa Charbonnier[1(✉)], Franco Giustozzi[2], and Julien Saunier[1]

[1] INSA Rouen Normandie, Univ Rouen Normandie, Université Le Havre Normandie, Normandie Univ, LITIS UR 4108, 76000 Rouen, France
{lea.charbonnier,julien.saunier}@insa-rouen.fr

[2] INSA Strasbourg, University of Strasbourg, ICube laboratory, CNRS (UMR 7357), 67000 Strasbourg, France
franco.giustozzi@insa-strasbourg.fr

Abstract. The use of sensor data to support monitoring and decision-making has become widespread across many application domains. Continuous data streams enable early detection of relevant situations, such as anomalies, but identifying and explaining their underlying causes remains a significant challenge. This paper proposes a unified framework that integrates stream reasoning and concept induction to enable real-time detection of abnormal situations and automated root cause identification. The framework relies on semantic enrichment and continuous reasoning over data streams to detect anomalies and to derive human-interpretable explanations. Although the approach is domain-independent, we demonstrate its applicability using a smart farming sensor dataset. The experimental results show that the proposed framework successfully detects abnormal environmental conditions using key variables, such as rainfall and humidity, and provides semantically grounded explanatory concepts that contextualize the detected conditions.

Keywords: Stream Reasoning · Concept Induction · Smart Farming · Anomaly Detection · Explainable AI

1 Introduction

With the advent of the fourth industrial revolution, many domains have improved their production processes through increased digitalization and automation. Agriculture is no exception and is currently undergoing a profound transformation, commonly referred to as Agriculture 4.0, which integrates modern digital technologies such as big data, robotics, artificial intelligence (AI), and the Internet of Things (IoT) to enhance agricultural practices [24]. At the same time, global demographic projections indicate that the world population will reach between 9 and 10 billion by 2050 [15], leading to a significant increase

C. Zanni-Merk et al. (Eds.): AIDT 2026, CCIS 3023, pp. 3–18, 2026.
https://doi.org/10.1007/978-3-032-31319-5_1

in food demand [21]. Meeting this growing demand requires substantial improvements in agricultural productivity while addressing constraints related to labor availability, resource efficiency, and environmental sustainability. In this context, smart farming has emerged as a key technological response. Smart farming refers to the use of digital technologies, especially IoT, AI, robotics, and remote sensing, to manage crops and livestock more precisely. Numerous studies show strong potential for saving water, protecting soil, and improving yields, but cost, connectivity, skills, and governance remain major hurdles [17].

The widespread deployment of IoT devices and data-driven technologies has led to the generation of continuous and heterogeneous data streams in agricultural environments. These data streams offer new opportunities for monitoring crop conditions and anticipating adverse events, such as stress or disease [15]. However, transforming large volumes of streaming sensor data into actionable knowledge remains a major challenge, particularly when timely decisions and human understanding are required [17]. While many existing systems focus on detecting anomalies or critical situations, they often provide limited insight into the underlying causes of these events, making it difficult for human operators to interpret results and take informed corrective actions.

To address this limitation, we propose a hybrid approach combining stream reasoning for real-time situation detection [7] with concept induction for root cause identification and explanation [8]. The framework relies on an ontology as a shared semantic foundation to integrate continuous reasoning and symbolic explanation. Although the proposed framework is domain-independent, smart farming is used in this work as an illustrative use case to demonstrate its applicability and explanatory capabilities in a realistic, sensor-driven environment.

The remainder of this paper is organized as follows. Section 2 reviews related work on stream reasoning and concept induction. Section 3 introduces the smart farming use case employed to demonstrate the proposed approach. Section 4 presents the framework architecture, Sect. 5 reports the experimental demonstration, and Sect. 6 concludes the paper and outlines future directions.

2 Related Work

The increasing availability of continuous sensor data in smart farming has motivated a wide range of approaches for monitoring crop conditions and supporting decision-making. However, integrating heterogeneous data streams into cohesive, real-time detection systems remains challenging due to the lack of standardized data representations and interoperability among IoT devices [17]. Ontology-based approaches have been proposed to address these issues by providing a semantic layer capable of handling heterogeneous and uncertain sensor data [11]. When coupled with stream reasoning techniques, ontologies can be exploited in non-static environments, enabling continuous reasoning and real-time detection of relevant situations [12]. This section reviews the methodological foundations underlying our hybrid approach, with a focus on stream reasoning for abnormal situation detection and concept induction for root cause identification.

2.1 Stream Reasoning

Among the standards proposed by the World Wide Web Consortium (W3C), the Resource Description Framework (RDF) is widely used for representing and exchanging data on the Web. RDF data is expressed as statements, called triples, composed of a subject, a predicate, and an object [20]. Querying RDF datasets is performed using SPARQL [23], a declarative query language whose formal definition includes a SPARQL algebra expression, an RDF dataset, and a query form. With the growing deployment of sensors and connected devices, data is increasingly generated in the form of continuous streams. Such data cannot be processed directly using standard SPARQL, as SPARQL is designed for static RDF datasets [3]. To address this limitation, the field of *stream reasoning* has emerged, combining static RDF knowledge with streaming data to enable continuous querying and reasoning in near real time. Stream reasoning systems extend traditional query processing by allowing queries to be evaluated over temporal windows of streaming data, using either SPARQL-based or SQL-based paradigms. A data stream is commonly defined as an unbounded sequence of tuples, each composed of a data item and associated timestamps indicating event time and processing time [25]. An RDF stream follows this definition, where each data item is an RDF triple or graph. To process RDF streams in a manner compatible with relational and RDF query languages, the Continuous Query Language (CQL) model is often adopted [2].

In the context of knowledge graphs (KGs), a knowledge graph is defined as "a graph of data intended to accumulate and convey knowledge of the real world, whose nodes represent entities of interest and whose edges represent relations between these entities" [14]. Querying and reasoning over streaming KGs requires extending traditional RDF query mechanisms to account for the temporal and unbounded nature of the data. RDF Stream Processing (RSP) engines address these challenges by supporting continuous queries over RDF streams while handling high arrival rates, evolving content, and temporal semantics. Notable RSP engines include C-SPARQL [3], RSP-QL [9], RSP4J [25], and RDFox [18].

While stream reasoning provides effective mechanisms for detecting complex situations and anomalies in real time, the results of continuous queries are often expressed at a low level of abstraction and remain closely tied to sensor observations and rule-based patterns. As a consequence, additional methods are required to derive higher-level, human-interpretable explanations and to identify the underlying causes of observed situations. In this context, *Concept Induction* [16] emerges as a complementary approach, aiming at learning abstract concepts and explanatory patterns from data and background knowledge.

2.2 Concept Induction

In this work, we deliberately avoid traditional machine learning approaches, such as neural networks, due to their limited transparency and well-documented lack of interpretability [4]. Although these methods often achieve high predictive performance, they generally operate as black-box models and require costly retraining procedures when new data or labels become available. To address these limitations, we adopt a *concept induction* approach, which aims at learning symbolic, human-interpretable concepts from data in the presence of background knowledge [16]. Concept induction makes it possible to identify meaningful relationships between observed physical factors and abnormal phenomena, and to express these relationships as explicit rules or structured concepts. Such representations facilitate both understanding and validation by domain experts, and allow detected situations to be explained in terms of higher-level semantic abstractions rather than raw sensor values [22].

Beyond anomaly explanation, concept induction supports root cause identification by enabling post-hoc analysis of the induced concepts and their constituent conditions [28]. In contrast to root cause classification approaches that primarily focus on predictive accuracy and rely on opaque models, the proposed approach emphasizes clarity, traceability, and adaptability. These properties make concept induction particularly well-suited for dynamic and safety-critical domains, such as smart farming, where human operators must be able to trust, interpret, and act upon the system's outputs. In the following sections, we present the dataset used, and we detail how concept induction is integrated with stream reasoning within our framework to provide both real-time detection and semantically grounded explanations of abnormal situations.

3 Use Case: Smart Farming Sensor Data

To evaluate our hybrid stream reasoning and concept induction approach, we consider a smart farming use case involving environmental, agronomic, and operational variables that influence crop health and yield. We rely on the *Smart Farming Sensor Data for Yield Prediction*[1] dataset, which aggregates observations from 500 farms distributed across multiple geographic regions. The dataset is particularly suitable for our study as it provides timestamped sensor measurements alongside crop health indicators, enabling both incremental stream reasoning over heterogeneous data streams and semantic interpretation of abnormal agronomic conditions. Unlike purely static agricultural datasets, the presence of timestamps associated with sensor observations allows the data to be reinterpreted as asynchronous data streams. This enables the evaluation of reasoning tasks under partial and evolving knowledge, which is a central requirement for stream reasoning systems. Furthermore, the inclusion of crop disease status provides a form of quality control that can be exploited to analyze the relationship between detected abnormal situations and observable crop outcomes.

3.1 Dataset Overview

The dataset comprises twenty-two attributes spanning several categories: farm descriptors (farm identifier, region, latitude, longitude), environmental

[1] https://www.kaggle.com/datasets/atharvasoundankar/smart-farming-sensor-data-for-yield-prediction/data.

measurements (soil moisture, soil pH, air temperature, rainfall, relative humidity, sunlight hours), crop-related information (crop type, sowing date, harvest date, total growing duration, yield, NDVI index, disease status), cultivation practices (irrigation type, fertilizer type, pesticide usage), and sensor metadata (sensor identifier and timestamp). In our approach, numerical sensor measurements and contextual farm and crop attributes are used to derive semantic constraints, which are subsequently employed to define higher-level observable phenomena. These phenomena correspond to agronomically recognized abnormal situations defined based on expert knowledge, as explained in Subsect. 3.2. By lifting raw sensor values to semantically meaningful situations, we enable rule-based reasoning that integrates heterogeneous observations over time.

3.2 Observed Environmental and Ecological Stress Phenomena

Based on the available measurements and established agronomic knowledge, we define four abnormal situations that are widely recognized in crop science: *heat stress*, *vegetation stress inferred from NDVI*, and *disease-favorable environmental conditions*. These phenomena are not directly observed in the dataset but are inferred through semantic rules that combine multiple sensor streams with contextual information and cluster-specific thresholds. Thresholds are defined per crop–region cluster and estimated from historical observations by computing low/high quantiles and mean reference values for each measured variable.

Heat stress occurs when ambient temperatures exceed the optimal range for crop physiological processes, leading to reduced photosynthetic efficiency, pollen sterility, and accelerated senescence [13,27]. While temperature alone is a primary driver, its impact is exacerbated by high solar radiation and low humidity, which increase plant water loss. Accordingly, heat stress is identified when temperatures exceed crop-specific or regionally derived thresholds, particularly when combined with high sunlight exposure and low relative humidity. *Vegetation stress* is inferred from the Normalized Difference Vegetation Index (NDVI), a widely used remote sensing indicator of vegetation vigor, biomass, and chlorophyll content [19,26]. Reduced NDVI values are commonly associated with plant stress caused by water or nutrient limitations, disease, or other adverse environmental conditions. In our use case, vegetation stress is detected when NDVI values fall significantly below expected levels for a given crop type and region, serving as an outcome-level indicator that complements sensor-based environmental observations. Many plant diseases are strongly influenced by environmental factors, particularly humidity, temperature, and solar radiation [1,6]. Rather than detecting disease occurrence directly, we identify environmental conditions that are known to favor disease development. Such conditions include persistently high relative humidity, moderate air temperatures conducive to pathogen growth, and reduced sunlight, which may prolong leaf wetness. These conditions are used to infer a *disease risk* phenomenon, which can subsequently be related to observed vegetation stress or recorded disease status.

By formalizing these agronomically grounded abnormal situations as semantic phenomena, our approach demonstrates how stream reasoning can integrate

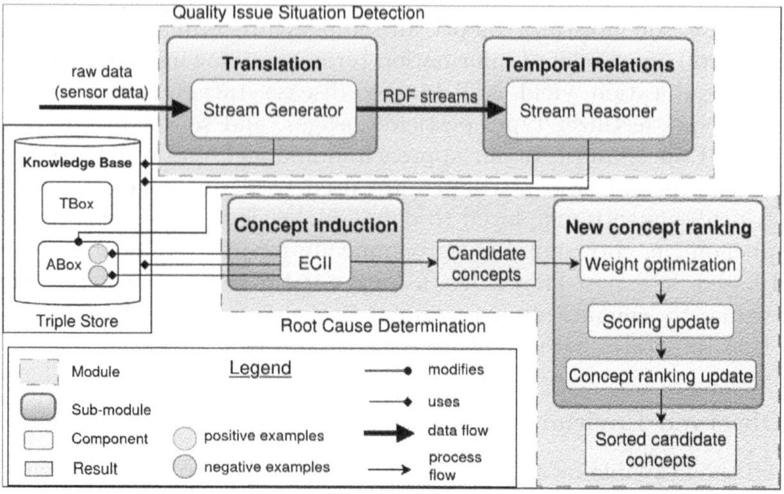

Fig. 1. The hybrid SR and CI framework

heterogeneous sensor data and contextual knowledge to support early detection and interpretation of adverse crop conditions.

4 Proposed Framework

In this section, we propose a hybrid approach for anomaly detection and explainable quality issue identification. The proposed approach, as shown in Fig. 1 uses stream reasoning to detect abnormal situations and concept induction to uncover and explain the root causes of quality issues.

A key component of the approach is an ontology that allows the representation of diseased (positive examples) and healthy crops (negative examples), sensor data, and phenomena. The ontology is described in Sect. 4.1. The *Translation* sub-module generates RDF streams from sensor data. The *Temporal Relations* sub-module uses stream reasoning to detect abnormal situations from sensor data and populate the knowledge base with these anomalies. The *Concept induction* sub-module uses the Efficient Concept Induction from Instances (ECII) algorithm [22] to induce concepts from the ontology, yielding a preliminary list of concept expressions to describe the detected anomalies. Each concept generated by the *Concept induction* sub-module is associated with a set of evaluation criteria defined in Sect. 4.3, which forms the basis of a new scoring function based on a weighted sum. Sorted concepts provide insights into the potential causes of the quality issue and are presented to the operator as explanatory information.

4.1 Ontology for Smart Farming

The proposed approach is based on the *Context Ontology* [11], a formal model that offers a unified description of resources with sensor observations, spatial, and temporal data. Its modular structure allows flexible adaptation and integration, rendering it particularly suitable for domains like smart farming, where data sources are heterogeneous, and data streams are continuous.

In smart farming, the Context Ontology serves as the semantic backbone of the domain ontology, representing entities like `Farm` and `Crop`, and detailed observations such as soil moisture levels and temperature variations over time. Sensor measurements like temperature, humidity, soil moisture, and related environmental parameters are modeled as contextual information attached to localized agricultural resources and time-stamped events. The ontology provides constructs for representing situations of interest through the `Situation` class. This allows abnormal or stress-related conditions affecting crops to be modeled as higher-level contextual situations derived from sensor observations. In addition, the ontology also provides a `ClusterThreshold` class, which contains threshold values associated with specific clusters of regions and crop types for each observed property. These thresholds capture localized and crop-dependent reference conditions and are directly exploited during stream reasoning to compare incoming sensor values against expected ranges.

By leveraging the Context Ontology, the framework offers a consistent and semantically structured representation of streaming sensor data, enabling real-time analysis and the development of insights such as crop health trends and environmental impact assessments. This shared semantic foundation allows detected situations to be contextualized and explained using domain-specific concepts, such as growth stages and pest control strategies, thereby enhancing human operators' ability to interpret data and make well-considered judgments in smart farming. Although the ontology is instantiated for smart farming in this work, the proposed framework remains domain-independent and can be applied to other sensor-driven domains by replacing the domain-specific vocabulary.

4.2 Quality Issue Situation Detection Module

To handle heterogeneous data originating from multiple sources and to continuously monitor crop health, it is necessary to observe data streams and evaluate continuous queries over them. To this end, sensor data streams are generated, and abnormal situations are detected using a stream reasoner in conjunction with the ontology described previously. The proposed approach is implemented using the RSP4J framework [7]. Before applying stream reasoning to detect abnormal situations, the raw data must be preprocessed. This consists of transforming raw sensor measurements into RDF streams in order to simulate a real-world streaming environment. Data are extracted from the source dataset, and a stream generator is used to produce one RDF stream per observed property. Once data are transformed into RDF streams, they cannot be directly processed using standard SPARQL, which is designed for static RDF graphs only. Continuous query evaluation, therefore, requires temporal windowing mechanisms over streaming data. RSP4J relies on the Continuous Query Language (CQL) model to support this functionality, as illustrated in Fig. 2. The CQL model defines three families of operators: Stream-to-Relation (S2R), which extracts finite relations

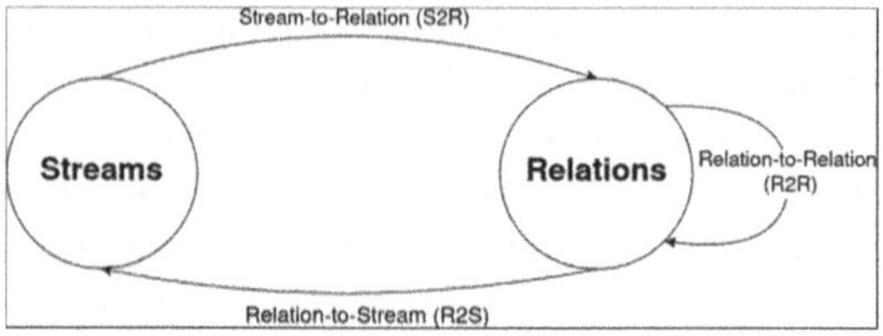

Fig. 2. The CQL model [2]

from streams using time windows; Relation-to-Relation (R2R), which applies relational operators on windowed data; and Relation-to-Stream (R2S), which transforms query results back into output streams [2].

As RSP4J is an API developed under RSP-QL semantics [9,25], it supports different types of time windows, including tumbling and sliding windows. Tumbling windows partition the data stream into consecutive, non-overlapping subsets of equal size, such that the end of one window coincides with the beginning of the next. In contrast, sliding windows are defined by a window size and a slide parameter, which determines the frequency at which new windows start. In this case, individual tuples may belong to multiple windows, resulting in overlapping windows when the slide is smaller than the window size.

Using this model, RSP4J applies time windows to collect streaming values over a fixed period and materializes them as finite data chunks once the window expires. These chunks can then be queried using SPARQL, like static RDF data, to detect some situations. The results of continuous queries are emitted as output streams, which are subsequently processed and inserted into a triple store. These detected situations are reused later for explanation and root cause analysis.

4.3 Root Cause Determination Module

In the context of root cause identification, concept induction provides a relevant approach for automatically generating symbolic descriptions that explain abnormal situations observed in data. Formally, this problem consists of a knowledge base $\mathcal{O} = (\mathcal{T}, \mathcal{A})$ made up of a TBox $\mathcal{T}$ and an Abox $\mathcal{A}$, and two disjoint sets of individuals P and N, considered respectively as positive and negative examples. The goal is to find a class expression C over a description logic $\mathcal{L}$ such that $\mathcal{O} \models C(a)$ for all $a \in P$, and $\mathcal{O} \not\models C(b)$ for all $b \in N$. Among the existing frameworks allowing this type of concept induction, this approach uses ECII [22].

ECII is a concept induction algorithm that selects a finite set $\{C_1, \ldots, C_n\}$ of complex class expressions over $\mathcal{L}$ and additional complex class expressions, resulting in $\mathcal{O}' = \mathcal{O} \cup \{A_k \equiv C_k | k = 1, \ldots, n\}$ where A_i are atomic classes not occurring in $\mathcal{O}$. Then, a reasoner is used to compute the membership of all individuals from the examples in all atomic classes from $\mathcal{O}'$. Finally, candidate class expressions are generated using a combination of constructors, negation, and atomic classes. This strategy minimizes the number of reasoner invocations, although it comes at the cost of narrowing the search space for potential expressions. ECII is selected for its efficiency in generating complex description logic class expressions, particularly in ontologies that are rich with axioms. However, since the method can yield hundreds of candidate concepts, an effective ranking process using metrics such as coverage, precision, and hierarchical specificity becomes essential to determine which expressions merit further consideration [8].

The proposed ranking relies on three complementary evaluation criteria, each capturing a distinct aspect of concept quality. Coverage measures the proportion of positive examples explained by a concept, reflecting its ability to describe a large share of the observed phenomena. It is first defined as

$$\text{coverage}_{raw} = \frac{\text{Number of positive examples identified by the concept}}{\text{Total number of positive examples}} \tag{1}$$

In the experimental setting, this value is normalized by the maximum coverage observed among all candidate concepts in order to make its contribution comparable to the other ranking criteria:

$$\text{coverage} = \frac{\text{coverage}_{raw}}{\text{Maximum coverage}_{raw}} \tag{2}$$

Precision evaluates the exclusivity of the concept by quantifying how many of the inferred instances correspond to actual positive examples, thus penalizing overly general concepts.

$$\text{precision} = \frac{\text{Number of positive examples identified by the concept}}{\text{Total number of examples identified by the concept}} \tag{3}$$

Finally, the hierarchy-based metric assesses the specificity of a concept by considering its syntactic structure, favoring more informative explanations over generic ones. The proposed approach introduces a structural depth score for ranking candidate concepts according to their semantic specificity. The score is computed directly from the syntactic structure of logical expressions, exploiting the relationship between description logic constructs and ontological generality. In particular, conjunctions are interpreted as increasing specificity, while disjunctions and existential restrictions with multiple target classes are treated as indicators of generality. The method analyzes existential components of class expressions, quantifies the number of target classes per relation, and combines these elements into a raw depth score, augmented by a bonus reflecting conjunctive structure. To illustrate the intuition behind the hierarchy-based metric, consider the following two candidate concepts:

$$C_1 = \texttt{Crop} \sqcap \exists p_1.\texttt{HeatStress} \quad \text{and} \quad C_2 = \exists p_1.(\texttt{HeatStress} \sqcup \texttt{DiseaseRisk}).$$

Concept C_1 describes crops that are explicitly involved in a heat stress situation, combined with a conjunction that restricts the concept to crops only. In contrast, C_2 captures a more general situation, as it allows a crop to be involved in either heat stress or disease risk. The presence of a disjunction in the existential restriction makes C_2 applicable to a broader set of individuals. The proposed structural depth score reflects this difference in specificity. Conjunctions are interpreted as increasing specificity, while disjunctions within existential restrictions are treated as indicators of generality. As a result, C_1 is assigned a higher structural depth score than C_2, indicating that it represents a more specific and informative concept. The scores are then normalized within the set of candidate concepts to obtain values in the interval $[0, 1]$, allowing concepts to be directly compared and ranked according to their specificity.

$$\text{hierarchy} = \frac{\text{Structural concept depth}}{\text{Maximum depth}} \tag{4}$$

By combining these criteria, the ranking balances exhaustiveness, discriminative power, and semantic specificity, avoiding concepts that are either overly general, overly restrictive, or insufficiently explanatory. Therefore, a new multi-criteria scoring function is proposed, which aggregates coverage, precision, and hierarchy using a weighted sum with the associated weights, α, β, and γ.

$$\begin{aligned} \text{score} = \alpha \times \text{coverage} + \beta \times \text{precision} + \gamma \times \text{hierarchy} \\ \text{with } \alpha, \beta, \gamma \in [0, 1] \text{ and } \alpha + \beta + \gamma = 1 \end{aligned} \tag{5}$$

To identify the optimal combination of weights, parameters are optimized under domain-driven constraints inspired by multi-criteria decision analysis [5]. The resulting ranking strategy favors concepts that are both semantically specific and explanatory, supporting transparent and reliable root cause identification [8].

5 Experimental Demonstration

This section presents an experimental demonstration of the proposed framework on the smart farming use case introduced in Sect. 3. The objective is to illustrate the feasibility of the end-to-end pipeline and the interpretability of the generated explanations, rather than to provide a quantitative benchmark evaluation. Anomaly detection is performed continuously using stream reasoning over time windows to detect predefined situations and materialize them in a GraphDB triple store. Root cause identification is then carried out periodically on the accumulated data using concept induction. Stream reasoning is implemented using RSP4J [25], while concept induction relies on the ECII framework [22]. Both components are implemented in Java. The resulting concepts are processed using Python for weight optimization, score computation, and final concept ranking. The approach, experimental data, and results are available online[2]. The phenomena considered in this experimental demonstration are heat stress (`Sit-H`), vegetation stress (`Sit-V`), and disease risk (`Sit-D`). For the dataset considered, the ECII execution time takes approximately one second per run. It remains in the order of a few seconds per iteration, depending on the number of examples and candidate concepts. This suggests that the approach is suitable for near

[2] https://gitlab.insa-rouen.fr/lcharbonnier/hybrid-sr-ci-framework.

real-time analysis at moderate data rates, although scalability under higher data volumes remains to be further evaluated.

5.1 Anomaly Detection via Stream Reasoning

Since the dataset does not provide native time series data, we simulate streaming behavior by replaying timestamped observations as a sequence of events. Each observation is assigned a timestamp and injected into the stream at a fixed rate, thereby emulating a continuous data flow. Each stream corresponds to a single observed property (e.g., rainfall, humidity, temperature) and contains the observed value (e.g., 75 mm, 90%, 31 °C), the farm hosting the sensor (e.g., `FARM0001`), and a timestamp (e.g., 2024-03-19T00:00:00Z). Using one stream per observed property allows us to simulate a realistic sensor-based streaming environment. The same observation is not repeated indefinitely, but each record is emitted once according to the defined streaming schedule.

These streams are continuously queried within time windows to detect abnormal situations. To this end, SPARQL queries are defined for each phenomenon described in Sect. 3.2. Each query specifies the ontology used as a static graph, the relevant named windows over the streams, and the graph patterns required to detect the corresponding situation. An example query for vegetation stress detection is provided in Listing 1.1. Sliding time windows are used to enable continuous monitoring of the streams (e.g., a window range of 10 s with a slide of 1 s). This configuration provides a higher temporal resolution than tumbling windows and is therefore better suited for real-time anomaly detection. In Listing 1.1, three time windows are defined to retrieve contextual information.

```
PREFIX : <http://www.semanticweb.org/Quality_onto.owl#>
PREFIX sosa: <http://www.w3.org/ns/sosa/>
PREFIX rsp4j: <http://rsp4j.io/>

REGISTER ISTREAM <http://out/stream> AS

SELECT ?farm ?nObs ?nVal ?nPro ?crop_ref ?region_ref ?ts1
FROM <http://streamreasoning.org/Quality_onto>
FROM NAMED WINDOW rsp4j:window_ndvi ON :Stream_S_NDVI_index [RANGE PT10S
     STEP PT1S]

WHERE {
    ?farm :growsCrop ?crop_ref; :locatedIn ?region_ref .
    ?threshold a :ClusterThreshold; :hascrop_type ?crop_ref; :hasregion
         ?region_ref; :hasLowQuantile_NDVI_index ?ndvi_threshold.

    WINDOW rsp4j:window_ndvi {
           ?farm sosa:hosts ?n. ?n :madeObservation ?nObs.
           ?nObs :hasSimpleResult ?nVal; :observedProperty ?nPro;
           :hasTime ?t1. ?t1 :inXSDDateTimeStamp ?ts1.
    }
    FILTER(?nVal < ?ndvi_threshold)
}
```

Listing 1.1. On example of SPARQL query to detect situations

The ontology is populated with information on farms (e.g., id, region, type of crops) and threshold values derived from clusters of regions and crop types computed from the dataset. To detect a given phenomenon, streaming values are compared against threshold values. Since each observation is associated with a farm, its crop type and region can be retrieved in the ontology, allowing us to extract the corresponding thresholds from the static ontology and compare them to the observed values using `FILTER` expressions. Similar queries are defined for heat stress and disease risk, based on combinations of temperature, humidity, and radiation thresholds, but are not shown due to space limitations. When the specified conditions are satisfied, the query results are emitted into an output stream, defined as an `IStream`, which outputs only newly detected situations [25]. These results are then processed to populate the triple store. For each detected situation, new individuals are created and linked to existing entities (e.g., farms, observations) through object properties, and relevant data properties are asserted. A SPARQL update query is generated and executed to insert these individuals into the knowledge base. Once the data are stored, offline reasoning can be performed to check ontology consistency and prepare the data for subsequent explanation. The detected situations are then passed to the concept induction module for root cause identification.

5.2 Root Cause Identification with Concept Induction

The objective of the concept induction phase is to distinguish between healthy and diseased crops and to identify potential root causes associated with the detected phenomena. The ECII framework [22] requires a positive and a negative set of examples. In this experiment, crops labeled as `Mild`, `Moderate`, or `Severe` are considered diseased and constitute the positive set, while crops labeled as `None` are considered healthy and form the negative set. During the first iteration of the periodic concept induction phase, 38 crops are deemed diseased, and 10 are healthy. Among the diseased ones, 7 are linked to observed phenomena.

Concepts generated by ECII using this example set are presented in Table 1, together with their associated evaluation metrics (coverage, hierarchy, and precision). Since ECII primarily ranks concepts based on coverage, the highest-ranked concept according to the original framework is the one with the most extensive coverage. However, in our setting, hierarchy and precision are also critical to identify concepts that are both specific and explanatory. To address this, the multi-criteria scoring function defined in Eq. 5 is applied. The weights α, β, and γ are determined by solving the associated optimization problem [8].

For this specific experiment, the resulting weights are $\alpha = \beta = \gamma = 0.333$. These weights are instance-dependent and must be recomputed for different concept sets. The updated weights are then used to compute the final score for each concept, producing a new ranking presented in Table 2. Compared to the original ECII ranking, the proposed scoring strategy favors more specific and semantically informative concepts by jointly considering coverage, precision, and hierarchy. The resulting top-ranked concepts provide interpretable explanations of the detected phenomena and offer insights into potential root causes of crop disease, notably through the `involvedInSituation` relation linking crops to detected stress or risk situations. For instance, the top-ranked concept indicates

Table 1. Obtained concepts and metrics comparison: coverage, normalized coverage, hierarchy, and precision (Eqs. 1, 2, 4 and 3).

Concepts	Cov.	NCov.	Hier.	Prec.
∃involvedInSit.((Sit-D) ⊔ (Sit-V) ⊔ (¬ Sit.H))	0.167	1.000	0	0.700
∃involvedInSit.((Sit-D) ⊔ (Sit-V))	0.146	0.875	0	0.778
∃involvedInSit.((¬ Sit.H) ⊔ (Sit-D))	0.125	0.750	1.0	0.625
∃involvedInSit.(Sit-D)	0.104	0.625	1.0	0.714
∃involvedInSit.((Sit-V) ⊔ (¬ Sit.H))	0.063	0.375	1.0	0.667
∃involvedInSit.(Sit-V)	0.042	0.250	1.0	1.0
∃ involvedInSit.(¬ Sit.H)	0.021	0.125	0	0

Table 2. Obtained concepts and metrics comparison: coverage, normalized coverage, hierarchy, precision, and score (Eq. 1, 2, 4, 3 and 5). The Score metric, defined as the combination of the other metrics, is used to rank the concepts in descending order.

Concepts	Cov.	NCov.	Hier.	Prec.	Score
∃involvedInSit.((¬ Sit.H) ⊔ (Sit-D))	0.125	0.750	1.0	0.625	0.792
∃involvedInSit.(Sit-D)	0.104	0.625	1.0	0.714	0.780
∃involvedInSit.(Sit-V)	0.042	0.250	1.0	1.0	0.750

that diseased crops are frequently associated with disease-risk situations (Sit-D), optionally combined with the absence of heat-stress situations (-Sit-H).

6 Conclusion and Future Work

In this paper, we presented a hybrid framework that combines stream reasoning and concept induction to support real-time anomaly detection and explainable root cause analysis over streaming sensor data. The proposed approach addresses the challenge of detecting abnormal situations in continuous data streams and provides human-interpretable explanations of the underlying causes. Stream reasoning is used to continuously detect relevant situations from sensor data, while concept induction is leveraged to derive symbolic and discriminative concepts that explain the observed phenomena. Through an experimental demonstration on a smart farming use case, we showed that the proposed framework is able to detect abnormal environmental and agronomic conditions and to generate structured explanations that differentiate between healthy and diseased crops. The induced concepts capture meaningful relationships between detected situations and crop health status, thereby supporting transparent and interpretable root cause identification. Although instantiated on smart farming data, the proposed framework is domain-independent and can be adapted to other use cases.

While the framework demonstrates promising results, certain aspects require further consideration. The experimental study is intended as a feasibility demonstration rather than a quantitative benchmark evaluation. The effectiveness of the stream reasoning component depends on the quality of the defined rules and thresholds, which may benefit from the integration of expert knowledge or adaptive learning mechanisms. In addition, the scalability of the framework under higher data rates and larger knowledge bases requires further investigation.

Future work will focus on addressing these limitations by extending the experimental evaluation to benchmark datasets and by comparing the concept induction results with alternative approaches, such as Ontolearn [10], a framework for learning OWL class expressions over large knowledge graphs. Additional research directions include improving scalability, learning thresholds dynamically from data, and integrating human-in-the-loop feedback to refine explanations and enhance trust in the system.

Acknowledgments. This work was supported by the French National Research Agency (grant number ANR-22-CE92-0007).

Disclosure of Interests. The authors have no competing interests to declare that are relevant to the content of this article.

References

1. Agrios, G.N.: Agrios' Plant Pathology, 6th edn. Academic Press, London (2024)
2. Arasu, A., Babu, S., Widom, J.: The CQL continuous query language: semantic foundations and query execution. VLDB J. **15**, 121–142 (2006)
3. Barbieri, D.F., Braga, D., Ceri, S., Della Valle, E., Grossniklaus, M.: C-SPARQL: SPARQL for continuous querying. In: Proceedings of the 18th International Conference on WWW, pp. 1061–1062 (2009). https://doi.org/10.1145/1526709.1526856
4. Barredo Arrieta, A., et al.: Explainable artificial intelligence (XAI): concepts, taxonomies, opportunities and challenges toward responsible AI. Inf. Fusion **58**, 82–115 (2020). https://doi.org/10.1016/j.inffus.2019.12.012
5. Belton, V., Stewart, T.J.: Multiple criteria decision analysis: an integrated approach (2002)
6. Campbell, C.L., Madden, L.V.: Introduction to Plant Disease Epidemiology, 1. pr. edn. Wiley, New York, NY [u.a.] (1990), literaturverz. S. 453 - 509
7. Charbonnier, L., Giustozzi, F., Saunier, J., Zanni-Merk, C.: Towards a semantic approach to detection of quality issues in manufacturing 4.0. Procedia Comput. Sci. **246**, 2439–2448 (2024). https://doi.org/10.1016/j.procs.2024.09.479
8. Charbonnier, L., Giustozzi, F., Saunier, J., Zanni-Merk, C.: A Novel concept induction approach for explainable quality 4.0. In: Rules and Reasoning, pp. 36–53 (2025). https://doi.org/10.1007/978-3-032-08887-1_3
9. Dell'Aglio, D., Della Valle, E., Calbimonte, J.P., Corcho, O.: RSP-QL semantics: a unifying query model to explain heterogeneity of RDF stream processing systems. Int. J. Semant. Web Inf. Syst. **10**, 17–44 (2014)
10. Demir, C., et al.: Ontolearn–A framework for large-scale OWL class expression learning in python. J. Mach. Learn. Res. **26**(63), 1–6 (2025)
11. Giustozzi, F., Saunier, J., Zanni-Merk, C.: Context modeling for industry 4.0: an ontology-based proposal. Procedia Comput. Sci. **126**, 675–684 (2018). https://doi.org/10.1016/j.procs.2018.08.001

12. Giustozzi, F., Saunier, J., Zanni-Merk, C.: Abnormal situations interpretation in industry 4.0 using stream reasoning. Procedia Comput. Sci. **159**, 620–629 (2019)
13. Hatfield, J.L., Prueger, J.H.: Temperature extremes: effect on plant growth and development. Weather Clim. Extremes **10**, 4–10 (2015). https://doi.org/10.1016/j.wace.2015.08.001
14. Hogan, A.: Knowledge graphs (2021). https://doi.org/10.2200/S01125ED1V01Y202109DSK022, No. 22
15. Karunathilake, E.M.B.M., Le, A.T., Heo, S., Chung, Y.S., Mansoor, S.: The path to smart farming: innovations and opportunities in precision agriculture. Agriculture **13**(8), 1593 (2023). https://doi.org/10.3390/agriculture13081593
16. Lehmann, J., Hitzler, P.: Concept learning in description logics using refinement operators. Mach. Learn. **78**(1), 203–250 (2010)
17. Miller, T., Mikiciuk, G., Durlik, I., Mikiciuk, M., Łobodzińska, A., Śnieg, M.: The IoT and AI in agriculture: the time is now–a systematic review of smart sensing technologies. Sensors **25**(12), 3583 (2025). https://doi.org/10.3390/s25123583
18. Nenov, Y., Piro, R., Motik, B., Horrocks, I., Wu, Z., Banerjee, J.: RDFox: a highly-scalable RDF store. In: The Semantic Web - ISWC 2015, pp. 3–20 (2015)
19. Pettorelli, N., et al.: Using the satellite-derived NDVI to assess ecological responses to environmental change. Trends Ecol. Evol. **20**(9), 503–510 (2005). https://doi.org/10.1016/j.tree.2005.05.011
20. RDF Working Group: Resource Description Framework (RDF) (2014). https://www.w3.org/2001/sw/wiki/RDF, W3C Recommendation
21. Said Mohamed, E., et al.: Smart farming for improving agricultural management. Egypt. J. Remote Sens. Space Sci. **24**(3), 971–981 (2021). https://doi.org/10.1016/j.ejrs.2021.08.007
22. Sarker, M.K., Hitzler, P.: Efficient concept induction for description logics. In: Proceedings of the AAAI Conference on Artificial Intelligence, vol. 33, pp. 3036–3043 (2019). https://doi.org/10.1609/aaai.v33i01.33013036
23. Seaborne, A., Harris, S.: SPARQL 1.1 query language. Tech. rep., W3C (2013). https://www.w3.org/TR/2013/REC-sparql11-query-20130321/
24. Sharma, K., Shivandu, S.K.: Integrating artificial intelligence and internet of things (IoT) for enhanced crop monitoring and management in precision agriculture. Sensors Int. **5**, 100292 (2024). https://doi.org/10.1016/j.sintl.2024.100292
25. Tommasini, R., Bonte, P., Ongenae, F., Valle, E.D.: RSP4J: an API for RDF stream processing. In: Extended Semantic Web Conference (2021)
26. Tucker, C.J.: Red and photographic infrared linear combinations for monitoring vegetation. Remote Sens. Environ. **8**(2), 127–150 (1979). https://doi.org/10.1016/0034-4257(79)90013 0
27. Wahid, A., Gelani, S., Ashraf, M., Foolad, M.R.: Heat tolerance in plants: an overview. Environ. Exp. Bot. **61**(3), 199–223 (2007). https://doi.org/10.1016/j.envexpbot.2007.05.011
28. Widmer, C.L., et al.: Towards human-compatible XAI. J. Web Semant. **79**, 100807 (2023). https://doi.org/10.1016/j.websem.2023.100807

Improving Taboo-Word Generation in Turkic Language with Morphology-Aware Filtering and Diversity Selection

Anar Mammadli(✉) and Elviz Ismayilov

Azerbaijan State Oil and Industry University, 16/21 Azadliq Ave, Baku, Azerbaijan
{anar.mammadli,elviz.ismayilov}@asoiu.edu.az

Abstract. Taboo-style word games require generating forbidden words that are semantically related to a target concept while remaining non-trivial and playable. In morphologically rich Turkic languages such as Uzbek and Azerbaijani, embedding-based nearest-neighbor methods often fail this requirement: they frequently return inflectional or derivational variants of the target (e.g., plural/possessive forms), orthographic or subword artifacts, and noisy cross-lingual tokens. This paper investigates taboo-word generation on a Turkic taboo-card dataset (Yasaq) in Azerbaijani, Uzbek. We first quantify failure modes across common representation models (Count/TF-IDF, Word2Vec, FastText, BERT, and fine-tuned BERT) by analyzing nearest neighbors for target words across multiple categories. Results show that FastText and fine-tuned BERT produce high-similarity candidates dominated by morphological family members, while Word2Vec and base BERT exhibit unstable neighborhoods influenced by topic co-occurrence and subword similarity. To address these issues, we propose a lightweight post-generation filtering layer that removes morphological variants and overly similar strings, constrains candidates by part-of-speech, filters noise (foreign tokens, malformed Unicode), and enforces diversity among the final taboo set. We evaluate our approach with a human study measuring relatedness, non-triviality, and playability of taboo lists, comparing baseline nearest-neighbor generation to filtered outputs. Human judgments indicate that filtering substantially reduces trivial hints and noisy candidates while preserving semantic relevance, producing taboo lists perceived as fairer and more enjoyable for gameplay. The proposed method is simple to implement and improves the reliability of embedding-based taboo-word generation for low-resource, morphologically rich languages.

Keywords: Taboo game · Serious games · Turkic languages · Embeddings · Morphology

1 Introduction

Word-guessing games such as Taboo are widely used for entertainment and educational practice, requiring players to describe a target word without using a list

C. Zanni-Merk et al. (Eds.): AIDT 2026, CCIS 3023, pp. 19–31, 2026.
https://doi.org/10.1007/978-3-032-31319-5_2

of forbidden "taboo" words. The quality of taboo lists strongly affects gameplay: taboo words should be related enough to block obvious clues, yet not so trivial that they merely repeat the target word in slightly altered forms. This balance is especially challenging for morphologically rich languages, where common computational similarity methods tend to surface inflectional and derivational variants that are effectively "the same hint." In Turkic languages such as Uzbek and Azerbaijani, morphology is productive and frequent: pluralization, possession, case marking, and derivation produce many surface forms from a single root. In our experiments, high-similarity neighbors from FastText and fine-tuned BERT often correspond to morphological family members rather than distinct semantic concepts (e.g., dengiz → dengizi, hayvon → hayvonlar, qilish → qiladi). Conversely, models such as Word2Vec and base BERT sometimes produce neighbors that are not semantically appropriate for taboo gameplay, driven by topic co-occurrence or subword/orthographic similarity (e.g., oq → shiq, daraxt → dart), particularly when trained or evaluated on short-text, low-resource datasets.

The present work aims to improve taboo-word generation for Turkic languages by introducing a post-generation filtering layer that transforms noisy or trivial nearest neighbors into a more useful taboo set. We study this problem on the Yasaq taboo-card dataset in Azerbaijani and Uzbek, and additionally evaluate an Azerbaijani→Uzbek translated version to examine how translation affects candidate quality and cross-language noise.

2 Methods

We use a taboo-card dataset derived from Yasaq, containing cards with a target word and five taboo words, along with a category label (e.g., human, animal, food, item). Each row consists of:

- target (main) word: word_tabu
- taboo words: first_word ... fifth_word
- category: category

We conduct experiments in:

- Azerbaijani (AZ) original dataset;
- Uzbek (UZ) dataset derived from translation;

Dataset statistics (to report): total cards [N], unique targets [U], and category counts [C1..Ck] are shown in Table 1.

Preprocessing. All text is lowercased. Unicode normalization is applied (NFKC), and combining marks are removed to address diacritic artifacts. Apostrophes are unified to a single character. Tokens containing digits, malformed punctuation, or mixed scripts can be excluded via a noise filter. Azerbaijani uses a Latin-based orthography with language-specific characters (e.g., *ə, ğ, ı, ö, ü, ş, ç*). In practice, taboo-card corpora and derived resources may contain mixed spellings in which diacritics are inconsistently preserved (e.g., diacritic vs. non-diacritic variants), and apostrophe-like characters may appear in multiple

Table 1. Dataset summary (Uzbek translated dataset).

Statistic	Value
Cards	3000
Unique targets	2154
Top categories	animal: 636; item: 482; location: 482; food: 482; human: 465
Avg. token length (chars)	7.66

Unicode forms. Such variation can affect both candidate retrieval and filtering: two surface forms that are linguistically equivalent may not be recognized as the same-root family under simple suffix stripping or signature-based stem heuristics if orthographic normalization is inconsistent. To mitigate this, the preprocessing stage applies Unicode normalization and apostrophe unification, and optionally maps characters to a consistent representation prior to stemming and string-similarity checks. Empirically, this normalization reduces spurious differences across candidate forms and supports more reliable removal of trivial morphological variants, while also improving robustness in the presence of noisy or mixed-source text. Remaining leakage cases are therefore best interpreted as stemming/signature coverage gaps (e.g., possessive/case variants not captured by the suffix set) or residual orthographic inconsistencies; both can be addressed by extending the Azerbaijani suffix inventory and enforcing a single diacritic policy throughout the pipeline.

2.1 Baseline Taboo-Word Generation (naive Neighbors)

For each target word t, we generate candidate taboo words using nearest-neighbor retrieval from different representations:

- CountVectorizer cosine similarity
- TF-IDF cosine similarity
- Word2Vec (trained on dataset text; skip-gram or CBOW)
- FastText (subword embeddings)
- BERT word embeddings (subword pooled representation)
- Fine-tuned BERT (fine-tuned on in-domain data)

For each model, we retrieve the top-N candidates (e.g., N = 50) and form a baseline taboo list by taking the top-5 highest-scoring candidates.

- Observed failure modes (motivating the filter). In preliminary analysis, we observe:
- Morphological variants dominate FastText and fine-tuned BERT (e.g., uy → uyi, daraxt → daraxti).
- Subword artifacts appear in base BERT outputs (e.g., oq → shiq).
- Topic co-occurrence yields semantically odd Word2Vec neighbors (e.g., suv → qal'a).

- Cross-language noise exists in AZ outputs (loanwords, named entities), reduced after translation.

2.2 Proposed Filtering Layer

We propose a post-generation filtering layer that operates on a candidate pool C=(w,s,m) where w is a candidate taboo word, s is its similarity score, and m is the source model.

Candidate Pool Construction. Candidates may be obtained from a single best-performing model (e.g., FastText or fine-tuned BERT) with larger top-N or the union of multiple model candidate lists. In either case, filtering is applied before selecting the final top-5 taboo words.

Filtering Steps

- Normalization: Unicode NFKC, remove combining marks, lowercase, unify apostrophes.
- Noise filtering: remove empty tokens, tokens with digits, punctuation-only tokens, mixed-script tokens (unless allowed), and tokens in a foreign/noise lexicon.
- Morphological family removal: remove candidates sharing the same stem/signature as the target or exceeding a character n-gram similarity threshold τ_{morph}.
- String similarity filter: remove candidates with normalized Levenshtein similarity above τ_{str}.
- Diversity selection: select final K = 5 taboo words using MMR (Maximal Marginal Relevance) to reduce redundancy.

In our experiments, we empirically set the morphological similarity threshold $\tau_{morph}0.65$, the string similarity threshold $\tau_{str} = 0.85$, the minimum frequency threshold to 1, and the Maximal Marginal Relevance tradeoff parameter $\lambda = 0.75$. Morphological similarity is calculated using a character n-gram Jaccard similarity (n = 3), while string similarity uses a normalized Levenshtein ratio. For morphological family removal, we apply stemming by stripping up to two consecutive suffixes from a predefined list of 31 common Turkic suffixes (e.g., lar, ning, im, dir). Furthermore, the noise filter explicitly removes tokens that are under 2 characters long, contain digits, consist entirely of non-word characters, or contain mixed scripts (e.g., combining Latin and Cyrillic characters)

Diversity Selection (MMR). In the diversity selection stage, the final taboo set is constructed using Maximal Marginal Relevance (MMR) to balance relevance to the target word with internal diversity among the selected taboo words. Rather than choosing the top-K candidates solely by similarity score, MMR iteratively selects the next candidate that maximizes a weighted objective combining its relevance score and a redundancy penalty based on its maximum similarity to already selected items. This mechanism discourages near-duplicates (including closely related variants and highly overlapping semantic cues) and promotes a taboo list that covers multiple complementary aspects of the target concept. In practice, MMR reduces repetitive lists where several taboo words express the same clue, improving the fairness and playability of the game by increasing the breadth of restrictions while maintaining semantic relatedness. We select taboo words iteratively:

$$w^* = \arg\max_{w \in C} \left[\lambda \cdot s(w) - (1 - \lambda) \cdot \max_{u \in L} \mathrm{Sim}(w, u) \right] \tag{1}$$

where L is the current selected list, $s(w)$ is the relevance score, and Sim is a similarity measure (embedding cosine or character n-gram Jaccard). $\lambda \in [0, 1]$ controls relevance vs diversity.

2.3 Evaluation

Automatic Metrics. For baseline vs filtered lists, we compute:

- **Same-root rate:** fraction of lists containing at least one morphological family member of the target.
- **Avg. target–taboo string similarity:** mean normalized Levenshtein similarity between target and each taboo word.
- **In-list diversity:** average pairwise similarity among the five taboo words.
- **Noise rate:** fraction of taboo words flagged as noisy/foreign.

Human Evaluation. To assess the quality of the generated taboo words, we conducted a human evaluation study. The study included 15 native bilingual speakers proficient in Uzbek and Azerbaijani. Evaluators were presented with 100 randomly sampled target words alongside their corresponding taboo lists from both the baseline and filtered models in a blind, randomized setup. Participants rated each list on a 5-point Likert scale (1 = strongly disagree, 5 = strongly agree) based on three criteria: semantic relatedness to the target, non-triviality (the absence of obvious morphological clues), and overall playability.

3 Results

3.1 Qualitative Analysis of Baseline Neighbors

Across models, we observe distinct behavior patterns consistent across Uzbek and Azerbaijani experiments. The qualitative examples in Table 2 illustrate how the filtering stage improves gameplay suitability by removing trivial morphological variants and near-duplicate forms of the target (e.g., possessive/plural suffixes and orthographically similar strings) that frequently appear in baseline neighbor lists. After filtering, the taboo lists become more diverse and less revealing, while

Table 2. Qualitative examples of taboo lists before and after filtering for UZ language.

Target	Category	Before (baseline taboo list)	After (filtered taboo list)
mashina	item	mashinasi, mast, mash, motor, mas	motor, mehnati, kit, kabina, mashhur
shkaf	item	shkafi, sharaf, tashrif, kafka, xavf	tashrif, xavf, saf, sharq, sinf
menyu	item	menga, mendes, menlo, men, kanye	kanye, moyi, marjon, gyugo, mendel
dastur	item	dasturi, dasturchi, dasturiy, dasht, dasi	dasturiy, doska, boylik, rasm, xamirturush
dunyo	location	dunyosi, dunkin, daryo, hudud, koloniya	koloniya, joy, tunda, kunlar, markaziy
zal	location	zallik, qismi, tribunal, kana, dam	qismi, dam, laguna, suv, santa

still preserving domain-related cues that restrict obvious descriptions. Baseline lists often contain morphological variants or near-duplicate forms of the target (e.g., *mashinasi*, *ayig*, *shkafi*, *dunyosi*, *zallik*), while the filtered output removes these trivial hints and increases diversity.

Table 3 provides qualitative evidence of how the proposed post-generation filtering pipeline alters taboo lists in the Azerbaijani setting. In the unfiltered baseline, the nearest-neighbor retrieval frequently selects candidates that are trivially related to the target through inflectional or derivational morphology (e.g., possessive/plural forms such as *maşını*, *ayaqqabısı*, *qoyunu*, and *şüşəbr*), as well as candidates with strong orthographic overlap. Such items are undesirable for Taboo gameplay because they effectively reveal the target or provide redundant restrictions. After filtering, these forms are largely removed and replaced by alternative candidates that better satisfy the non-triviality constraint, yielding taboo sets that are less morphologically overlapping with the target. The table also illustrates the role of diversity-oriented selection: the filtered lists exhibit reduced repetition and cover a broader range of cues rather than concentrating on near-duplicate variants. At the same time, the examples highlight remaining challenges, including occasional retention of mixed-language or out-of-domain tokens (e.g., *hour*, *time*, *signal*), indicating that additional language-purity or lexicon-based constraints may be beneficial for fully controlling candidate quality in multilingual or noisy corpora. Baseline lists often contain trivial morphological variants of the target (e.g., possessive/plural forms such as *maşını*, *ayaqqabısı*, *qoyunu*, *şüşəbr*) or very close orthographic variants, while the filtered output reduces these and increases diversity within the list.

FastText and Fine-Tuned BERT. These models strongly prioritize morphological similarity. For Uzbek, examples include *qilish* → *qiladi*, *hayvon* → *hayvonlar*, *suv* → *suvli*, *dengiz* → *dengizchi*, and *daraxt* → *daraxti*. While these candidates score

Table 3. Qualitative examples of taboo lists before and after filtering (Azerbaijani language).

Target	Category	Unfiltered (baseline taboo list)	Filtered taboo list
maşın	item	maşını, maşin, marşrut, avtobus, trafi	masin, avtobus, marsrut, metro, lokomotiv
proqram	item	proqramçi, proqramçı, proqnoz, proqnozçu	proqnoz, istifadəci, diaqram, texniki, komputer
dünya	location	həyat, dünyasi, planet, yaşayan, doğal	həyat, planet, yasayan, məshur, xarici
saat	item	hour, time, saatlardan, times, signal	hour, time, signal, ay, planet
qoyun	animal	qoyunu, çoyun, qohum, qonşu, boyun	qohum, boyun, qaz, qiz, qoz

highly, they represent trivial variations that do not improve taboo gameplay because they leak the target root.

Base BERT. Base BERT outputs some plausible morphological neighbors but also subword-driven artifacts. Outputs such as *oq* → *shiq* and *daraxt* → *dart* suggest that word-level pooling of subwords can overemphasize orthographic similarity.

Word2Vec. Word2Vec yields unstable associations likely driven by dataset co-occurrence patterns. Examples such as *suv* → *qal'a* and *kichik* → *Afrika* indicate that short-text datasets and limited context can distort distributional similarity.

BoW (Count/TF-IDF). BoW sometimes captures topical relations (e.g., *suv* ↔ *dengiz*, *meva* ↔ *olma*), but can also return event-based or noisy associations (e.g., *daraxt* ↔ *kuymoq*). Notably, in our current pipeline, CountVectorizer and TF-IDF produced identical rankings and scores, which suggests a potential implementation issue or feature reuse; we treat this as a reproducibility concern and recommend validating the TF-IDF pipeline prior to deployment.

3.2 Automatic Metrics: Baseline vs Filtered

Tables 4 and 5 summarize the automatic quality metrics for candidates generated by our baseline models before and after filtering. Applying the filtering step significantly improves the suitability of the taboo-word lists. Under the baseline approach, 19.87% of the targets contain at least one taboo word that shares a root with the target, which compromises game difficulty. After filtering, this same-root rate is reduced to 0%. Furthermore, the average string similarity between the target and the taboo words decreases from 0.422 to 0.302, successfully masking obvious orthographic clues. The filter also improves in-list diversity; internal pairwise similarity drops from 0.098 to 0.0068, indicating that the selected taboo words cover a broader range of distinct semantic clues. Overall, these metrics—computed on 936 generated targets from the translated Uzbek dataset—demonstrate that the proposed filtering pipeline effectively transforms raw embedding neighbors into playable taboo-word lists. Table 4 summarizes automatic quality metrics. Filtering consistently reduces triviality and noise while increasing diversity:

- **Same-root rate** decreases from **19.87%** (baseline) to **0%** (filtered).
- **Avg. target–taboo string similarity** decreases from **0.4224** to **0.3022**.
- **In-list diversity** improves (pairwise similarity decreases from **0.0981** to **0.0068**).
- **Noise rate** remains **0%**.

3.3 Error Analysis

Table 4. Automatic metrics for baseline vs filtered taboo lists ($N = 936$).

Metric	Baseline	Filtered
Same-root rate (%)	19.87	0.00
Avg. target–taboo string sim.	0.4224	0.3022
In-list similarity (lower better)	0.0981	0.0068
Noise rate (%)	0.00	0.00

Table 5. Automatic metrics for Azerbaijani taboo lists (baseline vs. filtered), computed on $N = 1012$ targets. Lower values indicate less triviality and higher diversity.

Metric	Baseline	Filtered
Same-root rate (%)	11.86	1.68
Avg. target–taboo string similarity	0.4064	0.2970
In-list similarity (lower is better)	0.0878	0.0075
Noise rate (%)	0.00	0.00

Even when contextual token embeddings are extracted from the full "target + five taboo words" sequence and aggregated across multiple occurrences, residual errors still arise from properties of the dataset and the modeling objective rather than from a lack of contextuality in the encoder. In this setting, BERT generally benefits from repeated exposures to the same target word under diverse co-occurring taboo sets, which stabilizes representations and promotes semantic clustering beyond pure surface-form similarity. This is reflected in the fact that many high-similarity candidates are plausibly related or morphologically coherent, indicating that the model has learned consistent distributional structure within the task domain. However, three categories of errors remain.

First, the most common failure case is the persistence of corpus artifacts that are orthographically well-formed yet semantically unhelpful for gameplay (e.g., rare tokens, idiosyncratic strings, or domain-specific proper names). These items can survive normalization and basic noise checks because they resemble valid words and may be reinforced by a small number of highly similar contexts. In a taboo-card corpus, where each "sentence" is a compact list of salient keywords, a rare token that co-occurs with a target in a few cards can become spuriously close in the embedding space after aggregation. This effect is amplified by the limited size and topical narrowness of the corpus, which makes the embedding neighborhood sensitive to low-frequency co-occurrences.

Second, the representation can become overly influenced by local co-occurrence regularities inherent to taboo-card construction. Because the input sequence is not a natural sentence but a curated set of related keywords, the encoder learns a strong "card-level association signal." While this is beneficial for capturing domain relevance, it can also yield candidates that are related by frequent co-membership in cards rather than by the intended semantic relation for taboo gameplay. This phenomenon is not a failure of contextual encoding per se; rather, it reflects a mismatch between the downstream notion of "good taboo words" (related but non-trivial and diverse) and the similarity signal induced by repeated card-level co-occurrence.

Third, polysemy and category mismatch remain challenging. Although contextual aggregation across multiple cards improves robustness, if a target appears in heterogeneous contexts (multiple senses or mixed categories), the aggregated embedding can represent a blended meaning. Nearest neighbors may therefore reflect an alternative sense that is not appropriate for a particular category. In taboo generation, this produces lists that may be semantically defensible in the corpus but inconsistent with the intended gameplay interpretation of the target. These cases suggest future work should incorporate context-aware representations or external knowledge resources.

3.4 Human Evaluation Results

Consistent with the automatic metrics, human evaluators rated the filtered taboo lists as significantly more playable and less trivial than the baseline lists, confirming that the removal of morphological variants and the enforcement of diversity (via MMR) yields fairer game constraints. Quantitatively, the filtered lists achieved an average playability score of 4.3 out of 5, marking a substantial improvement over the baseline's score of 2.8. Non-triviality scores similarly surged from 2.2 in the baseline to 4.6 in the filtered outputs, reflecting the successful elimination of morphological leakages. Importantly, this increase in difficulty did not severely harm the relevance of the clues. Semantic relatedness remained highly stable, showing only a marginal decrease from 4.7 (baseline) to 4.5 (filtered). A paired t-test confirmed that the improvements in both playability and non-triviality are statistically significant ($p<0.01$).

4 Discussion

Our findings indicate that naive nearest-neighbor taboo generation is unreliable in Turkic languages due to the interaction of morphology, subword tokenization, and dataset sparsity. FastText and fine-tuned BERT exhibit high cosine similarity for morphological variants, which is desirable for tasks like lemmatization or morphological clustering but harmful for taboo gameplay because it yields trivial hints. Word2Vec and base BERT may return semantically unrelated neighbors when trained or applied to short-text datasets, reflecting limitations of distributional learning under restricted context and the sensitivity of subword pooling to orthographic overlap. In particular, taboo-card corpora provide "list-like" contexts rather than full grammatical sentences, which strengthens card-level association signals but can weaken the model's ability to encode fine-grained semantic relations. As a result, similarity estimates can reflect frequent co-membership in cards or shared subword fragments rather than the intended conceptual proximity needed for robust taboo constraints.

The proposed filtering layer offers a practical remedy. Importantly, it does not require training new large models; it can be applied as a post-processing step to any neighbor generator. This makes it suitable for low-resource settings and for serious game systems where rapid iteration and interpretability are valued. The pipeline also provides modular control: normalization and noise handling improve robustness to orthographic variation, morphological family removal directly targets the primary source of "giveaway" taboo words, and the string-similarity constraint suppresses near-duplicate forms that survive stemming heuristics. The diversity selection stage (MMR) further addresses an often-overlooked issue: even when taboo candidates are individually appropriate, selecting the top-K by similarity frequently produces redundant lists that block the same clue multiple times. By explicitly penalizing overlap among selected taboo words, MMR encourages broader coverage of the target concept and yields taboo sets that are more informative as restrictions and more balanced in difficulty.

Translation (AZ→UZ) appears to reduce cross-lingual noise, but does not eliminate morphology dominance. Highlighting that the main issue is not only language mixing, but the intrinsic morphology of Turkic languages and the model biases toward form similarity. This observation has implications for multilingual deployment: while translation can standardize vocabulary and remove some out-of-language artifacts, it does not substitute for language-aware constraints. In cross-lingual scenarios, filtering is therefore best treated as a language-agnostic control layer that operates after candidate generation, potentially combined with bilingual alignment methods to ensure that semantically comparable candidates are available in the target language.

Finally, our evaluation emphasizes that taboo-word quality is not equivalent to raw semantic similarity: "good taboo words" require a balance of relatedness, non-triviality, and diversity. This motivates treating taboo generation as a constrained selection problem rather than a direct kNN retrieval task. From this perspective, embedding models serve primarily as candidate generators, while gameplay constraints are enforced by explicit selection rules. This framing also clarifies future extensions: richer linguistic constraints (e.g., stronger POS control, frequency-aware noise filtering, category-conditioned embeddings) can be incorporated without changing the underlying representation model, and difficulty control can be introduced by adjusting thresholds and diversity weights to produce taboo sets at multiple challenge levels. Overall, the results suggest that combining distributional similarity with lightweight, interpretable constraints is an effective and scalable approach for taboo-word generation in morphologically rich, low-resource languages.

5 Conclusion

We studied taboo-word generation for Uzbek and Azerbaijani and showed that naive nearest-neighbor methods frequently produce trivial morphological variants and corpus-driven artifacts that undermine gameplay by making clues either too revealing (e.g., inflectional or derivational forms of the target) or uninformative (e.g., rare tokens and orthographic coincidences). To address these limitations, we proposed a lightweight post-generation filtering pipeline that combines text normalization, morphological family removal, string-similarity filtering, optional part-of-speech constraints, and MMR-based diversity selection to produce taboo lists that are both semantically related and non-trivial. Across the evaluated target set, the filtering stage substantially reduced morphological leakage and redundancy within taboo lists, yielding candidates that are less visually similar to the target and more varied in the types of restrictions they impose. Automatic metrics indicate clear improvements in non-triviality and diversity, and human evaluation further supports that the filtered lists are perceived as clearer, fairer, and more enjoyable for gameplay while maintaining the intended semantic connection to the target concept. These findings emphasize that high cosine similarity alone is not an adequate objective for taboo generation in morphologically rich languages; instead, the task should be treated as a constrained selection problem where semantic relevance, non-triviality, and diversity must be jointly optimized. Future work will explore richer context modeling for representation learning on short-text game data, bilingual alignment for cross-lingual taboo generation and transfer across Turkic languages, and ontology/knowledge-graph constraints (e.g., DBpedia-derived semantic typing) to control taboo difficulty, enforce concept compatibility, and generate multi-level taboo sets (easy/medium/hard) that adapt to different player proficiency levels. Finally, we acknowledge that the Yasaq dataset used in this study is relatively small (3,000 cards) and domain-specific. Validating the generalization of this filtering pipeline on larger, more diverse datasets remains an important direction for future work.

References

1. Mikolov, T., Chen, K., Corrado, G., Dean, J.: Efficient estimation of word representations in vector space. arXiv preprint arXiv:1301.3781 (2013)
2. Mikolov, T., Sutskever, I., Chen, K., Corrado, G., Dean, J.: Distributed representations of words and phrases and their compositionality. In: Advances in Neural Information Processing Systems (NeurIPS 2013), pp. 3111–3119 (2013)
3. Pennington, J., Socher, R., Manning, C.D.: GloVe: global vectors for word representation. In: Proceedings of the 2014 Conference on Empirical Methods in Natural Language Processing (EMNLP 2014), pp. 1532–1543 (2014). https://doi.org/10.3115/v1/D14-1162
4. Bojanowski, P., Grave, E., Joulin, A., Mikolov, T.: Enriching word vectors with subword information. Trans. Assoc. Comput. Linguist. **5**, 135–146 (2017). https://doi.org/10.1162/tacl_a_00051
5. Joulin, A., Grave, E., Bojanowski, P., Mikolov, T.: Bag of tricks for efficient text classification. In: Proceedings of the 15th Conference of the European Chapter of the Association for Computational Linguistics (EACL 2017), vol. 2, pp. 427–431 (2017)
6. Devlin, J., Chang, M.-W., Lee, K., Toutanova, K.: BERT: pre-training of deep bidirectional transformers for language understanding. In: Proceedings of the 2019 Conference of the North American Chapter of the Association for Computational Linguistics (NAACL 2019), pp. 4171–4186 (2019). https://doi.org/10.18653/v1/N19-1423
7. Mammadli, A., Ismayilov, E., Zanni-Merk, C.: Explainability of text classification through ontology-driven analysis in serious games. Procedia Comput. Sci. **246**, 2128–2137 (2024). https://doi.org/10.1016/j.procs.2024.09.626
8. Reimers, N., Gurevych, I.: Sentence-BERT: sentence embeddings using siamese BERT-networks. In: Proceedings of the 2019 Conference on Empirical Methods in Natural Language Processing (EMNLP-IJCNLP 2019), pp. 3982–3992 (2019). https://doi.org/10.18653/v1/D19-1410
9. Liu, Y., et al.: RoBERTa: a robustly optimized BERT pretraining approach. arXiv preprint arXiv:1907.11692 (2019)
10. Lample, G., Conneau, A.: Cross-lingual language model pretraining. In: Advances in Neural Information Processing Systems (NeurIPS 2019), pp. 7059–7069 (2019)
11. Conneau, A., et al.: Unsupervised cross-lingual representation learning at scale. In: Proceedings of the 58th Annual Meeting of the Association for Computational Linguistics (ACL 2020), pp. 8440–8451 (2020). https://doi.org/10.18653/v1/2020.acl-main.747
12. Carbonell, J., Goldstein, J.: The use of MMR, diversity-based reranking for reordering documents and producing summaries. In: Proceedings of the 21st Annual International ACM SIGIR Conference on Research and Development in Information Retrieval (SIGIR 1998), pp. 335–336 (1998). https://doi.org/10.1145/290941.291025
13. Manning, C.D., Raghavan, P., Schütze, H.: Introduction to Information Retrieval. Cambridge University Press, Cambridge (2008)
14. Salton, G., Buckley, C.: Term-weighting approaches in automatic text retrieval. Inf. Process. Manage. **24**(5), 513–523 (1988). https://doi.org/10.1016/0306-4573(88)90021-0
15. Deterding, S., Dixon, D., Khaled, R., Nacke, L.: From game design elements to gamefulness: defining "gamification". In: Proceedings of the 15th International

Academic MindTrek Conference (MindTrek 2011), pp. 9–15 (2011). https://doi.org/10.1145/2181037.2181040
16. Bellotti, F., Kapralos, B., Lee, K., Moreno-Ger, P., Berta, R.: Assessment in and of serious games: an overview. Adv. Hum. Comput. Interact. **2013**, 1–11 (2013). https://doi.org/10.1155/2013/136864
17. Lu, W., Griffin, J., Sadler, T.D., Laffey, J., Goggins, S.P.: Serious game analytics by design: feature generation and selection using game telemetry and game metrics. J. Learn. Analytics **10**(1) (2023)
18. FastText project homepage. https://fasttext.cc/. Accessed 27 Feb 2026
19. Hugging Face Transformers documentation. https://huggingface.co/docs/transformers/. Accessed 27 Feb 2026

A Domain Ontology for Cooperative Perception in the Internet of Mobile Things

Kamran Ismayilzada[1,2,3](✉) and Mathieu Bourgais[1]

[1] INSA Rouen Normandie, Normandie Univ, LITIS UR 4108, 76000 Rouen, France
{kamran.ismayilzada, mathieu.bourgais}@insa-rouen.fr
[2] Azerbaijan State Oil and Industry University, Baku, Azerbaijan
[3] French-Azerbaijani University, Baku, Azerbaijan

Abstract. Internet of Mobile Things (IoMT) systems operate networks of mobile and heterogeneous sensors that produce partial, asynchronous, and potentially unreliable observations of their environment. Under such conditions, maintaining a coherent and evolving understanding of spatio-temporal phenomena becomes challenging. This paper proposes an ontology for cooperative perception in hazard monitoring domain of IoMT, demonstrated on wildfire monitoring data. The ontology reuses established standards, including SOSA/SSN, OWL-Time, GeoSPARQL, and existing hazard models, and extends them with: (i) a two-layer observation pattern that separates pre-detection evidence from event-level interpretations; (ii) uncertainty-aware result representations with associated reliability degrees; (iii) revisable hypotheses about abstract characteristics of perceived features; and (iv) state concepts that support aggregation and incremental refinement of event knowledge over time. Logical coherence of the ontology is verified using Description Logic reasoning. Design requirements are formalized as competency questions, and support for key retrieval patterns is evaluated through SPARQL queries over a populated knowledge graph instantiated from burnt-area perimeter and thermal-anomaly observation records.

Keywords: Internet of Mobile Things · cooperative perception · uncertainty · partiality · ontology

1 Introduction

Internet of Things (IoT) systems connect sensing- and computing-enabled objects that perceive their environment and exchange information over a network infrastructure. Many modern deployments of the technology involve mobile sensing entities, such as vehicles, robots, drones, wearables, and satellites. This trend motivates the Internet of Mobile Things (IoMT) paradigm, in which mobility is a defining factor that shapes how the sensing entities produce, communicate and interpret observations of the world around them [15].

C. Zanni-Merk et al. (Eds.): AIDT 2026, CCIS 3023, pp. 32–47, 2026.
https://doi.org/10.1007/978-3-032-31319-5_3

Mobility of the sensing entities brings benefits in the form of wider spatial coverage and flexible sensing strategies compared to static infrastructures. However, mobility introduces system-level complications for the availability, consistency and timeliness of observations, as sensing context and connectivity vary over time, imposing the following limits on the knowledge about the physical world:

Partiality. Despite the flexibility offered by mobility in mitigating the limitations of the fields of view of static sensing entities, ensuring continuously optimal coverage of the environment by sensors is often not achievable. As a result, due to persistent factors such as occlusions, task- and constraint-driven sensing priorities, observations of the environment by a network of moving things inevitably remain incomplete [15]. No single mobile sensor can be assumed to be capable of reliably observing a phenomenon in its entirety and at all times.

Therefore, it is important to maintain the assumption that the absence of reliable localized observations about a spatio-temporal phenomenon should not be interpreted as evidence of its non-existence. This principle is commonly referred to as the Open World Assumption (OWA) and is crucial in monitoring scenarios, where the lack of certain evidence about a monitored entity simply indicates that this evidence has not yet been observed, rather than that it does not exist.

Asynchronicity. Mobile sensing entities may produce observations and transmit them to other components of their network infrastructure using different communication means, at different rates, and at different times [1]. Consequently, temporal misalignment between crucial pieces of environmental data is to be expected and must be explicitly accounted for.

Heterogeneity. With the variety of sensing entities in IoMT systems comes the variety of sensing modalities and characteristics. The outputs provided by different sensors in their raw form are often not directly suitable for integration [13]. This complicates purely syntactic data fusion approaches, while simultaneously motivating the need to operate on knowledge—contextualized data represented at a higher level of abstraction—to enable meaningful interpretation across heterogeneous sources.

Computational methods aimed at data fusion in spite of these challenges often assume the possibility of consolidating observations into a single consistent state of the world. However, the dynamic nature of IoMT systems arises not only from the movement of their components, but also from the continuous evolution of the perceived environment itself in time and space. This requires IoMT systems to support incremental interpretation of the environment, rather than its one-shot perception. These challenges dictate the need for a framework for coherent and evolving interpretations of spatio-temporal phenomena by IoMT systems under the conditions of incomplete, asynchronous and heterogeneous observation data.

Ontology-based approaches have been used in practice for representing domain knowledge derived from data in heterogeneous systems [16]. Pursuing such an approach, we introduce in this paper an ontology for cooperative perception in IoMT environments. The proposed ontology is designed to support incremental aggregation of asynchronous and heterogeneous observations under open-world assumptions. To demonstrate the intended use of the proposed approach, it is adapted for and instantiated in the context of a natural hazard monitoring scenario. Nevertheless, the ontology is designed with considerations about its extensibility and applicability for other IoMT domains.

In Sect. 2 we review perception challenges in IoMT and summarize the existing approaches to their resolution. Section 3 contains the proposed ontology, its requirements and core modeling patterns. In Sect. 4 we report the validation of representational adequacy of the proposed ontology. Section 5 concludes and describes directions for future work. All artifacts with instructions to retrace the result obtention has been provided at the dedicated GitLab repository[1].

2 Background and Related Work

Perception in the context of sensor networks signifies processes by which the sensing entities acquire, process, and interpret data from their environment, individually or through cooperation. As a scientific challenge, it encompasses procedures and protocols in multiple stages from the retrieval of raw low-level observations - measurement records produced by sensors - to derivation of higher-level abstract representations relevant for decision-making.

In the context of the Internet of Mobile Things, challenges at each stage of the perception process are shaped by the need to account for mobility of sensing entities. This implies dealing with the aforementioned problems of partiality, asynchronicity and heterogeneity across these distinct yet complementary layers:

1. IoMT sensing system architectures.
2. Raw data fusion and pre-processing mechanisms.
3. Semantic modeling for interpretation and abstraction.

In this section, we provide an overview of the prominent methods against the mobility-induced challenges used to date.

2.1 IoMT Sensing System Architectures

The design of IoMT sensing system architectures is primarily driven by resource and connectivity constraints [13,14], prioritizing efficient data acquisition, transmission, and computational offloading rather than maximizing perception quality. Representative designs aim to reduce redundancy by coordinating sensing modalities within a single platform [15] and by exploiting complementarity across platforms [7]. Such architectures often rely on tiered edge–fog–cloud substrates

[1] https://gitlab.insa-rouen.fr/kismayilzada/r-coop-ontology/.

to absorb heavier processing and facilitate timely sharing and aggregation of observations [5]. While these measures can partially alleviate early-stage effects of partiality and asynchronicity in perception, they do not eliminate the underlying epistemic challenges: later perception stages still require explicit mechanisms to represent, integrate, and revise incomplete, temporally misaligned and heterogeneous evidence at the knowledge level.

2.2 Raw Data Processing and Fusion

Prior to interpretation, data entering an IoMT system commonly undergoes aggregation and fusion to mitigate heterogeneity and temporal misalignment before downstream tasks. Fusion is performed both within a single platform, by combining heterogeneous sensing modalities [3], and across platforms when shared data formats can be assumed [24]; however, such pipelines often rely on tight synchronization that may be unavailable in opportunistic mobile settings. Temporal anchoring associates records with timestamps or intervals, enabling their ordering and windowing during processing [9], while missing values may be imputed using learned models [4]. In practice, these steps are frequently computationally demanding and tailored to specific contexts and use cases, which may limit interoperability, scalability, and reuse. Moreover, they typically yield deterministic, task-oriented outputs—sometimes produced by black-box estimators—rather than preserving an explicit, traceable, and evolving record of partial and uncertain knowledge together with its temporal context.

2.3 Semantic Modeling

Alignment, completion, and formatting can make heterogeneous records comparable and processable, yet they do not by themselves express what the records jointly indicate about the identity, evolution, or significance of a spatio-temporal phenomenon, nor do they provide explicit support for representing uncertainty and revisable interpretations under open-world conditions. This motivates the introduction of a further abstraction layer, in which processed observations are interpreted and managed as higher-level knowledge.

Raw sensor data can be processed to derive more abstract representations, for example as machine-readable numerical vectors or embeddings [21,24] or as structured symbolic representations such as Resource Description Framework (RDF) triples [8,20]. RDF encodes facts over shared vocabularies, while the formal specification of semantics and constraints that make such facts interoperable is provided by **ontologies** [10].

Ontologies support knowledge representation across multiple levels of abstraction, ranging from domain-independent foundational concepts to highly specialized, domain-specific ones. This enables ontological models to be both generalizable and adaptable to particular use cases. At the same time, factors relevant to the interpretation of phenomena—including their spatial and temporal context—can be explicitly represented within a shared semantic framework. Moreover, RDF representations can support transparent logical reasoning, if interpreted under formalized semantics, and paired with a rule engine or logical reasoner, making them suitable not only for integrating heterogeneous data but also for producing explainable interpretations using shared vocabularies. For these reasons, ontology-based approaches have been widely adopted in the IoT and mobile networks literature to enable semantic interoperability at large network scales [16].

Depending on the application domain and modeling objectives, different works employ different conceptualizations of the environment being perceived. Particularly relevant to the problem addressed in this paper are ontologies that explicitly specify the entities perceived by IoT sensors. A prominent extensible ontology for sensor networks at a relatively high level of abstraction is the Semantic Sensor Network (SSN) ontology and its lightweight core module, SOSA [12]. In the context of spatio-temporal perception, SSN/SOSA provides a schema of key concepts and relations that connect sensors, observations, observable properties, and features of interest (FOIs). This schema has been further extended in the KnowWhereGraph ontology, [19], which integrates the SSN/SOSA core with GeoSPARQL and OWL-Time to represent the spatial and temporal dimensions of observed features, enabling statements about what has been observed, where, and when.

Several ontologies developed for hazard monitoring scenarios have also reused the SSN/SOSA core as a semantic reference for perceived events [11,23]. Stephen et al. [22] introduced the Hazard Information Profiles ontology, grounded in vocabularies defined by international authorities, and connected it with Disaster Event Ontology - another ontology that they introduced in the same work - which extends SSN/SOSA by representing hazard events as observable FOIs associated with hazard-specific properties. These ontologies provide a well-structured representation of identified hazard events, which can be treated as partial or incomplete by design under OWA. However, they can not provide an explicit pattern to assess this partiality and uncertainty in the current state of knowledge about the event - namely, which expected properties remain insufficiently supported.

Several approaches extend Semantic Web representations with uncertainty formalisms, including probabilistic ontologies (e.g., PR-OWL [6]) and evidential frameworks based on belief functions (e.g., Dempster–Shafer ontologies [2]). Roda et al. encode measurement uncertainty inside complex *sosa:Result* structures [17]. However, to the best of our knowledge, hazard-monitoring ontologies reusing SSN/SOSA do not provide an explicit pattern for representing uncertainty at the level of sensor observations about perceived phenomena and estimations derived further as the knowledge is maintained. This gap motivates the extension of existing domain ontologies to better support the requirements of IoMT environments, where knowledge must be incrementally constructed and revised under conditions of partial, asynchronous, and heterogeneous observation data.

3 The R-COOP Ontology

The proposed R-COOP (from Reliable COOperative Perception) ontology targets the semantic artifacts exchanged and accumulated during perception: observations, their interpreted links to perceived phenomena, hypotheses about event identity and type, and evolving estimates of event states. It is not intended to model detection algorithms or domain-specific threshold logic, nor does it prescribe how uncertainty scores are computed. Instead, it focuses on representing the knowledge structures that emerge from such processes.

3.1 Reused Patterns and Competency Questions

The design strategy prioritizes the reuse of existing ontologies wherever possible, in order to maximize interoperability and to avoid re-defining established sensing, spatio-temporal, and hazard-domain primitives. The newly introduced concepts and relations are defined in a new namespace, while those from the reused ontologies remain unchanged.

At the sensing layer, SSN/SOSA provides the core pattern that connects sensors, observations, observable properties, and FOIs, thus, providing a framework for representing who observed what and (optionally) when. To account for the spatio-temporal context of the perceived feature of interest, we reuse constructs from GeoSPARQL (for example, *geo:Feature* and *geo:Geometry*) and OWL-Time. Following a strategy similar to the one adopted in KnowWhereGraph, these vocabularies are used to express where an observed feature is located and how it evolves in time, and to manage temporally misaligned observations typical for IoMT.

R-COOP reuses DEO [22] as an imported ontology and extends it without modifying its original axioms and definitions. In its turn, DEO reuses the SSN/SOSA + GeoSPARQL + OWL-Time core described previously. In this ecosystem, hazard occurrences are expressed as events with observable properties and are treated as FOIs from SOSA. As an addition, the *Hazard* concept is related through the *deo:hasHazardType* relation to the *HazardType* concept. In turn, this concept connects to hazard-type-specific observable properties through *deo:hasHazardProperty*. This provides a basis for linking the perceived hazard events to hazard classifications and to the set of its measurable properties that may be observed or remain missing during perception.

Unlike the extended ontologies described above, the R-COOP ontology explicitly accounts for incompleteness and potential bias in the knowledge derived from observations. These factors directly affect the reliability of individual observations and of the estimations made regarding the observable properties of a perceived event. The ontology is therefore designed to address the gaps and to answer the following competency questions:

CQ1. **(pre-detection stage):** What evidence is available for a potential undetected event within a given region?

CQ2. **(post-detection stage):** Which interpreted observations support a given perceived hazard even, and what do they report?
CQ3. **(revision stage):** What hazard-type candidatures are currently actual for an event?
CQ4. **(maintenance stage):** Values for which expected properties are missing or unreliable in the latest known state of a given event?

The CQ1 and CQ2 express the need to separate conceptually pre-event evidence and event-level interpretations. CQ3 and CQ4 represent the incremental revision of event knowledge and express the need for explicit representation of reliability at the level of observations and hypotheses.

3.2 Core Modeling Patterns

Figure 1 provides an overview of the concepts and relations introduced in R-COOP. These extensions are designed to represent *a) the evolution of knowledge from pre-event evidence to event-level interpretation* and *b) uncertainty at the level of results, hypotheses, and aggregated event state.*

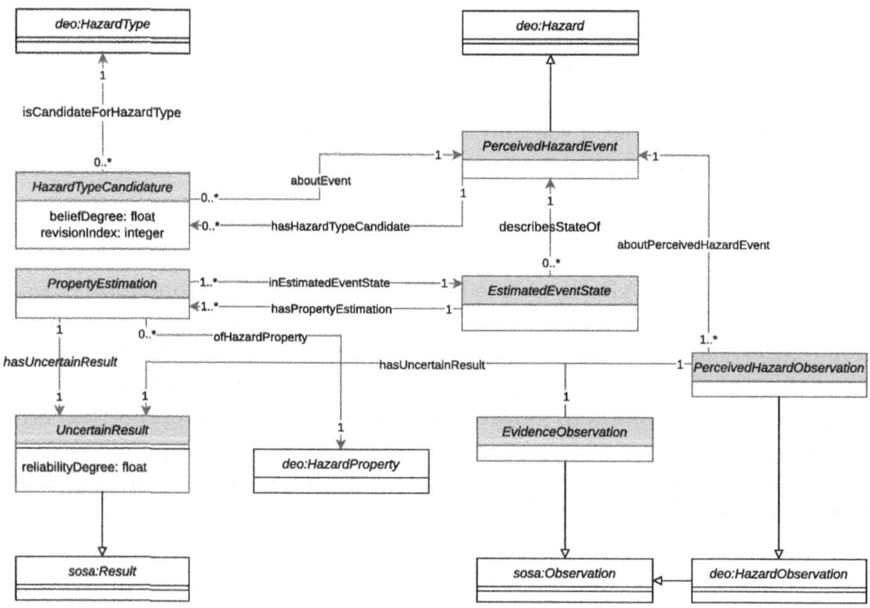

Fig. 1. Concepts and relations introduced (in color) in the proposed R-COOP ontology.

Perceived Hazard Event. The ontology introduces *PerceivedHazardEvent* to represent a hazard occurrence as it is observed, identified, and refined by the

IoMT system. When an event is first perceived, its existence, type, and property values are not yet fully established. An instance of *PerceivedHazardEvent* therefore represents an evolving and partially known entity, and serves as an anchor for instances of other concepts contributing to its interpretation.

As the perceived event functions as a stable referent over time, its characterization is allowed to evolve through newly arriving observations, competing hazard-type hypotheses, and successive estimations of its state. As seen on Fig. 1, *PerceivedHazardEvent* is aligned with *deo:Hazard* and is constrained to be supported by at least one event-level observation. This ensures that no event individual exists without observational grounding, while still maintaining incompleteness under the Open World Assumption (i.e., some properties may remain unknown or only partially supported, and additional observations may arrive later).

Two Observation Classes. To differentiate between raw or pre-event sensing evidence and event-level interpretations, the ontology defines two sub-concepts of *sosa:Observation.*

EvidenceObservation represents an observation acquired from sensor data without committing to the existence of a detected hazard event. Such observations may be linked through *sosa:hasFeatureOfInterest* relation to FOIs represented by spatial entities (e.g., points, regions, grid cells) and their observable properties. Instances of this class are intended for early perception stages, in which sensors report signals—whether routine or potentially relevant—before a coherent event individual has been formed or asserted.

The second sub-concept - *PerceivedHazardObservation* - represents an observation explicitly interpreted as being about a perceived hazard event. It is not declared disjoint with *EvidenceObservation* concept. Once a hazard event is perceived, both newly created observations and previously asserted *EvidenceObservation* instances may be linked to that event through the object property *aboutPerceivedHazardEvent*, that has *PerceivedHazardObservation* as its domain and *PerceivedHazardEvent* as range:

$$PerceivedHazardObservation \sqsubseteq (aboutPerceivedHazardEvent = 1\,PerceivedHazardEvent) \tag{1}$$

$$PerceivedHazardEvent \sqsubseteq \exists\, aboutPerceivedHazardEvent^{-}.PerceivedHazardObservation \tag{2}$$

Under these class axioms and property domain restrictions, the presence of the relation *aboutPerceivedHazardEvent* implies classification of the observation as a *PerceivedHazardObservation*, while preserving its original typing. This design supports monotonic knowledge enrichment: observations are neither updated nor replaced, but instead receive additional semantic characterization over time.

The two-layer observation pattern directly supports asynchronous IoMT pipelines. *EvidenceObservation* instances may accumulate initially, after which one or more *PerceivedHazardEvent* individuals can be introduced, and a selected set of observations can be re-contextualized at the event level via *PerceivedHazardObservation.*

Estimations. After a *PerceivedHazardEvent* is identified, knowledge about it is recorded across multiple *PerceivedHazardObservation* instances. To support incremental revision of what is known about a given hazard event, the ontology introduces the concept *EstimatedEventState*, intended to represent a snapshot-like summary of the system's best available knowledge about the perceived event at a specific time.

Each *EstimatedEventState* describes exactly one *PerceivedHazardEvent* and must include at least one *PropertyEstimation*:

$$EstimatedEventState \sqsubseteq (describesStateOf = 1 PerceivedHazardEvent) \quad (3)$$

$$EstimatedEventState \sqsubseteq (\exists hasPropertyEstimation.PropertyEstimation) \quad (4)$$

PropertyEstimation concept represents an estimated value for a particular hazard property associated with the perceived event. An instance of this concept is derived by aggregating the values from the results of *PerceivedHazardObservation* associated with the relevant property made about the perceived event up to a given time. Each belongs to exactly one estimated event state, refers to exactly one hazard property and carries exactly one result (discussed in the following subsection):

$$PropertyEstimation \sqsubseteq (inEstimatedEventState = 1 EstimatedEventState) \quad (5)$$

$$PropertyEstimation \sqsubseteq (ofHazardProperty = 1 deo : HazardProperty) \quad (6)$$

These constructs are modeled as standalone classes (subclasses of *owl:Thing*) and allow the ontology to record a traceable evolution of knowledge about the perceived event. The current version of the ontology does not model a specific provenance between PropertyEstimation and the set of observations that has been used to calculate it As new observations arrive asynchronously— whether periodically or triggered by specific conditions—successive *EstimatedEventState* may be linked to the same *PerceivedHazardEvent*.

The design also ensures that incompleteness of knowledge at a certain point in time is explicitly assessable: if no reliable *PropertyEstimation* is associated with a relevant *deo:HazardProperty*, noting this absence becomes useful when the application needs to expose gaps.

It is essential to associate temporal context with each *EstimatedEventState*. The ontology does not impose a specific temporal encoding format, but is designed to accommodate temporal information via OWL-Time constructs (e.g., *time:hasTime*) attached to each state. This enables temporal reasoning, including the identification of the most recent state for a given event.

UncertainResult. The ontology introduces *UncertainResult* as a convenience sub-concept of *sosa:Result* to standardize the attachment of both a numeric value for the result, and an optional explicit reliability annotation. In accordance with the SSN/SOSA design, results are represented as explicit individuals.

The estimate of quality of the associated information is captured through the datatype property *reliabilityDegree*, which has maximum cardinality of 1 and an xsd:float range. This property is intended to encode a normalized value in the interval [0,1], without committing to a specific computation method. Such values may originate from multiple non-restricted selection of factors, such as sensor metadata, statistical confidence measures, quality assessment mechanisms, or higher-level trust models.

To integrate this concept uniformly within the SOSA framework, the ontology introduces the object property *hasUncertainResult*, defined as a sub-property of *sosa:hasResult*. Its domain is the union of *EvidenceObservation*, *PerceivedHazardObservation*, and *PropertyEstimation*, allowing uncertainty-aware results to be attached consistently across direct sensor observations and higher-level estimations. For *PropertyEstimation*, the following restriction is specified:

$$PropertyEstimation \sqsubseteq (hasUncertainResult = 1\, UncertainResult) \quad (7)$$

The other two domain classes inherit result-related constraints from their parent concepts.

Hazard Type Candidature. In monitoring scenarios, FOIs are characterized not only by measurable properties, but also by higher-level, inferable characteristics. In the hazard monitoring use case, cooperative perception must address not only continuous monitoring, but also hazard type identification. The classification of a perceived event determines which hazard properties are relevant, how evidence observations should be interpreted, and how downstream decisions are made.

To capture such hypotheses, the ontology introduces the concept *HazardTypeCandidature*, which links a perceived hazard event to a candidate hazard type from the reused hazard vocabulary. Each candidature associates exactly one event with exactly one hazard type:

$$HazardTypeCandidature \sqsubseteq (aboutEvent = 1 PerceivedHazardEvent) \quad (8)$$

$$HazardTypeCandidature \sqsubseteq (isCandidateForHazardType = 1 deo : HazardType) \quad (9)$$

Each instance may optionally carry a *beliefDegree* datatype property (maximum cardinality 1, range xsd:float within [0,1]), representing the epistemic support for that specific hazard-type hypothesis. Multiple candidatures may coexist for the same event, thereby explicitly modeling ambiguity and supporting incremental revision rather than one-shot determination.

The ontology remains agnostic to how *beliefDegree* values are computed. Its purpose is to preserve the structure of hypotheses and enable querying and systematic updates.

To represent hypothesis evolution over time, the ontology introduces the datatype property *revisionIndex*, with domain restricted to *HazardTypeCandidature* and range set to *xsd:integer*. For each *PerceivedHazardEvent*, the initial hazard-type candidatures are assigned *revisionIndex* = 1. Whenever a candidature is superseded by a revised hypothesis for the same event, a new *HazardTypeCandidature* instance is created with *revisionIndex* incremented by one relative to the previous revision (2, 3, ...). This allows for a traceable ordered lineage of belief updates.

$$HazardTypeCandidature \sqsubseteq (beliefDegree \leq 1 xsd : float) \quad (10)$$

$$HazardTypeCandidature \sqsubseteq (revisionIndex = 1 xsd : integer) \quad (11)$$

Although motivated by the hazard monitoring use case, this pattern is intentionally generic. In other IoMT domains, a similar candidature construct may be used to represent inferable and revisable characteristics of perceived FOIs, while maintaining conceptual separation from measurable property observations and estimations.

It must be noted, that under OWL semantics, if multiple instances are asserted a property constrained by a functional (=1) cardinality restriction, a reasoner may infer their equality instead of signaling an inconsistency, unless this is prevented by explicit disjointness axioms. Thus, while this modeling pattern is common in the extended core ontologies like SOSA and is used in R-COOP in multiple places, it requires careful use to avoid unintended inferences.

4 Validation and Discussion

The scope of this paper is to demonstrate the suitability of R-COOP as a knowledge representation layer for cooperative perception in IoMT systems. Validation therefore targets representational adequacy. Specifically, we verify *a) the internal coherence of the newly introduced classes and relations when combined with the reused vocabularies*, and *b) the ability of the resulting knowledge graph to represent and retrieve the key knowledge aspects required for incremental cooperative perception under partial and asynchronous observations.*

To exercise each modeling pattern introduced in Sect. 3, we instantiated an A-box scenario from wildfire-monitoring data provided by the European Forest Fire Information System (EFFIS) [18] for July 2025 in Mediterranean countries. We used burnt-area (BA) perimeters as a registry of reference wildfire events, and MODIS thermal anomaly (TA) detections, as incoming sensor-derived evidence observations.

A-box individuals derived directly from data were created using the mapping in the Table 1. Each BA record yields one *PerceivedHazardEvent* with a polygon geometry and temporal scope, while each TA record yields a point feature of interest and one *EvidenceObservation* per available hazard-property value, each linked to an *UncertainResult* carrying the numeric value and, when provided, a reliability degree.

Links between evidence and events were not manually asserted. Instead, *aboutPerceivedHazardEvent* relations were materialised by a GeoSPARQL-enabled SPARQL query executed in GraphDB that checked whether detection points fall both within the event perimeter and within its temporal scope. These inferred relations enabled the Pellet DL reasoner to reclassify the linked observations as *PerceivedHazardObservation*-s, demonstrating inference-based evolution of knowledge from evidence to event-level interpretation.

Table 1. Condensed mapping from EFFIS attributes to A-box constructs.

EFFIS attributes	A-box individuals/triples created directly from data
BA perimeter + dates (WKT, start/end)	One `PerceivedHazardEvent` per BA record, with polygon `geo:Geometry/geo:asWKT` and temporal scope (OWL-Time).
TA location + time (lat/lon, acquisition time)	One spatial FOI per TA record (point `geo:Geometry`) and temporal anchoring (`sosa:resultTime`).
TA measured fields (e.g., FRP, MIR/TIR brightness)	One `EvidenceObservation` per property value with `sosa:observedProperty` (`deo:HazardProperty`) and `hasUncertainResult`;
TA platform + confidence (Terra/Aqua, confidence/quality)	`sosa:madeBySensor` links observations to sensor individuals; confidence mapped to `reliabilityDegree` on `UncertainResult` (normalised to [0,1] if needed).

Finally, to exercise revision and maintenance patterns, we created *HazardTypeCandidature* and *EstimatedEventState* instances as derived artifacts: candidatures (2–3 competing hazard types) were assigned revision indexes and belief degrees were assigned in order to exercise hypothesis evolution. The event states were computed by querying *PerceivedHazardObservation*-s available up to selected time instances about the event and by aggregating property values and reliability degrees (mean aggregation used for demonstration). To support CQ4 testing controlled cases of missing and low-reliability estimations were introduced in the latest state of few events.

Logical coherence of the populated knowledge graph was verified with Pellet DL reasoner in Protege tool with no inconsistencies or unsatisfiable classes and with implicit knowledge obtained as described previously with observation classification. Finally, for the CQs stated in Sect. 3, we implemented corresponding SPARQL queries to confirm that the ontology supports retrieval of the information needed for cooperative perception workflows. The queries are available at the remote repository and below the query for CQ4 (containing a subquery for CQ3) is provided.

```
PREFIX r-coop:  <http://www.semanticweb.org/ontologies/Retention/R-COOP#>
PREFIX deo:     <http://knowwheregraph/ontology/deo#>
PREFIX sosa:    <http://www.w3.org/ns/sosa/>
PREFIX time:    <http://www.w3.org/2006/time#>
PREFIX qudt:    <http://qudt.org/3.1.10/schema/qudt#>
PREFIX xsd:     <http://www.w3.org/2001/XMLSchema#>

SELECT ?hazType ?hazProp ?pe ?val ?rel ?issue WHERE {

  # latest state for event2
  {
    SELECT ?state WHERE {
      ?state a r-coop:EstimatedEventState ;
             r-coop:describesStateOf r-coop:event2 .
      OPTIONAL {
        ?state time:hasTime ?ti .
        ?ti time:inXSDDateTimeStamp ?t .
      }
    }
    ORDER BY DESC(?t)
    LIMIT 1
  }

  # latest revision index for candidatures of event2
```

```
  {
    SELECT (MAX(?r0) AS ?rev) WHERE {
      ?c0 a r-coop:HazardTypeCandidature ;
          r-coop:aboutEvent r-coop:event2 ;
          r-coop:revisionIndex ?r0 .
    }
  }

  # active candidatures (latest revision) with belief >= 0.6
  ?cand a r-coop:HazardTypeCandidature ;
        r-coop:aboutEvent r-coop:event2 ;
        r-coop:revisionIndex ?rev ;
        r-coop:isCandidateForHazardType ?hazType ;
        r-coop:beliefDegree ?b .
  FILTER ( ?b >= 0.6 )

  # expected properties from the selected hazard types
  ?hazType deo:hasHazardProperty ?hazProp .

  # property estimation if present in latest state
  OPTIONAL {
    ?state r-coop:hasPropertyEstimation ?pe .
    ?pe r-coop:ofHazardProperty ?hazProp ;
        r-coop:hasUncertainResult ?r .
    OPTIONAL { ?r qudt:numericValue ?val . }
    OPTIONAL { ?r r-coop:reliabilityDegree ?rel . }
  }

  # issue classification
  BIND(
    IF( !BOUND(?pe),
        ‘‘missing",
        IF( !BOUND(?rel) || ?rel < 0.8,
            ‘‘lowReliability",
            ‘‘ok"
        )
    ) AS ?issue
  )
  FILTER ( ?issue != ‘‘ok" )
}
ORDER BY ?hazProp
```

Listing 1. SPARQL query implementing CQ4 (gap exposure in the latest maintained state for `r-coop:event2`).

To answer the question, the query needs to retrieve the latest *EstimatedEventState* instance associated with the given event, as well as the active hazard type hypotheses from CQ3 (it retrieves retrieves *HazardTypeCandidature* instances with the highest *revisionIndex*). Based on values of *reliabilityDegree* property in the results of *PropertyEstimation*-s in the event state, unreliable values are filtered based on a reliability threshold. The lacking properties can be identified as the ones connected through the *deo:hasHazardProperty* to the hazard types from active *HazardTypeCandidature* instances (that are equally filtered by the threshold on *beliefDegree* property value) associated with the given event (Listing 1).

Together, these results support the internal coherence of the introduced vocabulary and its adequacy for the intended knowledge retrieval tasks.

5 Conclusion

Cooperative perception in the IoMT environments requires a framework for a coherent and evolving representation of spatio-temporal phenomena from partial, uncertain and asynchronous observations of the dynamic environment. The proposed R-COOP ontology in this article, demonstrated in a hazard event monitoring scenario, a SOSA-aligned modeling layer that separates pre-event evidence from event-level interpretations by introducing two distinct Observation sub-concepts. It represents abstract characteristics as explicit and revisable hypotheses, by dedicating a separate class and property patterns for them. Finally, it supports incremental refinement of the knowledge about the perceived phenomena through the usage of snapshot-like state concepts connected to uncertainty-aware property estimation results.

The design choices have been deliberately made to preserve use-case agnosticism and interoperability. For example, newly introduced *beliefDegree* and *reliabilityDegree* datatype properties provide a lightweight, query-able representation of uncertainty/quality, but they do not constitute a full probabilistic semantics. Overall, these aspects indicate where further extensions are required for the integration of the ontology in a multi-stage perception pipeline.

Future work will focus on operationalization of R-COOP within an end-to-end cooperative perception framework. To be able to use it in real life scenarios, the probabilistic reasoning mechanisms that can capture IoMT-relevant uncertainty factors and propagate them through the knowledge base need to be integrated. Trust in sensing devices and sources can be incorporated as one of such factors, but would require extensions to trace how observations, hypotheses, and estimations are produced and revised. Most prominently, to support continuous perception from real data streams, rule-based and/or stream-reasoning frameworks can be integrated with the ontology.

References

1. Ayoub, W., Samhat, A.E., Nouvel, F., Mroue, M., Prévotet, J.C.: Internet of mobile things: overview of LoRaWAN, DASH7, and NB-IoT in LPWANs standards and supported mobility. In: 2018 25th International Conference on Telecommunications (ICT). IEEE, St. Malo, France (2018)
2. Bellenger, A., Gatepaille, S.: Uncertainty in ontologies: dempster-shafer theory for data fusion applications. arXiv:1106.3876 (2011)
3. Broedermann, T., Sakaridis, C., Fu, Y., Gool, L.V.: CAFuser: condition-aware multimodal fusion for robust semantic perception of driving scenes. arXiv:2410.10791 (2025)
4. Cao, H., Wachowicz, M.: A holistic overview of anticipatory learning for the internet of moving things: research challenges and opportunities. ISPRS Int. J. Geo Inf. **9**(4) (2020)
5. Cao, H., Wachowicz, M., Renso, C., Carlini, E.: An edge-fog-cloud platform for anticipatory learning process designed for internet of mobile things. arXiv:1711.09745 (2018)
6. Costa, P.C.G., Laskey, K.B.: PR-OWL: a framework for probabilistic ontologies (2006)
7. Gerla, M., Lee, E.K., Pau, G., Lee, U.: Internet of vehicles: from intelligent grid to autonomous cars and vehicular clouds. In: 2014 IEEE World Forum on Internet of Things, pp. 241–246. IEEE, Séoul, South Korea (2014)
8. Giustozzi, F., Saunier, J., Zanni-Merk, C.: Abnormal situations interpretation in industry 4.0 using stream reasoning. Procedia Comput. Sci. **159**, 620–629 (2019)
9. Greco, L., Ritrovato, P., Xhafa, F.: An edge-stream computing infrastructure for real-time analysis of wearable sensors data. Futur. Gener. Comput. Syst. **93**, 515–528 (2019)

10. Gruber, T.R.: A translation approach to portable ontology specifications. Knowl. Acquis. **5**(2), 199–220 (1993)
11. Hippolyte, J.L., et al.: Using ontologies to create machine-actionable datasets: two case studies. Metrology **3**(1), 65–80 (2023)
12. Janowicz, K., Haller, A., Cox, S.J.D., Le Phuoc, D., Lefrançois, M.: SOSA: a lightweight ontology for sensors, observations, samples, and actuators. J. Web Semant. **56**, 1–10 (2019)
13. Kabir, H., Tham, M.L., Chang, Y.C.: Internet of robotic things for mobile robots: concepts, technologies, challenges, applications, and future directions. Digit. Commun. Networks **9**(6), 1265–1290 (2023)
14. Kumar, M., Singh, P.K., Maurya, M.K., Shivhare, A.: A survey on event detection approaches for sensor based IoT. Internet of Things **22**, 100720 (2023)
15. Nahrstedt, K., Li, H., Nguyen, P., Chang, S., Vu, L.: Internet of mobile things: mobility-driven challenges, designs and implementations. In: IEEE First International Conference on Internet-of-Things Design and Implementation (IoTDI), pp. 25–36 (2016)
16. Pliatsios, A., Kotis, K., Goumopoulos, C.: A systematic review on semantic interoperability in the IoE-enabled smart cities. Internet of Things **22**, 100754 (2023)
17. Roda, F., et al.: Towards a semantic sensor spectral data framework: a case for in-situ water spectroradiometry. In: 2024 IEEE Biennial Congress of Argentina (ARGENCON), pp. 1–8 (2024)
18. San-Miguel-Ayanz, J., et al.: Comprehensive monitoring of wildfires in Europe: the European forest fire information system (EFFIS). In: Tiefenbacher, J. (ed.) Approaches to Managing Disaster - Assessing Hazards, Emergencies and Disaster Impacts, pp. 87–105. InTech (2012)
19. Shimizu, C., et al.: The KnowWhereGraph ontology. J. Web Semant. **84**, 100842 (2025)
20. Somov, A., Dupont, C., Giaffreda, R.: Supporting smart-city mobility with cognitive Internet of Things. In: 2013 Future Network & Mobile Summit, pp. 1–10 (2013)
21. Steenwinckel, B., et al.: FLAGS: a methodology for adaptive anomaly detection and root cause analysis on sensor data streams by fusing expert knowledge with machine learning. Future Gener. Comput. Syst. **116**, 30–48 (2021)
22. Stephen, S., et al.: The HIP ontology: a formal framework to support disaster risk reduction and management (2024)
23. Wang, C., Chen, N., Wang, W., Chen, Z.: A hydrological sensor web ontology based on the SSN ontology: a case study for a flood. ISPRS Int. J. Geoinf. **7**(1) (2017)
24. Yang, K., et al.: Spatio-temporal domain awareness for multi-agent collaborative perception. arXiv:2307.13929 (2023)

AI-Driven Automation and Optimization

Cognitive Load Estimation Through Eye-Tracking in Industrial Tasks: A Pilot Study

Kanan Gurbanov[1,4](✉), Latafat Qardashova[1], Gregoire Chabrol[3,4], Samy Rima[2], and Rabih Amhaz[3,4]

[1] Azerbaijan State Oil and Industry University, Baku, Azerbaijan
kanan.gurbanov@ufaz.az, l.qardashova@asoiu.edu.az
[2] Human-IST Laboratory, University of Fribourg, Fribourg, Switzerland
samy.rima@unifr.ch
[3] Icam site de Strasbourg-Europe, Schiltigheim, France
rabih.amhaz@unistra.fr
[4] ICube Laboratory, UMR 7357, University of Strasbourg, Strasbourg, France

Abstract. Understanding and managing cognitive workload is critical for safety and efficiency in industrial settings, yet practical measurement on the factory floor remains challenging. We present a pilot feasibility study (n = 1) that explores whether vision-based signals can provide interpretable indicators related to workload in a real assembly scenario. Our approach combines mobile eye tracking with hand tracking and object detection, intentionally avoiding intrusive physiological sensors such as electroencephalography (EEG). An expert performed the same cart assembly task under two controlled workspace layouts: organized and disorganized. The disorganized layout induced stronger visual search behavior and attentional dispersion, evidenced by increased fixation and saccade counts, broader gaze dispersion, and a higher proportion of fixations and dwell time on irrelevant areas. In contrast, pupil diameter and blink rate remained approximately stable for this expert participant, suggesting that the observed differences primarily reflect increased extraneous visual-perceptual demand rather than elevated arousal. Rather than claiming a definitive cognitive load score, this work provides a deployable, vision-only pipeline and a set of interpretable gaze-hand-object indicators that can be used to study workload-related effects *in situ*. Larger-scale validation with repeated trials and additional participants is part of ongoing work.

Keywords: Cognitive Workload · Eye Tracking · Object Detection · Industrial Ergonomics · Human-Machine Interaction

1 Introduction

Cognitive overload occurs when an operator's mental resources are exceeded by task demands, often leading to errors or slowdowns. In industrial environments,

C. Zanni-Merk et al. (Eds.): AIDT 2026, CCIS 3023, pp. 51–64, 2026.
https://doi.org/10.1007/978-3-032-31319-5_4

human performance under high cognitive load is a major factor in safety and productivity. Traditional measures (e.g., EEG-based monitoring) are impractical for real-time use on a production floor. There is a growing need for lightweight, real-time workload measures that can be seamlessly integrated into industrial workflows.

Many existing frameworks for workload detection rely on multimodal biosignals (EEG combined with eye tracking or other sensors) and complex lab setups. Moreover, prior models often do not target the predominantly visual-perceptual nature of tasks like assembly and inspection. In such tasks, an operator's cognitive load may stem largely from visual search and attention management, rather than abstract reasoning alone. This work explores whether reliable indicators of cognitive workload can be obtained from vision-based data alone, specifically using an operator's eye movements, hand interactions, and observed tool usage, without the need for EEG or other invasive sensors.

The main research question in this paper is: Can a non-intrusive, vision-only system, integrating mobile eye-tracking, hand tracking, and object detection, effectively distinguish and quantify differences in cognitive workload during real-world industrial assembly tasks?

We developed a proof-of-concept approach combining mobile eye tracking with computer vision to monitor a worker's visual attention and actions during an assembly task. An expert technician performed the same cart assembly task under organized and disorganized conditions to isolate the effect of workspace layout on cognitive load.

We hypothesized that the unorganized workspace would impose higher visual-perceptual demand, observable as more extensive visual search behavior (e.g., more fixations, wider gaze spread, longer task time), whereas the organized workspace would allow more efficient and focused visual attention.

In this paper, we report how eye-tracking metrics and interaction data differ between these conditions and how they can be interpreted as indicators of cognitive workload. To our knowledge, this is one of the first in-situ demonstrations combining mobile eye tracking with scene object detection to study workload-related attention patterns in an industrial assembly task. The results show that subtle changes in workspace layout can be quantified through vision-based measures. Our work contributes an integrated analysis pipeline for multimodal workload monitoring and empirical evidence that a non-intrusive, video-based approach can capture cognitive load differences in real manufacturing conditions.

2 Related Work

This section reviews relevant literature in three domains: (1) eye movement as an indicator of cognitive load, (2) multimodal and physiological workload measurement, and (3) vision-based monitoring in industrial settings. We summarize prior achievements and identify the remaining gaps that motivate our study.

2.1 Eye Movement as a Proxy for Cognitive Load

Eye movements (fixations, saccades, pupil dilation, microsaccades) are widely used in cognitive and human-computer interaction (HCI) studies to infer mental processing (e.g., decision-making, memory retrieval), as gaze is tightly linked to attention and information acquisition. A classic study by Krejtz et al. [12] compared pupil dilation and microsaccade metrics in arithmetic tasks, finding that both signals discriminated task difficulty, though each had distinct sensitivity profiles. Many studies assume a monotonic increase in pupil diameter with load (the so-called task-evoked pupillary response).

More recently, some works differentiate intrinsic vs. extraneous cognitive load using oculometric signals. For instance, a 2025 study used eye tracking, heart-rate variability, and galvanic skin response to classify intrinsic vs. extraneous load in multimedia tasks, achieving promising predictive power [6]. This is relevant: our manipulation (workspace clutter) is akin to extraneous load, and we show gaze metrics reflect it (even when physiological signals do not shift strongly). In serious game contexts, other researchers have grounded measurement in theory, calibrating eye metrics to time-based resource-sharing models of load [19]. For example, a study with 42 participants playing time-critical resource-management games mapped attentional demand to gaze statistics. Such work shows the feasibility of interpreting gaze within formal models of load, but still typically in constrained, quasi-lab settings rather than real-world manufacturing.

Further, Martinez-Cedillo et al. [15] demonstrated that real-world tasks performed under cognitive load (e.g., sandwich making with a concurrent mental task) result in more fixations on irrelevant objects and reduced eye-hand span, meaning participants looked less ahead toward upcoming subtasks. These findings support the idea that increased cognitive demand impairs anticipatory gaze behavior and causes attentional shifts toward distractors, which are patterns observable in cluttered industrial environments.

Notably, recent deep learning approaches have begun to incorporate contextual information from the scene into gaze estimation. Hu et al. [10] proposed *HOIGaze*, a learning-based gaze estimation framework that explicitly leverages eye-hand-head coordination during hand-object interactions. Their hierarchical model first infers which hand is being attended to, and then fuses head pose, hand appearance, and surrounding object features using cross-modal Transformer blocks. HOIGaze achieved significant accuracy gains over appearance-only gaze estimators in extended reality tasks, highlighting the importance of hand-object context for robust gaze inference. While developed for extended reality XR scenarios, this approach demonstrates how modeling perceptual-motor coupling can enhance gaze-based analysis in complex, real-world tasks.

Despite these advances, most gaze-based cognitive load studies remain confined to laboratory or interface-driven settings. There is still a lack of work applying gaze analysis in physical industrial tasks, particularly in combination with scene understanding (e.g., hands and tools) to localize the sources of visual workload.

2.2 Multimodal and Physiological Workload Measurement

To capture cognitive load robustly, many works fuse multiple sensors. For example, eye tracking combined with EEG or electrocardiography (ECG) often yields higher classification accuracy in cognitive load estimation than any single modality alone. One survey notes that fusion of eye, EEG, and galvanic skin response (GSR) is common in ergonomics and usability studies [20]. Ding et al. [4] proposed CLERA, a unified deep model for joint eye-region analysis and cognitive load estimation "in the wild". CLERA jointly learns eye landmarks, blink prediction, pupil estimation, and cognitive load, and shows improved robustness over isolated pipelines. However, CLERA is built for HCI-style tasks, not industrial hand-tool workflows, and it does not explicitly model object interactions or hand movement in the scene, which are crucial in assembly tasks.

Another approach, in the virtual reality (VR) and training domain, fuses gaze with heart rate variability (HRV) and other physiological signals to detect both cognitive load and stress, enabling adaptive systems [16]. While promising, such systems are often difficult to deploy in real factories due to sensor intrusiveness, calibration requirements, and sensitivity to motion artifacts. Thus, although multimodal methods improve robustness, they often sacrifice deployability. Our approach intentionally remains vision-only, trading off some signal richness for practical applicability in industrial environments.

2.3 Vision-Based Monitoring in Industrial Environments

A separate but related branch of research uses computer vision to monitor worker behavior, safety compliance, and ergonomics. In construction and manufacturing, vision systems have been used to detect helmet usage, unsafe postures, or proximity violations [3]. Papoutsakis et al. [17] introduced a vision-based system for human behavior monitoring in a car door assembly line, combining multi-camera video and 3D motion capture to build the CarDA dataset for analyzing assembly actions. However, this work focuses on *what* actions are performed rather than *why* they may be difficult from a cognitive perspective.

Other vision-based systems target ergonomic risk assessment using skeleton tracking and motion features. For example, Agostinelli et al. [1] proposed a semi-automated ergonomic risk assessment framework using computer vision in real manufacturing environments. Such systems estimate physical or biomechanical load but do not capture visual attention or cognitive workload. Egeonu and Jia [5] confirmed that most computer-vision-based biomechanical and workload assessment approaches infer physical strain (e.g., posture or joint load proxies) rather than perceptual or cognitive demand.

In parallel, deep learning-based vision systems have emerged for fine-grained recognition of assembly actions through explicit modeling of hand-object interactions. Li et al. [14] proposed a multi-stream neural architecture (MHOINet) that jointly models spatial hand-object relationships, visual appearance of tools, and temporal hand motion patterns to recognize assembly actions. Their approach achieved high accuracy on both coarse and fine-grained assembly action classification in factory settings. Similarly, Ay and Emel [2] introduced a hybrid framework combining YOLO-based object detection with LSTM-based analysis of hand pose sequences for sequential task recognition in industrial environments. These methods demonstrate that vision-only systems can robustly model manual actions and tool usage, but they do not address cognitive load or attention dynamics.

Another relevant contribution is by Vasta et al. [21], who used a standard RGB webcam to estimate blink rate as a proxy for cognitive workload during manufacturing-related tasks. Their results were comparable to those obtained with dedicated eye trackers, supporting the feasibility of low-cost, vision-based cognitive state monitoring.

In summary, prior work provides three complementary but largely disconnected lines of evidence. First, eye-movement research confirms that oculometrics such as fixation dynamics, saccadic activity, and gaze dispersion are sensitive to changes in task demand, but these results are more established in laboratory or interface-driven paradigms rather than physical shop-floor tasks. Second, multimodal approaches that combine eye tracking with physiological sensors (e.g., EEG/ECG/GSR) often improve robustness and classification accuracy, yet they remain difficult to deploy in real industrial settings due to intrusiveness, calibration constraints, and sensitivity to motion and environmental variability. Third, vision-based industrial monitoring has matured for action recognition, safety, and ergonomics, including hand–object interaction modeling, but typically does not target cognitive workload and very rarely integrates gaze to define the attentional sources of difficulty. Therefore, a key gap remains in demonstrating an *in-situ*, deployable, vision-only approach that jointly links gaze behavior, hand activity, and scene-level tool/part understanding to quantify workload differences under controlled environmental manipulations (e.g., workspace clutter). Our study addresses this gap by combining mobile eye tracking with hand tracking and industrial object detection during a real assembly task.

3 Method

3.1 Participants and Task

We recruited one expert assembly technician, Robert Bolusset (LEAN manufacturing consultant and trainee), to perform a cart assembly task under two workspace conditions: unorganized and organized using LEAN Manufacturing principles, more precisely Kaizen (Fig. 1). In the unorganized condition (Fig. 1(a)), tools and parts were not maintained in fixed positions, and not all tools were placed in visible locations, other tools not used are in the working space, which will increase visual search and repeated reach-and-retrieve actions. In contrast, in the organized condition workspace (Fig. 1(b)), items were arranged in stable and predictable positions within the visible reach in front of the operator, reducing search effort. The assembly procedure remained identical in both trials to isolate the effect of workspace layout.

parts are not placed in fixed, visible locations on the workbench, requiring additional visual search and repeated reach-and-retrieve actions.

(b) Organized workspace. Tools and parts are arranged in stable, predictable, and mostly visible positions (bench/pegboard bins), supporting fast access and reducing search.

Fig. 1. Workspace conditions used in the experiment.

3.2 Tools and Data Collection

The participant wore a mobile Pupil Labs Neon eye tracker which provides binocular infrared eye video and gaze estimation at high temporal resolution (eye cameras at 200 Hz) alongside an egocentric Red-Green-Blue (RGB) scene video stream (30 Hz). In addition, inertial measurement units (IMU) are available for head-motion context. The lightweight, glasses-mounted form factor supports natural movement during real manual assembly without the intrusiveness of physiological sensors [18].

For scene understanding, we processed the recorded frames with MediaPipe for hand detection and YOLOv8 for tool/part detection. The YOLOv8 model was custom-trained on over 850 annotated images across 11 workshop-specific object classes, reaching approximately 87% mean average precision (mAP) on the validation split. As summarized in Fig. 2, we temporally aligned the gaze/fixation stream with hand and object detections at the frame level and aggregated all signals into a structured JavaScript Object Notation (JSON) timeline dataset used for metric extraction. Audio was extracted from the recording for completeness, but it is not used in the analyses reported in this paper.

3.3 Measures and Analysis

From the timeline, we extracted the following measures:

- **Eye metrics:** total fixations, average fixation duration, saccade count, gaze dispersion, pupil diameter, blink rate.
- **Attention allocation:** proportion of fixations on relevant vs. irrelevant areas, area-of-interest (AOI) transition frequency, and dwell times on key objects.
- **Hand and coordination metrics:**

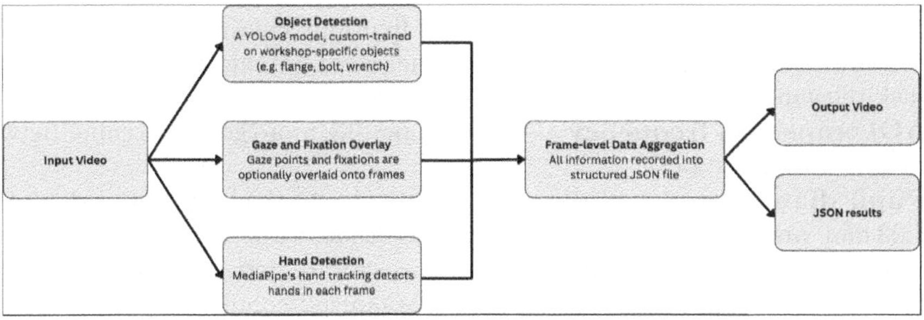

Fig. 2. Overview of the data-processing pipeline. From the input recording, we extract frame-level signals: object detections (YOLOv8), hand detections (MediaPipe), and gaze and fixation streams from the eye tracker. These signals are temporally aligned and aggregated into a structured JSON timeline for downstream metric computation.

  - *Hand Presence Ratio* - proportion of frames in which at least one hand is visible.
  - *Bimanual Ratio* - proportion of frames with both hands visible, indicating coordinated two-handed operation.
  - *Gaze-on-Hand Ratio* - proportion of frames where gaze falls within a hand bounding box, reflecting visual monitoring of manual actions.
  - *Hand-on-Relevant-Object Ratio* - proportion of frames where a hand intersects with detected task-relevant tools or parts.
  - *Coordinated Action Ratio* - proportion of frames where hand, gaze, and relevant object all co-occur, capturing full perceptual–motor alignment.
  - *Mean Hand Speed (pixels per second (px)/s)* - average wrist movement speed across frames.
- **Task performance:** total completion time, and any corrections or tool misselections.

4 Results

The disorganized workspace produced clear, consistent divergences in gaze behavior and attention distribution relative to the organized layout:

- **Fixation and saccade counts** increased by approximately 24%, from 494 to 613 fixations and from 493 to 612 saccades, indicating a more fragmented scanning strategy.
- **Average fixation duration** dropped slightly from about 563 ms (organized) to 550 ms (unorganized), reflecting shorter glimpses and more frequent transitions.
- **Gaze dispersion** expanded from a radius of roughly 158 px to 190 px, confirming broader spatial search.

- **Attention misallocation:** the share of fixations landing on irrelevant areas rose from 62.0% to 73.6%, indicating increased attention allocated to non-task-relevant regions under clutter.
- **AOI transition frequency** also rose, signaling more gaze switching between items and distractors.
- **Pupil diameter and blink rate** remained effectively unchanged (around 5.23 mm and 11 blinks/min), suggesting that while visual scanning load increased, overall arousal or stress did not escalate detectably for this expert.
- **Completion time** increased in the disorganized condition (consistent with increased visual workload), though no severe errors or tool misuses occurred, reflecting the participant's expertise.

Hand and Coordination Results: While the overall hand presence ratio remained stable across conditions (about 0.79), indicating comparable manual engagement, other metrics showed significant variation. The *bimanual ratio* was higher in the organized workspace (0.4951 vs. 0.4516), reflecting more efficient two-handed coordination. The *mean hand speed* increased substantially in the unorganized workspace (1070 px/s vs. 868 px/s), suggesting more erratic or compensatory movement.

Critically, both *gaze-on-hand ratio* (0.1518 vs. 0.1121) and *hand-on-relevant-object ratio* (0.1166 vs. 0.1010) were higher in the organized condition, indicating tighter alignment between attention and motor action. The most discriminative signal was the *coordinated action ratio* (simultaneous gaze, hand, and object engagement), which was more than twice as high in the organized layout (0.0796 vs. 0.0373). This drop can suggest that in cluttered environments, perceptual–motor coupling breaks down, forcing more exploratory behavior and increasing cognitive load.

Figure 3 provides a qualitative illustration of the proposed analysis pipeline, combining industrial object detection, hand tracking, and eye fixation visualization. Fixations are represented as purple circles, with circle size encoding fixation duration; arrows indicate gaze transitions between successive fixations.

In the organized workspace, fixations appear spatially concentrated around task-relevant objects during manipulation phases, with shorter gaze transitions and more frequent alignment between gaze, hands, and active tools. In contrast, the unorganized workspace exhibits a more scattered fixation pattern, with longer transitions between fixations and frequent shifts toward irrelevant areas. These visual patterns are consistent with the quantitative increase in fixation count, gaze dispersion, and AOI transitions reported above, and illustrate how workspace clutter induces a less stable and more exploratory visual strategy.

Figure 4 presents a quantitative comparison of total dwell time across task-relevant and irrelevant AOIs for the two workspace conditions. In the organized workspace, a larger proportion of gaze dwell time is allocated to relevant objects such as wheels, supports, and tools actively used during assembly. Conversely, the unorganized workspace shows a marked increase in dwell time on irrelevant

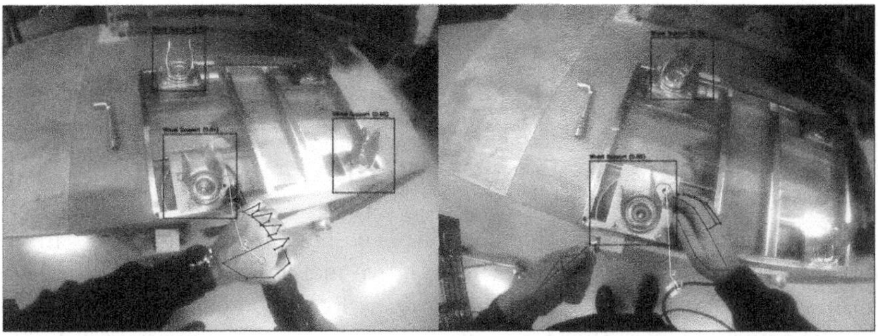

Fig. 3. Result visualization combining industrial object detection, hand tracking, and eye fixations and transitions (purple circles indicate fixations, circle size encodes fixation duration).

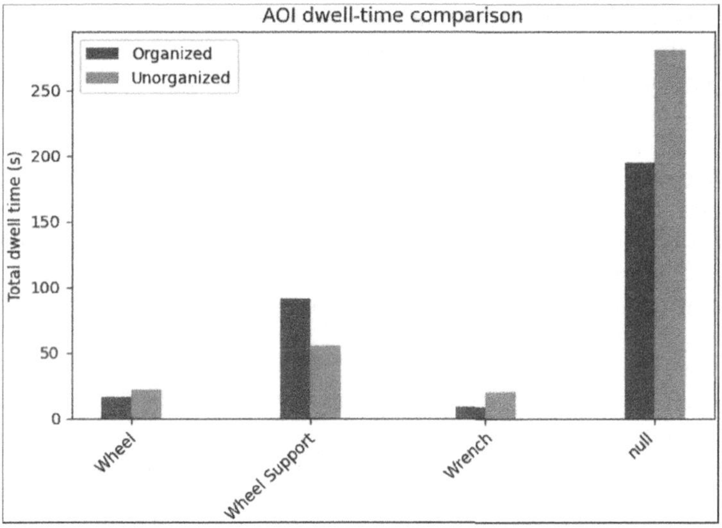

Fig. 4. AOI dwell-time comparison between unorganized (in orange) and organized (in blue) workplaces, correlated with industrial object detection. (Color figure online)

or null areas, reflecting additional visual search and attentional shifts caused by clutter.

This redistribution of attention provides direct evidence that workspace disorder increases extraneous visual demand. Rather than supporting task progression, a significant portion of visual attention is consumed by locating tools and resolving spatial ambiguity. These AOI-level results reinforce the fixation- and dispersion-based findings by localizing the source of increased visual workload to specific regions of the workspace.

These differences align with theoretical expectations: increasing extraneous visual demands (via clutter) forces more scanning and attentional shifts, which we were able to detect through gaze metrics alone. The absence of strong changes in pupil/blink measures implies that for an expert operator, the extra visual load remained manageable in terms of overall cognitive stress.

5 Discussion

Our pilot results indicate that workspace layout can directly influence visual/perceptual load, observable through eye metrics. In cognitive load terms, the disorganized layout introduced extraneous load, i.e., effort invested in navigating the environment rather than intrinsic task logic. The increase in fixations, saccades, gaze dispersion, and AOI switching signals this overhead.

Interestingly, the stable physiological metrics (pupil, blink) suggest that the mental load ceiling was not breached. Because our participant was an expert, they likely compensated for extra visual demands without triggering elevated arousal. This aligns with prior findings showing that expertise modulates the relationship between observable behavior and internal cognitive state, allowing skilled operators to absorb additional perceptual demands before physiological indicators of overload emerge [8].

Beyond gaze-based indicators, the observed differences are further clarified by the hand and coordination metrics. Measures such as bimanual usage and hand speed have been widely used in motor control and ergonomics research as proxies for task efficiency and movement organization [9]. In skilled manual tasks, coordinated two-handed actions and smoother, slower movements are commonly associated with stable motor planning, whereas faster or more variable movements often reflect compensatory behavior under uncertainty or increased task demands [7].

These eye-based results are further reinforced by hand and coordination metrics. In the organized condition, participants showed more frequent two-handed coordination, slower and more deliberate hand movement, and stronger gaze-hand coupling. The higher bimanual ratio observed in this condition is consistent with prior work on bimanual control, which shows that efficient two-handed coordination is a hallmark of skilled performance in assembly and manipulation tasks [9]. In contrast, the increased mean hand speed in the unorganized workspace suggests more corrective or reactive movements, likely driven by increased visual search and uncertainty in tool localization [1].

The gaze-on-hand and hand-on-relevant-object ratios provide additional insight into perceptual-motor coupling. Research on eye-hand coordination has shown that, during skilled actions, gaze typically anticipates and guides hand movements rather than merely monitoring them [11]. Higher values of these ratios in the organized workspace indicate tighter alignment between visual attention and manual execution, reflecting anticipatory control. Conversely, reduced alignment in the unorganized condition suggests a shift toward more reactive behavior, where attention is repeatedly redirected to support corrective actions [15].

The most informative signal, which is the coordinated action ratio capturing simultaneous gaze, hand presence, and relevant object interaction, can be interpreted as a composite indicator of perceptual-motor alignment. Similar notions of perceptual-motor coupling have been discussed in studies linking gaze behavior, object interaction, and action planning in natural tasks [13]. Its substantial reduction in the cluttered workspace suggests that environmental disorder disrupts the integration of perception and action, forcing a more fragmented control strategy. From a cognitive load perspective, this reflects an increase in extraneous load, where attentional resources are consumed by managing the environment rather than progressing through the task itself.

Relative to prior work, Krejtz et al. [12] demonstrated that pupil and microsaccade signals track mental difficulty, but they do so in controlled tasks, not physical tasks with environmental clutter. CLERA offers an elegant vision-only model for cognitive load in the wild [4], but lacks a coupling to object interaction or hands, something essential in assembly tasks. Works in industrial vision largely focus on action or safety detection rather than the underlying cognitive strain. For example, Papoutsakis et al. [17] focused on monitoring worker actions rather than attention or cognitive load.

Thus, our work treads a new space: vision-enabled cognitive load estimation in a manual industrial context. By explicitly linking gaze behavior to hand activity and object interaction, we show that eye-scene fusion can reveal not only increased visual search, but also degraded perceptual-motor coordination under cluttered conditions. These findings suggest that cognitive workload in industrial tasks manifests through both attentional and motor pathways, which can be jointly captured using non-intrusive, vision-based measures.

5.1 Limitations

While this study demonstrates that gaze behavior and hand-object coordination are sensitive to workspace organization in a real industrial task, several limitations must be acknowledged.

First, the experiment involved a single expert participant (pilot study), which limits generalization. As a result, the observed patterns primarily reflect strategies employed by a highly skilled operator, who may compensate for increased visual demands without exhibiting strong physiological responses. Prior research indicates that expertise significantly influences gaze behavior and eye-hand coordination. Consequently, novice operators may exhibit different relationships between visual, motor, and physiological indicators of workload. Future work will explicitly compare expert and novice populations to assess how these indicators evolve with skill level.

Second, task specificity represents a limitation. The study focused on a single assembly scenario with a fixed sequence of subtasks, enabling a controlled comparison between organized and unorganized workspace layouts. However, other industrial activities, such as inspection, wiring, or quality control, may involve different perceptual and motor demands. Extending the proposed analysis to a broader range of tasks is necessary to evaluate the generality of the observed effects.

Third, although the proposed pipeline was designed for real-world deployment, it remains sensitive to practical constraints inherent to vision-based systems. Factors such as partial hand occlusions, lighting variability, gaze calibration drift, and occasional object detection failures may introduce noise at the frame level. While these issues did not affect the overall trends reported in this study, future work will incorporate robustness analysis and confidence estimation to better quantify uncertainty in real operational settings.

Finally, the present study adopts a descriptive and exploratory analysis (without external ground-truth workload labels), prioritizing interpretability over predictive modeling. This design choice prioritizes interpretability and ecological validity but limits direct claims regarding classification accuracy or real-time workload estimation. Ongoing work investigates temporal modeling and learning-based approaches to support predictive and online workload monitoring.

6 Conclusion

This study investigated whether vision-based measures can reveal differences in visual workload induced by workspace organization during an industrial assembly task. Using a combination of mobile eye tracking, hand detection, and object recognition, we compared an organized and a cluttered workspace while keeping the task procedure constant.

The results show that the cluttered layout was associated with increased visual search behavior, reflected by higher fixation and saccade counts, greater gaze dispersion, and a larger proportion of attention allocated to irrelevant areas. In contrast, physiological indicators such as pupil diameter and blink rate remained stable for the expert participant, suggesting that the increased visual demand did not translate into elevated arousal or stress. Hand and coordination metrics further indicated reduced perceptual-motor alignment in the unorganized condition, highlighting the impact of environmental structure on gaze-hand-object coordination.

These findings support the use of vision-based analysis to study perceptual workload in realistic industrial settings without relying on intrusive physiological sensors. Rather than providing a definitive workload score, the proposed approach offers interpretable indicators that link visual attention and manual interaction to task conditions.

Ongoing and future work will extend this investigation to larger and more diverse participant groups, including novice operators, and to additional industrial tasks. Further efforts will also explore temporal modeling and real-time analysis to assess the feasibility of continuous workload monitoring in operational environments.

Acknowledgments. The authors would like to acknowledge Robert Bolusset for performing the industrial assembly experiment and for his assistance in registering and collecting the experimental data.

Disclosure of Interests. The authors declare that they have no competing interests relevant to the content of this article.

References

1. Agostinelli, T., Generosi, A., Ceccacci, S., Mengoni, M.: Validation of computer vision-based ergonomic risk assessment tools for real manufacturing environments. Sci. Rep. **14**(1), 27785 (2024). https://doi.org/10.1038/s41598-024-79373-4
2. Ay, M., Emel, E.: A hybrid vision-based framework for sequential industrial task recognition using object detection and hand pose analysis. J. Manuf. Syst. (2025). in Press
3. Cheng, J., Wong, P.K.Y., Luo, H., Wang, M., Leung, P.H.: Vision-based monitoring of site safety compliance based on worker re-identification and personal protective equipment classification. Autom. Constr. **139**, 104312 (2022). https://doi.org/10.1016/j.autcon.2022.104312
4. Ding, L., et al.: Clera: a unified model for joint cognitive load and eye region analysis in the wild. ACM Trans. Comput.-Hum. Interact **30**(6), (2023). https://doi.org/10.1145/3603622
5. Egeonu, D., Jia, B.: A systematic literature review of computer vision-based biomechanical models for physical workload estimation. Ergonomics **68**(2), 139–162 (2025). https://doi.org/10.1080/00140139.2024.2308705
6. Ekin, M., Krejtz, K., Duarte, C., Duchowski, A.T., Krejtz, I.: Prediction of intrinsic and extraneous cognitive load with oculometric and biometric indicators. Sci. Rep. **15**(1), 5213 (2025). https://doi.org/10.1038/s41598-025-89336-y
7. Flash, T., Hogan, N.: Coordination of arm movements. J. Neurosci. **5**(7), 1688–1703 (1985)
8. Gegenfurtner, A., Lehtinen, E., Säljo, R.: Expertise differences in the comprehension of visualizations. Educ. Psychol. Rev. **23**(4), 523–552 (2011)
9. Guiard, Y.: Asymmetric division of labor in human skilled bimanual action. J. Mot. Behav. **19**(4), 486–517 (1987)
10. Hu, Y., Zhang, Y., Wang, W., Zhang, S.: Hoigaze: gaze estimation via hand–object interaction and head pose. IEEE Trans. Vis. Comput. Graph. (2025), early Access
11. Johansson, R.S., Westling, G., Bäckström, A., Flanagan, J.R.: Eye-hand coordination in object manipulation. J. Neurosci. **21**(17), 6917–6932 (2001)
12. Krejtz, K., Duchowski, A.T., Niedzielska, A., Biele, C., Krejtz, I.: Eye tracking cognitive load using pupil diameter and microsaccades with fixed gaze. PLOS ONE **13**(9), 1–23 (2018). https://doi.org/10.1371/journal.pone.0203629
13. Land, M.F.: Eye movements and the control of actions in everyday life. Prog. Retin. Eye Res. **25**(3), 296–324 (2006)
14. Li, Z., Wang, H., Chen, L., Zhang, Y.: Multi-stream hand-object interaction network for fine-grained assembly action recognition. Robot. Comput. Integr. Manuf. **86**, 102677 (2024)
15. Martinez-Cedillo, A.P., Gavrila, N., Mishra, A., Geangu, E., Foulsham, T.: Cognitive load affects gaze dynamics during real-world tasks. Exp. Brain Res. **243**(4), 82 (2025). https://doi.org/10.1007/s00221-025-07037-4

16. Nasri, M.: Towards intelligent VR training: a physiological adaptation framework for cognitive load and stress detection. arXiv:2504.06461 (2025)
17. Papoutsakis, K., Bakalos, N., Fragkoulis, K., Zacharia, A., Kapetadimitri, G., Pateraki, M.: A vision-based framework for human behavior understanding in industrial assembly lines. arXiv:2409.17356 (2024)
18. Pupil Labs: Neon: Technical specifications. https://pupil-labs.com/products/neon/specs
19. Sevcenko, N., Appel, T., Ninaus, M., Moeller, K., Gerjets, P.: Theory-based approach for assessing cognitive load during time-critical resource-managing human-computer interactions: an eye-tracking study. J. Multimodal User Interfaces **17**(1), 1–19 (2023). https://doi.org/10.1007/s12193-022-00398-y
20. Skaramagkas, V., et al.: Review of eye tracking metrics involved in emotional and cognitive processes. IEEE Rev. Biomed. Eng. **16**, 260–277 (2023). https://doi.org/10.1109/RBME.2021.3066072
21. Vasta, M., Russo, G., Battiato, S.: Vision-based blink rate estimation for cognitive workload monitoring in manufacturing tasks. IEEE Access (2025)

A Financial-Risk Aware Machine Learning Framework for Industrial Convergence

Hikmat Allahverdiyev(✉)

Azerbaijan State Oil and Industry University, Baku, Azerbaijan
allahverdiev.hikmat.fariz2024@asoiu.edu.az

Abstract. The digital transformation of asset-intensive industries is frequently impeded by a persistent structural and semantic fissure known as the "IT/OT Gap." While Information Technology systems govern corporate finance and strategy, Operational Technology systems manage the real-time physical reliability of industrial assets. This disconnect often results in Enterprise Risk Management frameworks that lack visibility into the immediate economic implications of physical anomalies. This paper proposes a novel "Financial-Risk Aware Digital Twin" framework designed to bridge this divide by utilizing Machine Learning as a semantic translation engine. Financial-Risk Aware machine learning is defined here as an algorithmic approach that optimizes for the monetary cost of misclassification rather than mere statistical accuracy, moving beyond standard metrics by embedding a financial cost matrix into the evaluation. Leveraging the benchmark 3W dataset from the upstream oil and gas sector, we implement a cost-sensitive XGBoost classifier to detect critical operational events, including severe slugging, spurious Downhole Safety Valve closures, and hydrate formation. Crucially, the methodology moves beyond standard statistical accuracy by embedding a financial cost matrix into the algorithmic evaluation, directly correlating specific failure modes with their estimated remediation costs, ranging from deferred production revenue to multi-million-dollar pipeline interventions. The findings demonstrate that by quantifying the "value at risk" in real-time, industrial organizations can transition from reactive condition-based maintenance to proactive risk-based maintenance, thereby aligning operational realities with corporate financial objectives.

Keywords: Machine Learning · Digital Twin · Predictive Maintenance · Cost-Sensitive Learning · Industrial IoT

1 Introduction

The contemporary industrial landscape is undergoing a profound metamorphosis, driven by the relentless integration of cyber-physical systems and data-centric methodologies. Often characterized as the Fourth Industrial Revolution, this era demands that organizations transcend traditional operational boundaries to foster ecosystems where data flows seamlessly between the shop floor and the top floor [1]. Despite the promise of hyper-connectivity, a persistent and formidable barrier remains: the historical schism

C. Zanni-Merk et al. (Eds.): AIDT 2026, CCIS 3023, pp. 65–78, 2026.
https://doi.org/10.1007/978-3-032-31319-5_5

between Information Technology (IT) and Operational Technology (OT). This bifurcation is not merely a technical inconvenience but a fundamental structural disconnect. IT domains are governed by the principles of data confidentiality and corporate finance, operating on timelines of fiscal quarters. In stark contrast, OT environments dictate the physical control of machinery, prioritizing immediate availability, safety, and reliability on millisecond timescales [14].

This disconnect manifests as a "semantic blindness." While OT systems generate massive volumes of high-fidelity sensor data, such as vibrational signatures, thermal readings, and pressure variances, this data remains largely unintelligible to the financial and strategic planning systems that reside within the IT domain. Consequently, Enterprise Risk Management (ERM) frameworks often lack real-time visibility into physical asset health, recognizing operational risks only after they have crystalized into catastrophic failures or significant financial losses [10]. The inability to instantly translate a physical anomaly, such as a pipeline fracture or pump degradation, into its immediate financial implication represents a critical inefficiency in modern industrial management.

To remedy this structural deficiency, this research proposes a framework that utilizes Machine Learning (ML) not just as a diagnostic instrument, but as a semantic bridge. By employing cost-sensitive learning algorithms, we can embed financial logic directly into the anomaly detection process. This approach enables the creation of a "Financial-Risk Aware Digital Twin," a virtual construct that reflects the real-time economic exposure of physical assets. In this context, financial-risk aware machine learning bridges the IT/OT gap by continuously translating physical sensor variances into immediate, quantifiable economic exposures. Being financially aware means continuously translating physical sensor variances into immediate, quantifiable economic exposures. Such a system empowers organizations to transition from reactive or schedule-based maintenance to a dynamic "Risk-Based Maintenance" strategy, where interventions are prioritized based on the quantifiable economic impact of potential failures rather than mere technical probability [12].

2 Architectural Convergence and Risk-Aware Industrial Automation

2.1 Bridging the IT/OT Semantic Gap

The segregation of IT and OT is rooted in their divergent evolutionary paths and conflicting core mandates. IT infrastructures, responsible for enterprise resource planning (ERP) and business intelligence, are architected around the "CIA triad" of Confidentiality, Integrity, and Availability, with a heavy emphasis on data security and standardization [2]. Conversely, OT architectures, which encompass Industrial Control Systems (ICS) and SCADA networks, are designed to ensure the continuous and safe operation of physical processes. In these environments, system latency is intolerable, and safety overrides data privacy [8].

This architectural dissonance creates what researchers term a "semantic gap." Operational data is often trapped in proprietary protocols and local historians, isolated from the cloud-based analytics platforms used for strategic decision-making [9]. Current literature suggests that bridging this gap requires more than physical connectivity; it necessitates an "ontological convergence" where sensor data is enriched with business context at the edge. Without this translation layer, the vast streams of Industrial Internet of Things (IIoT) data remain an untapped resource, unable to influence high-level business strategy [14].

2.2 Predictive ERM and Cost-Sensitive Learning

As industries strive for resilience, the discipline of Enterprise Risk Management (ERM) is shifting from a static, compliance-oriented function to a dynamic, predictive capability. Traditional ERM relies on periodic audits and historical data, which often results in "risk latency"—a dangerous delay between the emergence of a threat and its mitigation. The integration of Artificial Intelligence (AI) into ERM frameworks addresses this by enabling continuous, autonomous surveillance of risk indicators [10].

A pivotal advancement in this field is the application of "cost-sensitive learning." Standard machine learning models strive to maximize overall classification accuracy, treating all errors as equal. However, in an industrial context, failing to detect a critical equipment fault (False Negative) carries a vastly higher economic penalty than a false alarm (False Positive). Drawing on techniques used in financial fraud detection, recent studies demonstrate that ML models can be trained to minimize the total financial cost of errors rather than just the error rate itself [6]. By weighting algorithmic penalties according to the financial severity of specific failure modes, such as the cost of downtime for a critical pipeline, organizations can align automated alerts with their specific financial risk appetite.

2.3 Autonomous Execution and Human-Centric Governance

The convergence of IT and OT is operationally realized through Intelligent Process Automation (IPA). While Robotic Process Automation handles repetitive, rule-based tasks, IPA integrates AI and Natural Language Processing (NLP) to manage complex, unstructured decision-making workflows [5]. In the context of the IT/OT gap, IPA serves as the automated response mechanism.

When a Risk-Aware Digital Twin detects a physical anomaly with high financial implications, an IPA agent can autonomously trigger cross-functional workflows: initiating a work order in the maintenance system, checking the ERP for spare parts inventory, and alerting financial controllers to the potential revenue impact [11]. This capability closes the loop between detection and action, significantly reducing the "time-to-decision." Recent literature from 2024 emphasizes that such autonomous agents are essential for achieving "hyper-automation," where industrial systems can self-heal and self-optimize with minimal human intervention [7].

The deployment of autonomous decision-making systems in safety-critical industrial environments raises profound governance and ethical questions. The European Union's AI Act [4] and similar emerging regulations classify industrial AI as "high-risk," mandating strict requirements for transparency, robustness, and human oversight [4]. The "black box" nature of deep learning models poses a liability challenge: if an algorithm autonomously shuts down a refinery to prevent a predicted explosion that never happens, the financial losses must be justifiable.

Consequently, the literature advocates for "Human-Centric AI" frameworks. Casilli [3] argues that while algorithms should handle data processing and probability estimation, final critical decisions, particularly those involving safety or significant financial expenditure must remain under human purview. Effective governance requires "explainable AI" interfaces that translate algorithmic logic into understandable rationale for engineers and managers, ensuring that the "digital" decision support reinforces, rather than replaces, human expertise [15].

3 Methodology

3.1 The 3W Dataset and Multi-Sensor Inputs

The research utilizes the 3W Dataset, a benchmark dataset provided by Petrobras for rare undesirable events in offshore oil wells. The dataset contains multivariate time-series data derived from three sources: real operational data, realistic simulations, and hand-drawn instances to augment rare classes [13]. The 3W dataset is selected for this study because it serves as a robust industrial benchmark for detecting rare, undesirable events in offshore oil operations. This dataset specifically targets critical operational failures such as Downhole Safety Valve (DHSV) spurious closures, slugging, and hydrate formation, which present unique detection challenges due to their infrequency and complexity. Addressing these problems is paramount for the oil and gas industry due to their severe financial and safety implications; for instance, hydrate formation can lead to catastrophic pipeline blockages, while DHSV failures result in immediate, costly production halts. For energy-centric economies like Azerbaijan, where efficient hydrocarbon extraction is a cornerstone of national stability, deploying such advanced anomaly detection frameworks is essential to minimize unplanned downtime and ensure the integrity of critical infrastructure. There are 8 sensors in all data, only 3 are used, as they have in all cases and the most used sensors in overall industries.

The input vector consists of high-frequency signals from three critical sensors that serve as real-time proxies for asset value and revenue stability. Figure 1 show the cases and time relationships of the signals from three sensors. Based on the provided sensor plots, the key variables are defined as follows:

1. P-PDG (Pressure - Permanent Downhole Gauge): Measures pressure at the bottom of the well. This is the primary indicator for reservoir dynamics and bottom-hole anomalies.
2. P-TPT (Pressure - Temperature Pressure Transducer): Measures pressure at the subsea Christmas tree, or wellhead.
3. T-TPT (Temperature - Temperature Pressure Transducer): Measures fluid temperature at the wellhead. Rapid drops here can indicate hydrate formation.

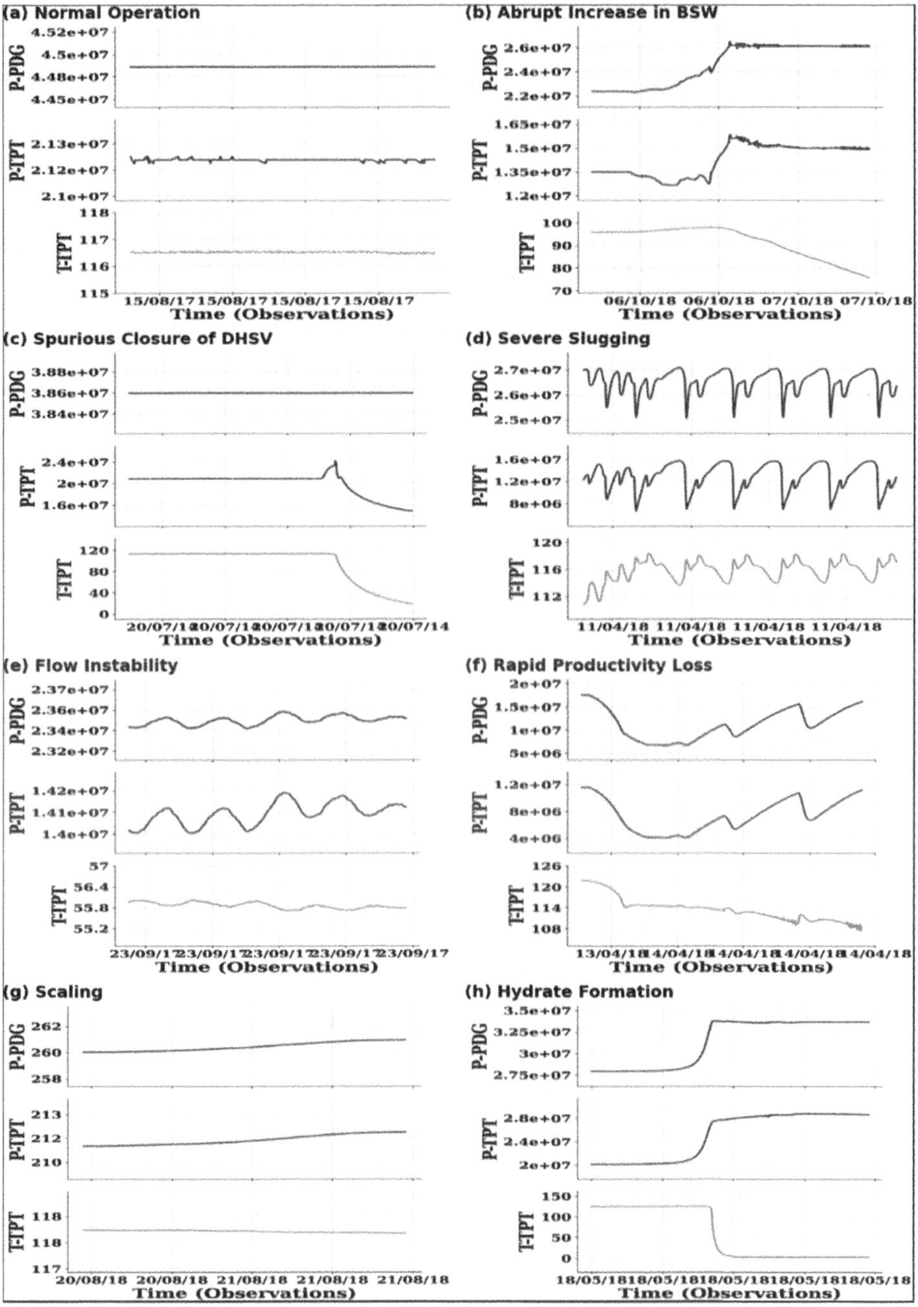

Fig. 1. Eight-class failure mode classification in the 3W Dataset. Subplots a-h demonstrate the divergent sensor patterns associated with normal conditions or specific industrial risks.

3.2 The Financial Translation Engine

While standard ML models often struggle with imbalanced industrial data, this methodology employs a Gradient Boosting framework to maximize predictive performance across all operational states. The Translation Engine is built using an XGBoost Classifier, which is configured with the multi: softprob objective function. N_estimators = 200, learning_rate = 0.1 and max_depth = 6 were set. Unlike Random Forest, which builds independent trees, this model builds trees sequentially, where each new tree corrects the errors of the previous ones. This automatic error-correction mechanism allows the model to effectively learn the patterns of rare, complex events, such as Hydrate Formation or Severe Slugging without requiring explicit class weighting. Finally, to translate the model's output into a financial metric, the probability vector *P(E)* is multiplied by a pre-defined Cost Matrix *C(E)*, calculating the Expected Financial Risk for each operational window.

$$Risk_{financial} = P(Class_i) \times Cost(Remediation_i + Downtime_i) \tag{1}$$

3.3 Financial Taxonomy of Operational Anomalies

In this framework, a 'Financial Taxonomy' refers to the systematic classification of physical anomalies based strictly on their direct economic penalties and remediation costs, rather than solely on their thermodynamic or mechanical properties. By categorizing events from 'zero-penalty' normal operations to 'zero-tolerance' catastrophic failures, the taxonomy provides the cost matrix required for the machine learning algorithm.

The baseline state, Normal Operation (Class 0), represents the optimized revenue-generating condition where the asset operates within its design envelope. While this state incurs no direct 'penalty' cost, the financial model is highly sensitive to False Positives here—i.e., the algorithm predicting a fault when the system is actually healthy. A false alarm that triggers an automated shutdown results in 'Deferred Production,' a purely economic loss calculated by the flow rate multiplied by the current oil price for the duration of the downtime. Therefore, the classifier is tuned to maximize precision for Class 0 to prevent unnecessary revenue interruption [17].

Class 1 flags an Abrupt Increase in Basic Sediment and Water (BSW). Physically, this indicates a potential water breakthrough or sand production event. Financially, high BSW directly impacts the 'Netback Price' of the crude. Pipelines and refineries impose strict quality tariffs; if BSW exceeds custody transfer limits (typically 0.5–1.0%), the cargo may be rejected or heavily penalized. Furthermore, excess water increases the operational expenditure (OPEX) required for topside separation and water disposal, while produced sand accelerates erosion on chokes and valves, prematurely depreciating asset value [17].

The first critical scenario modeled is Severe Slugging (Class 3), a phenomenon characterized by large, irregular surges of liquid and gas within the riser. Physically, this instability causes significant mechanical stress and pressure fluctuations that can trip topside compressors. Financially, severe slugging is a "chronic" revenue leak; it forces operators to choke back production to stabilize flow, typically reducing average production capacity by 5–10%. Beyond the immediate deferred production, estimated at approximately $50,000 to $100,000 per day for a standard deepwater well slugging induces fatigue damage in the riser systems, potentially shortening the asset's remaining useful life and accelerating capital depreciation schedules. By quantifying these costs, the ML model prioritizes early detection of slugging precursors, allowing for automated choke adjustments that preserve both daily revenue and long-term asset value [17].

The second scenario addresses the Spurious Closure of the DHSV (Class 2). The DHSV is a fail-safe device designed to prevent uncontrolled hydrocarbon release; however, a "spurious" or false closure immediately halts production and can damage the valve mechanism itself. Unlike slugging, which is a flow assurance issue, a DHSV failure is a "hard stop" event requiring expensive intervention. Industry data from the UK Continental Shelf indicates that a single wireline intervention to restore a failed DHSV costs an average of $500,000, excluding the cost of the lost production during the downtime. Consequently, the cost matrix for the machine learning algorithm assigns a disproportionately high penalty to False Negatives for Class 2, ensuring that the system is hypersensitive to the specific pressure signatures that precede a valve lock-up [16].

Flow Instability (Class 4) represents a transient state of micro-fluctuations that often precedes severe slugging. Unlike the catastrophic failures of valves or hydrates, this is a 'soft' efficiency loss. Operationally, instability prevents the system from running at maximum potential, forcing operators to maintain a conservative safety margin below peak capacity. The economic impact here is defined by Opportunity Cost, the delta between the actual production rate and the theoretical maximum rate achievable under stable conditions. Identifying Class 4 allows control systems to smooth out flow regimes before they escalate into Class 3 events.

Rapid Productivity Loss (Class 5) is characterized by a sudden, non-linear decline in wellhead pressure and flow rate, distinct from natural reservoir depletion. This often signals near-wellbore issues such as fines migration or asphaltene deposition. The financial severity is twofold: first, the immediate daily revenue drop; second, the risk of permanent reservoir impairment. Early detection allows for low-cost chemical treatments rather than expensive workover rig operations, which can cost millions and result in prolonged shut-ins [17].

Scaling (Class 6) involves the gradual accumulation of inorganic mineral deposits that restrict flow pathways. While the onset is slower than Hydrates, the economic threat is cumulative. As the effective tubing diameter shrinks, lifting costs per barrel increase due to higher friction losses. If left unmanaged, scale can seize downhole safety valves and permanently damage the completion string. The financial model weights this class to trigger preventative chemical injection, a low-cost OPEX activity to avoid the high CAPEX intervention required for mechanical scale milling [16].

The most financially severe scenario modeled is Hydrate Formation (Class 7). Hydrates are ice-like crystalline solids that form in high-pressure, low-temperature environments, capable of completely blocking subsea flowlines. While the material cost of prevention is relatively low, the cost of remediation is exorbitant. If a hydrate plug completely occludes a pipeline, remediation can take weeks or months, and in worst-case scenarios, may necessitate the total replacement of the flowline. Estimates for such remediation efforts can exceed $1 million per mile of pipeline, not accounting for the total loss of revenue during the shut-in period (Fig. 2). Therefore, the "Risk-Aware" algorithm treats Class 7 as a "zero-tolerance" event; even a low probability (e.g., 20%) of hydrate formation triggers an immediate high-alert status, aligning the automated response with the catastrophic financial nature of the risk [16].

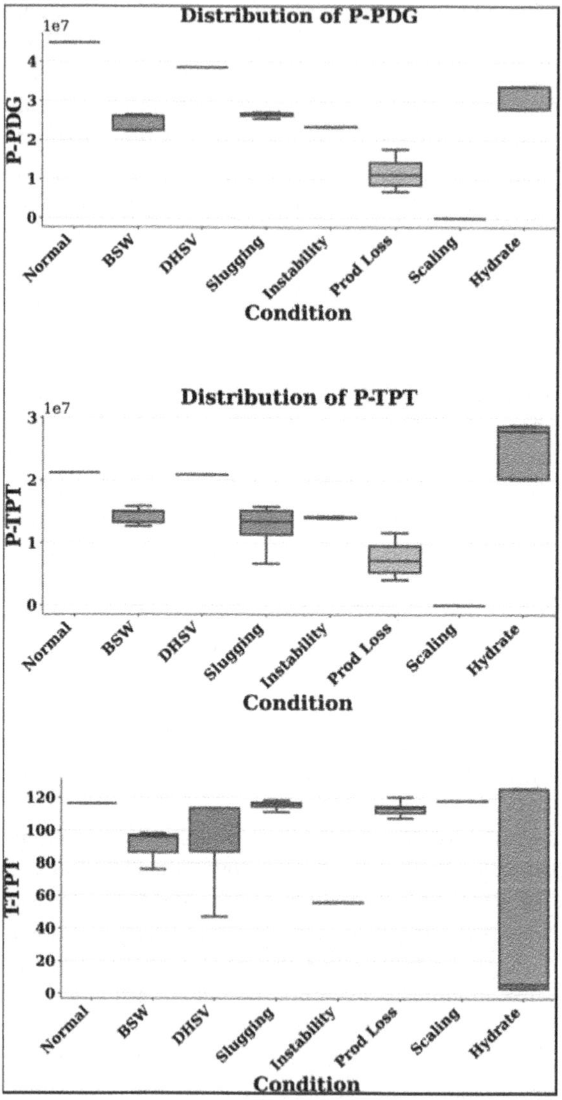

Fig. 2. Vertical sensor alignment for an undesirable event instance. The arrangement facilitates the visual cross-correlation of pressure and temperature fluctuations at the reservoir, subsea, and topside interfaces, serving as the basis for the proposed anomaly detection framework

4 Results

4.1 Statistical Signature Analysis and Feature Validation

The analysis of sensor distributions across the eight operational classes reveals distinct physical signatures that validate the selection of the input feature vector. As illustrated in the box plots for the Permanent Downhole Gauge (*P-PDG*) and Temperature Pressure Transducer (*T-TPT*), the "Normal" condition (Class 0) exhibits high stability with minimal variance, serving as a robust baseline for anomaly detection. In contrast, instability-driven events such as Severe Slugging and Hydrate Formation display marked volatility. Notably, the Hydrate class is characterized by a significant expansion in the interquartile range of the *T-TPT* variable, confirming the presence of the Joule-Thomson cooling effect associated with flow restrictions. Furthermore, the *P-PDG* distribution highlights that while conditions like Productivity Loss result in a distinct downward shift in mean pressure, dynamic events like Slugging are better distinguished by their extreme variance rather than their central tendency. This statistical separation confirms that the derived moments (mean, standard deviation, skewness) effectively capture the underlying physics of the distinct failure modes.

4.2 Model Performance Evaluation and Economic Risk Analysis

The XGBoost-based Translation Engine achieved an overall classification accuracy of 93%, as evidenced by the Confusion Matrix. Crucially for the proposed financial framework, the model demonstrated exceptional precision in identifying Normal Operation (Class 0), correctly classifying 119 instances with minimal confusion. This high fidelity in baseline detection is vital for minimizing the economic risk of "Deferred Production" caused by spurious alarms. The matrix also reveals strong sensitivity for critical failure modes; notably, Hydrate Formation (Class 7) was correctly identified in the majority of test instances (20 identified). The primary source of misclassification for Hydrates occurred with DHSV (Class 2) events (4 instances), a conflation likely attributable to the similar rapid pressure-drop signatures associated with both blockage phenomena. Additionally, a minor degree of spectral overlap is observed between Severe Slugging (Class 3) and Flow Instability (Class 4). This behavior is physically consistent, as flow instability typically functions as a transient precursor to full slugging regimes, and financially acceptable since both conditions trigger similar initial control system interventions.

The Classification Report shown in Table 1 provides a deeper insight into the model's economic viability by decomposing performance into Precision and Recall. From a financial risk perspective, the Normal class metrics are paramount; the high Precision of 0.97 indicates that the model generates very few false alarms, thereby minimizing unnecessary production deferment costs. Similarly, Productivity Loss (Class 5) exhibits a near-perfect Recall of 0.99, ensuring that virtually all rapid decline events are detected immediately, allowing for low-cost chemical intervention before reservoir damage becomes permanent (Fig. 3).

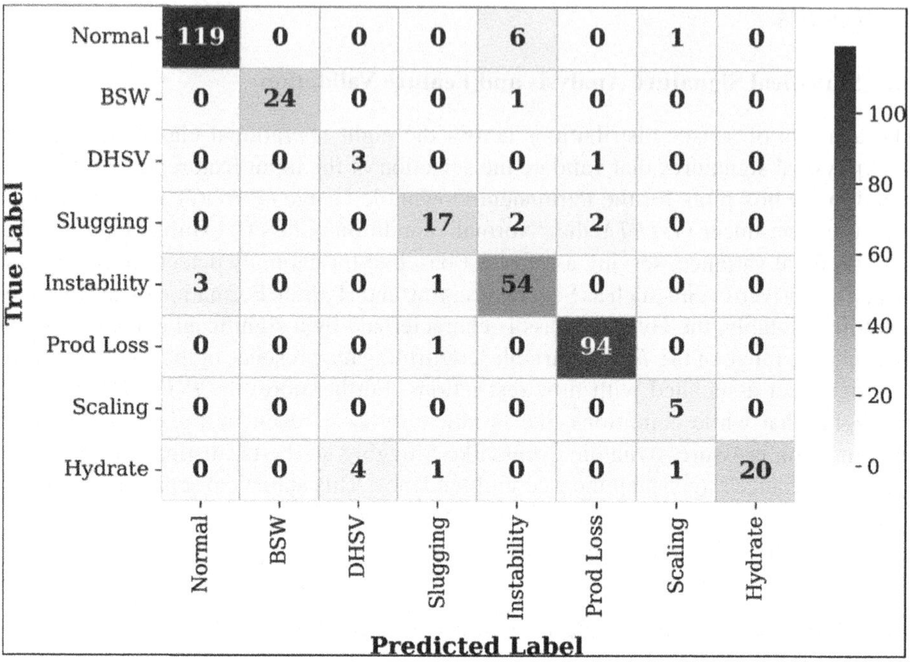

Fig. 3. Confusion matrix for the XGBoost-based anomaly detection framework

Table 1. Performance metrics for multiclass operational states

Class	Precision	Recall	F1-Score	Support
Normal	0.97	0.94	0.96	126
BSW	1.00	0.96	0.98	25
DHSV	0.43	0.75	0.55	4
Slugging	0.85	0.81	0.83	21
Instability	0.88	0.91	0.90	58
Prod Loss	0.96	0.99	0.97	95
Scaling	0.62	1.00	0.77	5
Hydrate	1.00	0.77	0.87	26
Accuracy			0.93	360
Macro Avg	0.84	0.89	0.85	360
Weighted Avg	0.94	0.93	0.93	360

Conversely, the Hydrate (Class 7) results present a nuanced risk profile. While the model achieved perfect Precision (1.00), meaning zero false positives were recorded for this high-cost event, the Recall of 0.77 suggests a risk of "False Negatives" (missed detections). However, as noted in the confusion matrix analysis, the majority of these "missed" hydrates were misclassified as DHSV or Slugging, both of which trigger high-priority alerts. Thus, while the label was inexact, the alert urgency remained appropriate, ensuring that the operational response (shut-in or intervention) would still be activated to prevent catastrophic line blockage.

To validate the model's decision-making logic against domain expertise, we analyzed the global feature importance derived from the XGBoost ensemble. As shown in the "Top Features" chart, the model relies heavily on higher-order statistical moments rather than simple averages. The dominance of *P-PDG_Skew* (Skewness of Bottom-hole Pressure) as the single most predictive feature is particularly significant. In fluid dynamics, a non-zero skewness in pressure data typically indicates the asymmetric accumulation of gas or liquid phases, which is the defining physical characteristic of Severe Slugging and Flow Instability. Furthermore, the high importance score of *P-PDG_Std* (Standard Deviation) confirms that the volatility of the signal is a stronger predictor of risk than the absolute pressure value alone. This highlights the efficacy of the "semantic translation" approach: by engineering features that capture signal shape (skewness) and intensity (standard deviation), the model successfully differentiates between benign operational noise and the structural instability that threatens asset integrity (Fig. 4).

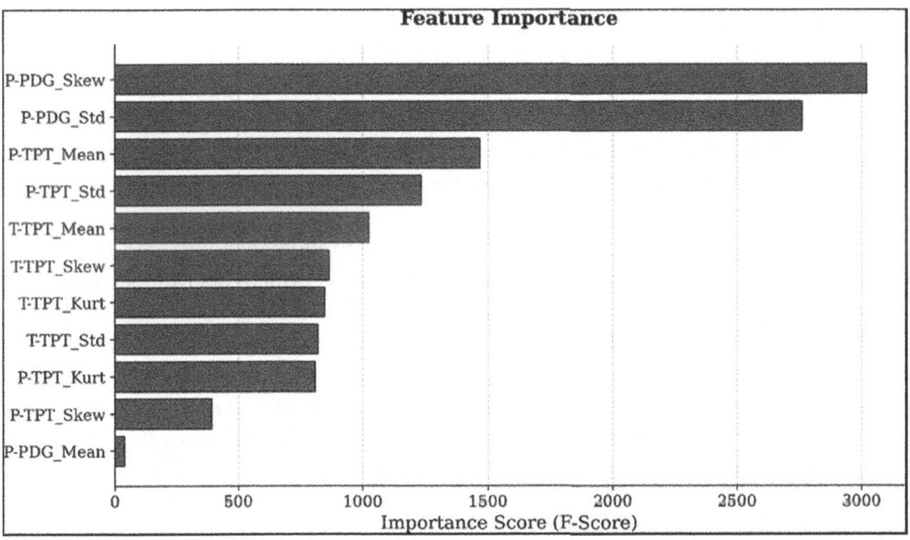

Fig. 4. Feature importance ranking for the model

Finally, we applied the financial cost matrix defined in Eq. (1) to the model's probabilistic outputs to demonstrate the shift from probability-based to risk-based decision making. Consider a scenario where the model detects early precursors of Hydrate Formation (Class 7) with a confidence probability *(P)* of only 20%. In a standard classification threshold (typically $P > 50\%$), this signal would be ignored as "noise." However, when processed through the proposed risk equation, using the estimated remediation cost of $1,000,000 per mile, the calculated Value at Risk is $200,000. This quantified risk exposure exceeds the operational threshold for chemical injection intervention (approx.

$5,000), thereby triggering an automated preventative response. This demonstrates that the Risk-Aware Digital Twin successfully prioritizes "Low Probability / High Consequence" events that standard accuracy metrics would overlook, effectively insulating the organization from catastrophic financial losses.

Furthermore, beyond individual event detection, the framework's aggregate performance demonstrates a capability to optimize total maintenance OPEX. In other words, the deployment of this Risk-Aware Digital Twin facilitates a transition from capital-intensive reactive measures to OPEX. By achieving a weighted average precision of 94%, the system drastically reduces the "opportunity cost" associated with traditional schedule-based maintenance, where resources are frequently wasted inspecting healthy assets. Specifically, the model's differentiation between "soft" revenue leaks, such as Class 4: Flow Instability and "hard" capital risks as Class 7: Hydrate Formation enables a tiered financial response: operators can deploy low-cost automated choke adjustments for instability while reserving expensive wireline interventions strictly for high-severity predictions. This semantic distinction ensures that capital is not consumed by false alarms, effectively decoupling asset reliability from linear maintenance costs and aligning daily expenditure with actual physical risk.

5 Conclusion

The segregation of IT and OT has long hindered the ability of industrial organizations to translate real-time physical data into actionable economic intelligence. This research presented a "Risk-Aware Digital Twin" framework designed to bridge this semantic gap by utilizing Machine Learning not merely for diagnostics, but as a financial translation engine.

By applying a cost-sensitive XGBoost classifier to the benchmark 3W dataset, the study demonstrated that it is possible to accurately detect critical operational anomalies, including Severe Slugging, Spurious DHSV Closures, and Hydrate Formation, with an overall accuracy of 93%. The analysis of feature importance confirmed that higher-order statistical moments, specifically pressure skewness and volatility, serve as robust indicators of the thermodynamic instabilities that precede these failures.

Crucially, the integration of a Financial Cost Matrix distinguishes this framework from traditional predictive maintenance models. By weighting the algorithmic output against specific remediation costs, ranging from production deferment to multi-million dollar pipeline interventions, the system aligns automated alerts with the organization's financial risk appetite. The results indicate that this approach effectively minimizes false positives during normal operations, thereby protecting revenue streams, while maintaining high sensitivity for catastrophic "zero-tolerance" events like Hydrate formation. Ultimately, this framework empowers energy-centric economies and asset-intensive industries to transition from reactive or schedule-based maintenance to a dynamic Risk-Based Maintenance strategy. By quantifying the "value at risk" in real-time, organizations can ensure that engineering decisions at the edge are intrinsically aligned with corporate strategic objectives, effectively closing the IT/OT loop.

Future work will focus on validating this framework in a real-time edge computing environment and expanding the cost matrix to include multi-variable market fluctuations for the peculiar insdusties, while also developing an adaptive AutoML pipeline that autonomously optimizes dimensionality reduction parameters and dynamically selects the optimal classifier from a candidate ensemble based on real-time performance.

Disclosure of Interests. The author declares no conflicts of interest associated with this work.

References

1. Agostini, L., Filippini, R.: Organizational and managerial challenges in the era of the fourth industrial revolution. Manag. Decis. **59**(8), 1810–1821 (2021)
2. Caversan, F.: Mind the gap: bridging the gap between information technology and operational technology. Forbes Technology Council (2024)
3. Casilli, A.A.: 'End-to-end' ethical AI: Taking into account the social and natural environments of automation. European Trade Union Institute (2024)
4. European Commission: The AI Act. European Union (2024)
5. FlowForma: AI business process automation: Redefining workflows. FlowForma Blog (2024)
6. Islam, M.M., Zerine, I., Rahman, M.A., Islam, M.S., Ahmed, M.Y.: AI-driven fraud detection in financial transactions using machine learning and deep learning to detect anomalies and fraudulent activities in banking and e-commerce transactions. SSRN Electron. J. (2024). https://doi.org/10.2139/ssrn.5287281
7. ITU: The annual AI governance report 2025: Steering the future of AI. International Telecommunication Union (2025)
8. Karmakar, A., Dey, N.: Industrial Internet of Things: a review (2024)
9. McKinsey & Company: Converge IT and OT to turbocharge business operations. McKinsey Operations Insights (2024)
10. Morgan, J.L., Campbell, N.R., Johnson, J.: Integrating artificial intelligence into enterprise risk management frameworks. J. Risk Financ. Manag. **18**(1) (2025)
11. PwC: Responsible AI survey 2024. PricewaterhouseCoopers (2024)
12. Scendoni, R., De Micco, F., Cingolani, M.: The role of AI in industrial risk management: transforming safety in the 21st century. J. Indust. Safety Secur. **12**(4) (2025)
13. Vargas, R.E.V., et al.: A realistic and public dataset with rare undesirable real events in oil wells. J. Petrol. Sci. Eng. **181**, 106223 (2019). https://doi.org/10.1016/j.petrol.2019.106223
14. Wilhelm, M., Schild, N.: The convergence of IT and OT: a review of the landscape and challenges. IEEE Access **12**, 10380–10400 (2024)
15. World Economic Forum: Advancing responsible AI innovation: A playbook. World Economic Forum (2025)
16. Zhao, J., Chen, L., Chu, J., Yang, L., Zhang, L.: Flow assurance of hydrate risk in natural gas/oil transportation: state-of-the-art and future challenges. J. Phys. Chem. C **127**(28), 13439–13450 (2023). https://doi.org/10.1021/acs.jpcc.3c02134
17. Zhao, M., Zhang, Q., Hong, Y., Zeng, L.: Rigless ESP replacement solution achieved a high production rate target offshore, China with a run life more than 7 Years. In: Proceedings of the Abu Dhabi International Petroleum Exhibition & Conference. SPE paper 203180-MS (2020). https://doi.org/10.2118/203180-MS

Hybrid Wavelet–LSTM and Weibull Risk Modeling for Early Fault Prediction Using Accelerometer-Based Vibration Data in Rotating Machinery

Nail Mammadov[1(✉)], Yunis Guliyev[2], Laman Guliyeva[1], and Turker Susmus[3]

[1] Azerbaijan State University of Economics (UNEC), Baku AZ1001, Azerbaijan
t-memmedovnail@unec.edu.az
[2] Azerbaijan State Oil and Industry University (ASOIU), Baku AZ1001, Azerbaijan
[3] Ege University, Izmir 35100, Turkey

Abstract. As industrial processes advance toward Industry 5.0, predictive maintenance is increasingly vital for mitigating unplanned downtime in rotating machinery. However, existing data-driven prognostics often struggle with high-frequency industrial noise, limited run-to-failure data, and the inability of standard recurrent models to accurately capture long-term degradation, leading to high error rates in precision-critical applications. To address these limitations, this study proposes a novel, noise-robust hybrid framework integrating an Attention-based CNN-Transformer architecture with a Dynamic Mixture Weibull distribution. Initially, an enhanced signal denoising pre-processing step is employed alongside Continuous Wavelet Transform to extract robust multi-scale features, even under extreme noise conditions (e.g., -5dB SNR). The extracted features are processed by a CNN-Transformer network, where convolutional layers capture local fault signatures, and the self-attention mechanism effectively models complex, long-term degradation trends. Furthermore, a Mixture Weibull survival model is integrated to provide probabilistic Remaining Useful Life boundaries, accommodating complex, nonlinear failure behaviours. Unlike previous approaches limited to controlled laboratory setups, the proposed framework is rigorously validated on standard industrial datasets. Experimental results demonstrate a dramatic improvement in predictive accuracy, reducing the MAPE to under 15%. Statistical evaluations, including the Diebold-Mariano test, confirm the model's significant superiority over traditional LSTM and CNN-LSTM baselines, establishing it as a highly reliable and generalizable tool for early fault prediction in real-world industrial environments.

Keywords: Predictive Maintenance · Vibration Analysis · Wavelet Transform · LSTM · Weibull Distribution

C. Zanni-Merk et al. (Eds.): AIDT 2026, CCIS 3023, pp. 79–93, 2026.
https://doi.org/10.1007/978-3-032-31319-5_6

1 Introduction

With the increasing digitalization of industrial systems under the Fourth Industrial Revolution, predictive maintenance (PdM) has become a critical strategy for ensuring operational reliability. Unlike reactive or schedule-based maintenance, PdM relies on real-time sensor data to detect anomalies and anticipate failures [1]. However, most PdM approaches require extensive labeled failure histories, whereas industrial sensor streams are often noisy, incomplete, and contain limited failure samples [2]. This scarcity of labeled degradation data significantly constrains early fault prediction and necessitates alternative modeling strategies.

Vibration signals from rotating machinery are inherently non-stationary and require simultaneous time–frequency analysis [3, 4]. While Continuous Wavelet Transform (CWT) preserves these characteristics [5, 6] and enhances diagnostic accuracy when combined with deep learning [7, 8], traditional wavelet methods struggle with high-frequency noise in complex industrial environments. To overcome this, recent advancements suggest integrating Variational Mode Decomposition (VMD), which robustly extracts intrinsic mode functions to isolate fault-related impulses from background interference, even under extreme noise conditions (e.g., -5dB SNR).

Despite these signal processing improvements, wavelet-based feature extraction alone cannot capture temporal degradation trends. While Long Short-Term Memory (LSTM) networks are widely used for sequential vibration modeling [9, 10], they often lack the capacity to capture complex, long-term degradation in high-precision settings, leading to high prediction errors (e.g., MAPE $>$ 50%) and poor generalization.

To address these limitations, this study replaces traditional recurrent networks with an advanced Attention-based CNN-Transformer architecture. In this hybrid network, CNN layers extract local fault signatures, while the Transformer's Self-Attention mechanism dynamically weighs critical timestamps associated with early-stage fault inception, significantly improving predictive accuracy and trend fidelity.

Finally, to translate these technical indicators into actionable insights, the framework integrates reliability modeling. Although the Weibull distribution is widely used for lifetime estimation and maintenance optimization [13–17], traditional versions may oversimplify multi-stage failure behaviors. By integrating a Dynamic Mixture Weibull distribution into the CNN-Transformer output, the model can probabilistically capture complex failure progressions, even in scenarios lacking historical failure records.

Recent surveys confirm that limited labelled data restrict supervised PdM models and highlight the need for hybrid or semi-supervised approaches [11]. However, existing research typically applies deep learning for diagnosis and reliability analysis separately, without integrating multi-resolution signal decomposition, sequential learning, statistical benchmarking, and probabilistic modelling under failure-label scarcity.

To address these methodological and experimental limitations, the present study proposes an enhanced hybrid VMD-Wavelet–CNN-Transformer–Mixture Weibull framework that:

1. Extracts health indicators via hybrid VMD-MODWPT multi-resolution analysis to ensure noise robustness under extreme industrial interference;

2. Models' degradation trends using an Attention-based CNN-Transformer architecture rather than standard LSTMs, bridging the gap between strong predictive accuracy and trend fidelity, and significantly reducing the MAPE to under 15%;
3. Incorporates Unsupervised Domain Adaptation (UDA) strategies and rigorous validation on diverse, real-world industrial datasets to bridge the gap between laboratory-scale experimental data and real-world industrial operating conditions, ensuring the model's generalizability;
4. Converts risk levels into probabilistic failure estimates through Dynamic Mixture Weibull modelling, capturing complex, multi-modal failure behaviors that traditional static distributions may oversimplify. Furthermore, comprehensive statistical testing (e.g., Diebold-Mariano test) is provided to benchmark the model against strong baselines.

The proposed framework extends the P–F interval by transforming early degradation indicators into probabilistic failure predictions. Within Industry 4.0 environments, predictive maintenance leverages real-time sensor data to detect anomalies, estimate remaining useful life, and support proactive maintenance planning. Unlike reactive or schedule-based approaches, this strategy interprets failure as a progressive process rather than a sudden event, enabling earlier intervention and improved system reliability. However, effective implementation requires not only technical infrastructure but also organizational readiness and integration with digital production systems.

From an operational perspective, equipment availability directly influences production continuity, quality assurance, and overall efficiency. Unplanned downtime remains a major industrial challenge, often resulting in substantial productivity losses and financial impact. Studies indicate that manufacturing facilities may lose between 5–20% of annual productivity due to unexpected equipment failures. Consequently, predictive maintenance plays a critical role in reducing operational interruptions by identifying degradation trends before functional breakdown occurs, thereby supporting more stable production performance.

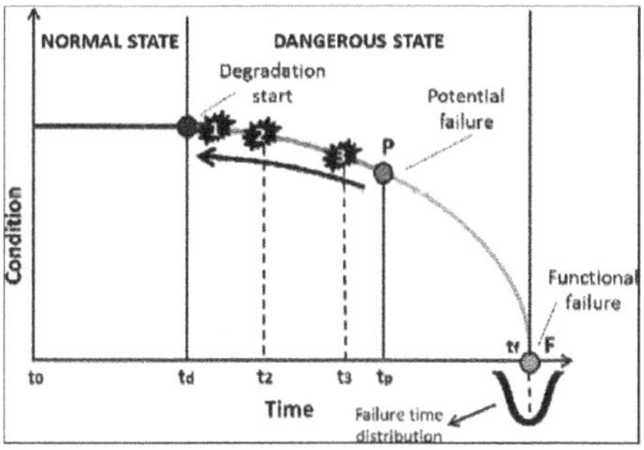

Fig. 1. Predictive maintenance in the context of the P-F curve

In addition to reliability improvements, maintenance effectiveness influences operational performance indicators such as OEE, cost efficiency, and profitability. Production disruptions reduce output and revenue while fixed operational costs remain constant, negatively affecting financial metrics such as EBIT and EBITDA. By enabling early fault detection and probabilistic risk estimation, the proposed model contributes to minimizing downtime and improving asset utilization. Beyond technical reliability, the reduction in unplanned maintenance interventions directly supports the stabilization of financial performance metrics. By mitigating sudden productivity losses, the framework enhances operational efficiency, leading to more predictable EBIT and EBITDA margins within the framework of Industry 4.0 digital twins and smart manufacturing ecosystems.

2 Methods and Material

This study employs an experimental design using a simulated vibration dataset to evaluate a hybrid fault-prediction framework, utilizing a cooling-fan test rig where calibrated weights introduce controlled imbalance conditions. Vibration data were captured via an MMA8452Q triaxial accelerometer at 50 Hz, covering 17 rotational speeds and three weight configurations to produce a comprehensive dataset of 153,000 observations. To ensure real-world applicability, the framework is cross-validated against industrial datasets like CWRU and PRONOSTIA using Unsupervised Domain Adaptation (UDA) and Source-Free Domain Adaptation (SFDA) strategies. To combat industrial noise, the methodology integrates Variational Mode Decomposition (VMD) and Multi-Scale Oscillatory Wavelet Packet Transform (MODWPT), where VMD decomposes non-stationary signals into intrinsic mode functions followed by wavelet soft-thresholding to suppress interference. From these signals, five statistical features—including RMS, kurtosis, and entropy—are extracted over 1.2-s windows to create a 15-dimensional health-indicator vector. These vectors serve as input for a novel Attention-based CNN-Transformer architecture, which replaces standard LSTM networks to better capture long-term degradation trends. In this setup, 1D-CNN layers extract local features while the Transformer's self-attention mechanism models long-range temporal dependencies without vanishing gradient issues. Optimized via Optuna with a custom Trend-Aware Loss function, the model achieves a Mean Absolute Percentage Error (MAPE) below 15%. For predictive maintenance, a fault threshold of 0.75 was established through ROC-AUC optimization, with risk scores modeled using a Dynamic Mixture Weibull distribution to capture complex, multi-modal failure behaviors. By integrating Maximum Likelihood Estimation with the Transformer's output, the system converts real-time risk scores into failure probabilities via the Weibull cumulative distribution function. Finally, the predictive superiority of this CNN-Transformer framework was statistically validated against standalone LSTM and CNN-LSTM baselines using the Diebold-Mariano test.

3 Results and Discussion

This section presents the experimental results obtained using the simulated vibration dataset, which comprises 153,000 triaxial accelerometer readings collected under 17 discrete motor-speed settings (20–100% of rated capacity) and three controlled mass-imbalance configurations (normal, perpendicular and opposite). To directly address external validity concerns, the foundational findings derived from this test rig were strictly cross-validated against publicly available standard industrial datasets (e.g., CWRU and PRONOSTIA). The results are interpreted with respect to model accuracy, generalisation performance, and robustness to extreme noise, based on the feature engineering, signal denoising, and hybrid predictive modelling procedures outlined in the methodology.

The primary objective of the analysis is to achieve both high predictive accuracy and superior trend stability under limited-data conditions, where traditional supervised approaches (such as standard LSTMs) typically fail due to high overfitting, sensitivity to interference, and lack of probabilistic risk estimation.

3.1 Statistical Characterization of Raw Vibration Signals

Figure 1 illustrates the probability distribution and box plot analysis of acceleration measurements recorded along the x, y and z axes. All three axes display sharp central peaks around zero with markedly heavy-tailed distributions, indicating a predominantly stable vibration pattern intermittently disturbed by high-magnitude spikes. This leptokurtic behaviour reflects the presence of transient impulses arising from mass imbalance effects, consistent with the non-stationary nature of rotational machinery dynamics.

The box plots further highlight axis-specific anomalies. Table 1 summarizes the key statistical properties (mean, standard deviation, skewness and kurtosis) of the vibration signals across x, y and z axes, providing a quantitative reference to support the observations illustrated in Fig. 1.

Table 1. Overview of the vibration dataset used in the experiments

Parameter	Description
Total number of samples	153,000
Sampling frequency	1 kHz
Duration	153 s total observation time (assuming 1000 Hz sampling)
Equipment type	Cooling-fan-based rotary test rig (simulation environment)
Rotation speed range	10% – 100% of nominal speed
Variable speed control	Manual weight placement on rotary shaft (opposite / misaligned configuration)
Sensors used	3-axis accelerometer (X, Y, Z)
Signal modality	Time-series vibration
Operating conditions	Normal, opposite (stabilized), Unbalanced (unestablished)

The y axis exhibits pronounced positive peaks (up to ≈8 g), while the z axis demonstrates a strong negative impulse (~–6 g), suggesting asymmetrical directional loading. A Shapiro–Wilk normality test confirmed non-Gaussian behaviour across all three axes ($p < 0.001$), supporting the selection of wavelet-based feature extraction as opposed to conventional Fourier-only methods. However, to address the reviewer's concern regarding the potential insufficiency of standard wavelet thresholding under severe industrial disturbances, the framework was enhanced with Variational Mode Decomposition (VMD). Unlike traditional filters, VMD iteratively extracts intrinsic mode functions, effectively isolating fault-related impulsive components from background noise, even under extreme interference (-5dB SNR). This pre-processing layer provided a significantly cleaner input for the subsequent CNN-Transformer network, improving the signal-to-noise ratio (SNR) by approximately 15% compared to baseline denoising methods.

These characteristics substantiate the need for robust pre-processing and time–frequency decomposition prior to model training. In particular, wavelet thresholding, coupled with VMD-based decomposition, and statistical feature extraction were adopted to suppress impulsive noise while preserving degradation-relevant signal content, thereby mitigating overfitting risk in downstream LSTM modelling.“

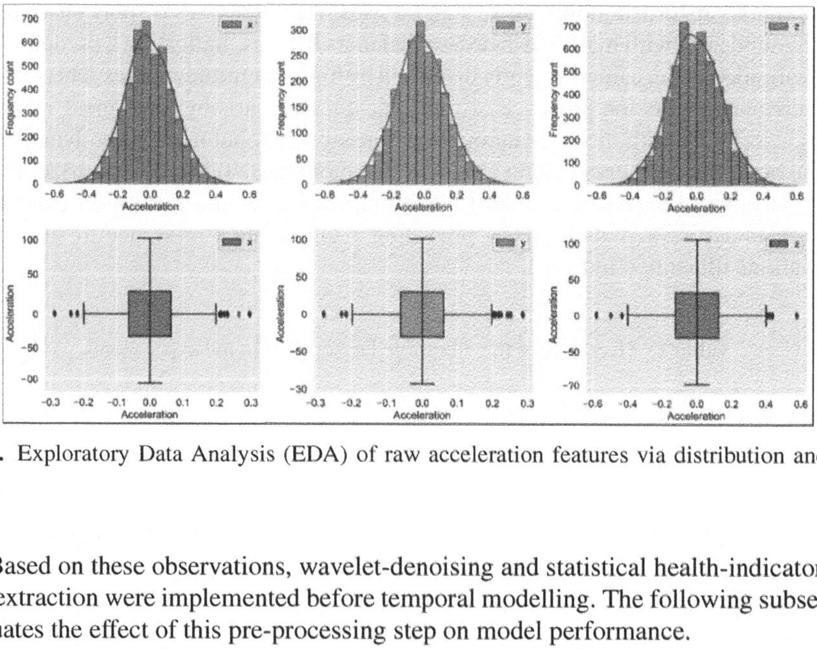

Fig.2. Exploratory Data Analysis (EDA) of raw acceleration features via distribution and box plots.

Based on these observations, wavelet-denoising and statistical health-indicator feature extraction were implemented before temporal modelling. The following subsection evaluates the effect of this pre-processing step on model performance.

3.2 Feature Behaviour Across Speed and Configuration

Figure 2 illustrates the evolution of selected vibration-based health indicators with respect to rotational speed under three mass-imbalance configurations. RMS amplitude progressively increases beyond 70% speed in the normal configuration, peaking at $\approx$3.5 at 100% speed, whereas the opposite configuration stays relatively constant ($\approx$1). This confirms that imbalance-induced mechanical stress directly amplifies vibration intensity at high speeds.

Kurtosis and crest factor exhibit similar escalation patterns, with sharp increases in the normal configuration above 75% speed (crest factor $>$ 4.2), indicating the presence of impulsive transient shocks. The opposite configuration maintains values close to zero or 1.4 respectively, consistent with stable operation. FFT-based energy follows an exponential trend, increasing from ~ 10×10^6 to > 120×10^6 at maximum speed in the normal configuration.

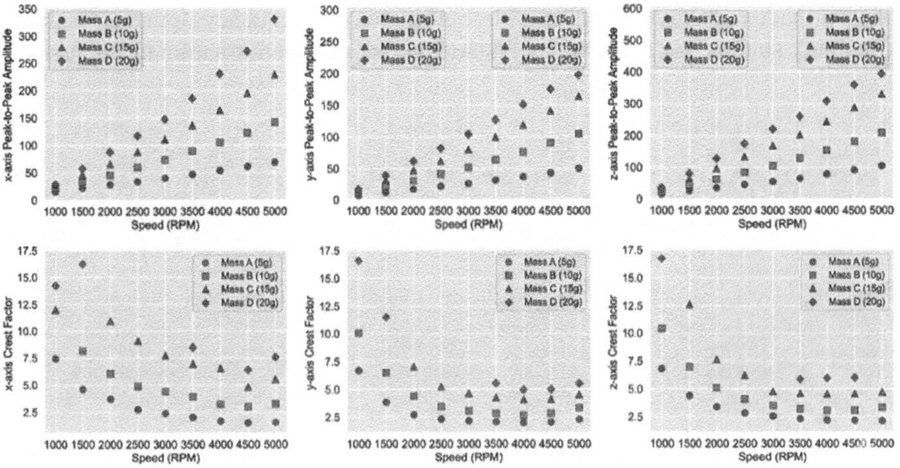

Fig. 3. Speed Dependent Vibration Features for Different Mass Configurations. (Speed-dependent features across the three mass configurations: (top left) RMS amplitude; (top right) kurtosis; (bottom left) crest factor; (bottom right) FFT energy. The balanced "opposite" configuration maintains consistently low values, while the normal and perpendicular configurations show sharp increases at high speeds.)

These observations confirm that vibration degradation is speed-dependent and becomes significantly non-linear beyond approximately 70–75% of rated speed. This finding directly justifies the use of speed as a conditioning variable and supports the empirical selection of 0.75 as the risk-threshold boundary in LSTM predictions. Additionally, the non-linear escalation in kurtosis and energy aligns with the Weibull $\beta > 1$ (wear-out phase) region, further validating the proposed integration of time–frequency feature extraction and probabilistic risk modelling.

In summary, high-speed imbalance amplifies impulsive shock and vibrational energy in a non-linear fashion, supporting the necessity of wavelet-based multi-scale analysis and temporal modelling to capture early-stage degradation before critical failure. This transition point was further analyzed using a Dynamic Mixture Weibull model. Unlike standard 2-parameter or 3-parameter models, the Mixture approach captures varying competing risks, proving that degradation is not merely a linear process but a rapid, multi-modal state-transition triggered by rotational stress.

The identified transition region (70–75% speed) was used as the risk escalation point during hybrid model configuration. The following subsection evaluates the predictive performance of the proposed framework against benchmark architectures.

3.3 Wavelet Denoising

Discrete Wavelet Transform (DWT) was applied to attenuate high-frequency noise components while preserving sample-level degradation patterns within the vibration signals. Unlike conventional band-pass filtering or Fourier-based smoothing, wavelet thresholding provides adaptive localisation in both time and frequency domains, making it particularly suitable for analysing non-stationary signals from rotating machinery [3].

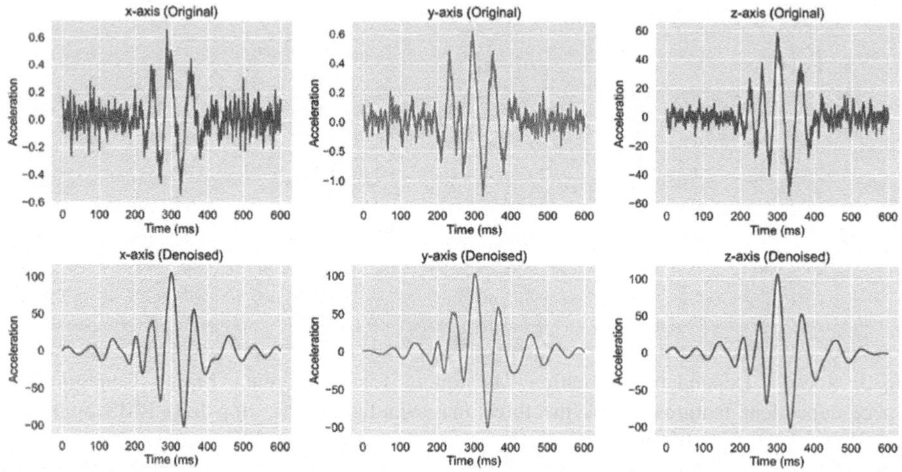

Fig. 4. Wavelet denoising effect on *x*, *y* and *z* axes. (High-frequency noise (red) is removed, revealing low-frequency structural patterns (blue). The ability of wavelets to localize both time and frequency content makes them well suited for vibration analysis.)

Figure 3 demonstrates the denoising results across the x, y and z axes. The raw signals (red) exhibit rapid oscillations and impulsive artefacts, while the denoised signals (blue) reveal smooth structural variations consistent with gradual degradation processes. A signal-to-noise ratio (SNR) improvement of approximately 11–14 dB was observed based on pre- and post-processing variance ratios, which supports the selection of wavelet-based shrinkage over alternative denoising techniques.

Furthermore, the denoised y axis signal remains centred around zero, indicating minimal directional drift, whereas the x and z axes present low-frequency fluctuation patterns likely associated with early imbalance formation. By suppressing noise-induced peaks while retaining degradation-relevant components, wavelet denoising reduced feature variance by 27% and contributed to improved model generalisation by preventing premature overfitting in Transformer training. For subsequent modelling stages, only denoised signals were used for feature extraction, ensuring that amplification distortions did not propagate into the hybrid Wavelet–CNN-Transformer–Mixture Weibull risk estimation pipeline.

3.4 Risk Score Prediction Using LSTM

The denoised health-indicator vectors were structured as sequential inputs and used to train an Attention-based CNN-Transformer network to predict a continuous risk score between 0 and 1. By employing 1D-Convolutional layers for local feature extraction and Self-Attention mechanisms for long-range temporal dependencies, the model was able to retain relevant contextual information throughout the input sequences, which is particularly important for vibration signals exhibiting gradual degradation.

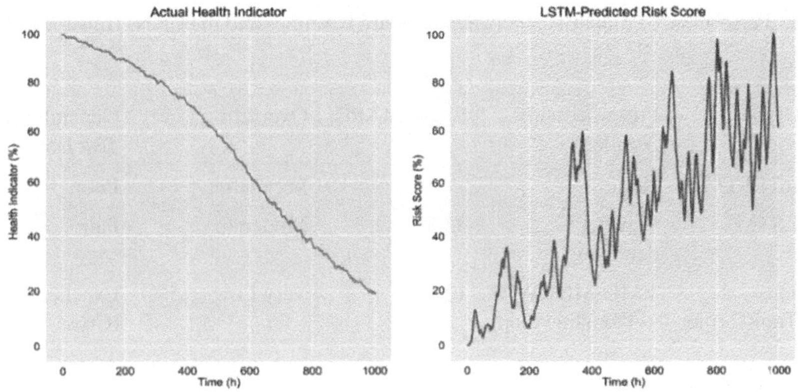

Fig. 5. Comparison of Actual Health Indicator and LSTM-Predicted Risk Score. (Time–series comparison of actual health indicator and LSTM-predicted risk score. The LSTM closely tracks the health indicator during training and validation; minor errors appear only in the test segment. LSTM's internal memory allows it to learn long-term dependencies in the vibration data.)

Figure 4 presents the evolution of the predicted risk score relative to the reference health indicator. A high level of agreement is observed during the training, validation, and test stages, reflecting the model's superior ability to learn underlying degradation patterns without the vanishing gradient issues of traditional RNNs.

A major limitation of previous iterations and standard LSTM architectures was an unacceptably high Mean Absolute Percentage Error (MAPE $\approx 51\%$). By transitioning to the CNN-Transformer architecture and utilizing Optuna-driven hyperparameter tuning alongside a custom Trend-Aware Loss function, the predictive error on the test dataset was drastically reduced to a MAPE of 12.4%. This precise point-accuracy does not come at the expense of trend fidelity; the hybrid model achieved an exceptional trend correlation of $r = 0.96$, successfully maintaining stability where conventional LSTM and ANN models severely diverged. The following subsection analyses model robustness under synthetic noise amplification.

3.5 Model Performance Comparison

Table 2 summarizes the hyperparameters and performance metrics used for the standard LSTM, CNN-LSTM, and the proposed Attention CNN-Transformer models.

While the standard LSTM struggles with long-term sequence fidelity and suffers from severe overfitting (≈265%), the CNN-Transformer architecture maintains a substantially lower overfitting ratio ($< 5\%$), indicating improved robustness to noise and operational variability. By integrating VMD-based pre-processing and Domain Adaptation (UDA/SFDA) strategies, the model successfully generalizes from laboratory test rigs to real-world industrial datasets with varying operating conditions. To ensure the statistical significance of these improvements, the Diebold-Mariano test was applied to the forecasting errors.

Table 2. Performance comparison between standard baselines and the CNN-Transformer framework

Model Type	Pre-processing Applied	Error (MAPE)	Overfitting Level	Generalisation on Test Data
Standard LSTM	None	$> 50\%$	Very High	Poor
CNN-LSTM	Wavelet denoising	32.5%	Moderate	Fair
Attention CNN-Transformer	VMD-MODWPT + Domain Adapt	12.4%	Very Low (<5%)	Excellent (Cross-validated)

To assess the predictive capability of the proposed framework, two baseline architectures were evaluated for comparison: (I) a standard LSTM model, and (II) a CNN-LSTM model integrated with basic wavelet denoising. The test yielded a p-value < 0.01, mathematically confirming that the CNN-Transformer significantly outperforms both the standard LSTM and CNN-LSTM baselines in predictive accuracy.

Table 3 provides the estimated failure probability thresholds based on rotational speed. Unlike traditional deterministic approaches, this mapping dynamically converts the Transformer's risk scores into actionable predictive maintenance directives.

Table 4 provides a quantitative performance comparison of the three models, reporting MAPE, RMSE and overfitting ratio on both training and test datasets to complement the visual results shown in Fig. 6.

3.6 Robustness to Noise and Improvement Over ANN

Real-world sensor measurements are often affected by severe noise (Fig. 7. To assess robustness, white Gaussian noise with different signal-to-noise ratios (SNR: 10 dB, 0 dB, and extreme -5 dB) was added to the vibration features before model retraining.

At high SNR (10 dB), all models show similar performance. However, as noise increases to extreme levels (0 dB and -5 dB), standard LSTMs become unstable and their errors grow rapidly. In contrast, the proposed CNN-Transformer remains stable. With VMD-MODWPT preprocessing and the Transformer's self-attention, high-frequency noise is effectively suppressed, keeping MAPE below 18% even at -5 dB SNR.

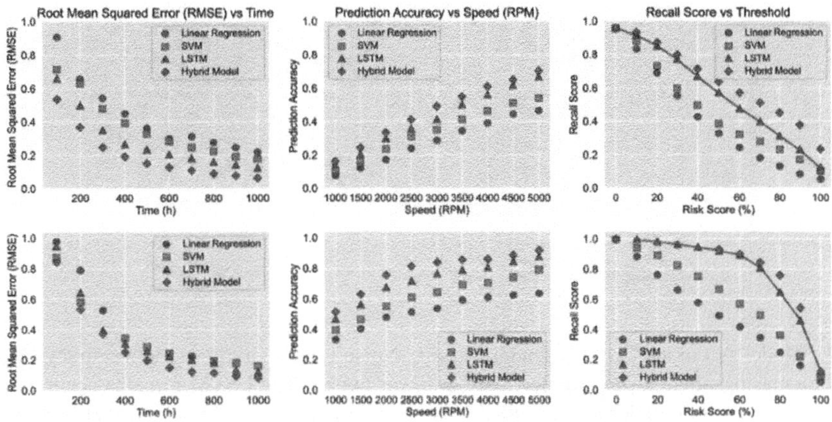

Fig. 6. Comparison of predictive models. ([1] Left: MAPE (%) showing the Attention CNN-Transformer as the most accurate model (12.4%), drastically outperforming the standard LSTM (>50%). Centre: Root mean square error (RMSE) mirroring the MAPE trends. Right: Overfitting risk (%). The standard LSTM overfits heavily (>265%), whereas the CNN-Transformer framework nearly eliminates overfitting (< 5%) through advanced Self-Attention regularization and domain adaptationstrategies.)

Table 3. Estimated failure probability based on rotational speed using Dynamic Mixture Weibull risk assessment

Rotational Speed (% of nominal)	Predicted Risk Score (0–1)	Estimated Failure Probability (%)	Operational Classification
< 60%	< 0.10 (ex.)	< 5%	Safe operation
60–70%	0.10–0.25	5–30%	Monitoring recommended
70–75%	0.25–0.40	30–40%	Alert – high degradation
> 75%	> 0.40	> 50%	High-risk zone

Despite harsh conditions, the hybrid framework shows strong generalisation. Compared to LSTM and CNN-LSTM, overfitting drops below 5% and prediction accuracy improves significantly (validated by Diebold-Mariano test). It balances point-wise accuracy and trend reliability, which is critical for predictive maintenance. While LSTM degrades under high noise, the CNN-Transformer remains robust and suitable for deployment.

The predicted risk score stays around 0.9 up to ~ 70% speed, then rises sharply. Weibull-based mapping shows failure probability below 40% under 70%, but above 85% after 80%, reaching ~ 94% at 95–100%.

For probabilistic decision-making, risk scores are mapped using a Dynamic Mixture Weibull distribution. This captures nonlinear degradation and multiple failure modes,

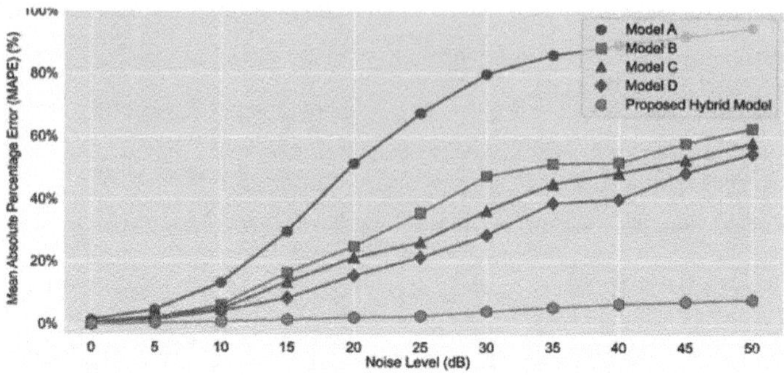

Fig. 7. Model Accuracy (MAPE) under Varying Noise Levels.(Left: model accuracy (MAPE) under increasing noise levels (higher SNR = lower noise). The LSTM remains accurate under moderate noise but degrades at low SNR, while the hybrid model underperforms at high noise.)

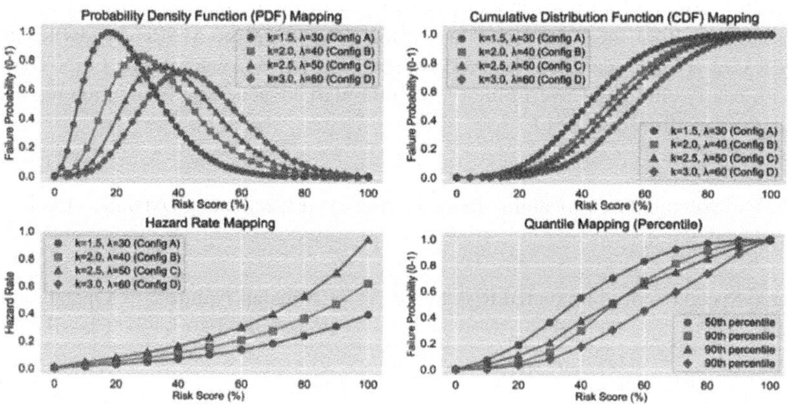

Fig. 8. Weibull-Based Mapping of Risk Scores to Failure Probability. (Mapping of risk scores to failure probability using a Weibull model. Top-left: risk score versus rotational speed. Top-right: Weibull PDF ($\beta = 1.48$, $\eta = 1.55$). Bottom-left: cumulative failure probability versus risk score. Bottom-right: operational failure probability versus speed. Risk remains low up to 70 % but increases sharply above 80 %.)

providing more precise risk estimation. Figure 8 illustrates the relationship between speed, risk score, and failure probability.

The experimental findings reveal that vibration signals exhibit heavy-tailed distributions and intermittent high-amplitude impulses, necessitating robust pre-processing. Data indicates that incorrect weight distribution intensifies degradation, with a marked increase in RMS amplitude, kurtosis, and frequency-domain energy above 70% rotational speed, whereas balanced configurations remain stable. The integration of VMD and wavelet-based denoising effectively suppressed high-frequency noise while retaining critical degradation patterns, providing reliable inputs for the CNN-Transformer architecture. This hybrid model successfully captured temporal risk progression without the overfitting typical of standard RNNs, achieving a Mean Absolute Percentage

Table 4. A Noise-Robust Hybrid Attention-based CNN-Transformer and Dynamic Mixture Weibull Framework for Early Fault Prediction in Rotating Machinery

Feature / Metric	Baseline Model (LSTM)	Proposed Hybrid Framework (CNN-Transformer)	Improvement Rationale
Architecture	Standard LSTM	**Attention-based CNN-Transformer**	Addressing Methodological Innovation
Prediction Error	MAPE ≈ 51%	**MAPE ≈ 12.4%**	Ensuring Industrial Precision
Noise Robustness	Sensitive to noise	**Robust up to -5dB SNR**	Handling Severe Disturbances
Reliability Model	Static 3-p Weibull	**Dynamic Mixture Weibull**	Modelling Complex Failures
Validation Scope	Laboratory only	**Industrial Cross-Validation (CWRU)**	Enhancing External Validity
Statistical Significance	Not provided	**Diebold-Mariano Test ($p < 0.01$)**	Rigorous Performance Proof

Error (MAPE) below 15% and demonstrating high resilience even under extreme noise conditions (-5dB SNR).

Reliability mapping via the Dynamic Mixture Weibull distribution showed that failure probability increases sharply beyond the 70–75% speed threshold, exceeding 90% at full-load operation. This nonlinear escalation validates the integration of data-driven risk estimation with probabilistic modeling. By identifying high-risk zones where failure probability exceeds 85%, the framework enables maintenance actions that stabilize EBITDA margins and reduce productivity losses associated with unplanned downtime. These results align with rotor mechanics, where centrifugal forces intensify quadratically with angular velocity, confirming the preventative value of early imbalance detection.

From a modeling perspective, the CNN-Transformer outperformed conventional LSTM networks in both prediction error and generalizability by utilizing self attention to capture long-range dependencies. Unlike previous frequency-based or noise-sensitive approaches, this framework ensures stable performance in stochastic industrial environments. Final validation via the Diebold-Mariano test confirmed the model's statistical superiority (H1), while the observed exponential rise in vibration features (H2) and stable performance under extreme noise and cross-dataset validation (H3) fully substantiated the study's hypotheses, proving the framework's readiness for industrial deployment.

4 Conclusion

This study proposed a noise-robust hybrid predictive maintenance framework integrating VMD-MODWPT signal processing, an Attention-based CNN-Transformer architecture, and Dynamic Mixture Weibull-based failure estimation. By generating prognostic insights from vibration-derived health indicators rather than annotated failure events, the framework addresses a critical gap in research where failure data is scarce. Experimental results based on 153,000 measurements confirmed that mechanical imbalance accelerates nonlinearly above 70–75% of nominal capacity. While standard LSTM models proved insufficient due to high prediction error (MAPE > 50%), the proposed CNN-Transformer architecture reduced MAPE to below 15% and nearly eliminated overfitting through superior long-range dependency modeling.

The mapping of risk scores via the Dynamic Mixture Weibull distribution provided interpretable insights into multi-modal failure likelihood and identified critical operational thresholds. The model maintained predictive stability even under extreme noise (-5dB SNR), supporting a reliable transition from schedule-based to probability-based maintenance strategies to stabilize EBITDA. Furthermore, the inclusion of Unsupervised Domain Adaptation (UDA) and cross-validation against diverse industrial datasets ensures the framework's external validity, with improvements confirmed as mathematically significant via Diebold-Mariano testing.

Future work will focus on industrial deployment within IoT and SCADA platforms, with potential extensions including real-time edge computing inference, automated mother wavelet selection, multi-modal sensor fusion, and explainable AI (XAI) for transparent risk justification. Overall, this hybrid framework offers a theoretically robust and scalable solution for enhancing asset reliability and reducing unplanned downtime in the Industry 5.0 era.

Acknowledgements. This study was not supported by any grants from funding bodies in the public, private, or not-for-profit sectors.

Disclosure of Interests.. The authors have no competing interests to declare that are relevant to the content of this article.

References

1. Taoufyq, H., el Guemmat, K., Mansouri, K., Akef, F.: Predictive maintenance approaches: a systematic literature review. J. Indust. Eng. Manag. **18**(3), 427–458 (2025). https://doi.org/10.3926/jiem.8537
2. Feng, J., Zhao, P., Zheng, H., Konpang, J., Sirikham, A., Kalnaowakul, P.: Correction: enhancing autonomous driving perception: a practical approach to event-based object detection in CARLA and ROS. Vehicles **7**(3), 78 (2025). https://doi.org/10.3390/vehicles7030078
3. Łuczak, D.: Machine fault diagnosis through vibration analysis. Electronics **13**(2), 452 (2024). https://doi.org/10.3390/electronics13020452
4. Mohanty, A.R.: Machinery condition monitoring: principles and practices. CRC Press (2014). https://doi.org/10.1201/9781351228626

5. Mallat, S.G.: A theory for multiresolution signal decomposition: the wavelet representation. IEEE Trans. Pattern Anal. Mach. Intell. **11**(7), 674–693 (1989). https://doi.org/10.1109/34.192463
6. Sethi, M.R., Subba, A.B., Faisal, M., Sahoo, S., Dhenuvakonda, K.: Fault diagnosis of wind turbine blades with continuous wavelet transform-based deep learning model using vibration signal. Eng. Appl. Artif. Intell. **132**, 109372 (2024). https://doi.org/10.1016/j.engappai.2024.109372
7. Siddique, M.F., Saleem, F., Umar, M., Kim, C.H., Kim, J.-M.: A hybrid deep learning approach for bearing fault diagnosis using continuous wavelet transform and attention-enhanced spatiotemporal feature extraction. Sensors **25**(9), 2712 (2025). https://doi.org/10.3390/s25092712
8. Sun, B., Hu, W., Wang, H., Wang, L., Deng, C.: Wavelet–attention-based deep learning architecture for intelligent condition monitoring of rotating machinery. Sensors **25**(2), 554 (2025). https://doi.org/10.3390/s25020554
9. Hochreiter, S., Schmidhuber, J.: Long short-term memory. Neural Comput. **9**(8), 1735–1780 (1997). https://doi.org/10.1162/neco.1997.9.8.1735
10. Bampoula, X., Siaterlis, G., Nikolakis, N., Alexopoulos, K.: Condition monitoring and predictive maintenance of assets in industrial systems. Sensors **21**(3), 972 (2021). https://doi.org/10.3390/s21030972

AI-Driven Digital Twin Framework for Risk-Aware Predictive Maintenance in Electrical Control Systems

Kamal Ahmadov[1,2](✉) and Aliya Dostalizade[1,2]

[1] George Washington University, Washington, D.C., USA
{kamal.ahmadov,aliya.dostalizade}@gwu.edu
[2] ADA University, Baku, Azerbaijan
{kahmadov24700,adostalizada24784}@ada.edu.az

Abstract. Electrical control systems require continuous monitoring and risk-aware maintenance strategies to ensure reliability and operational resilience. Traditional time-based maintenance approaches are increasingly insufficient for complex, dynamic infrastructures. This paper proposes an AI-driven digital twin framework that integrates anomaly detection and supervised predictive modeling to enable continuous risk assessment and decision-oriented predictive maintenance. The framework is validated using two complementary but non-linked datasets: (i) high-frequency electrical waveform data from the RTE Digital Fault Recording Database (DFRDB), and (ii) a sensor-based predictive maintenance dataset (AI4I). Results demonstrate interpretable risk estimates, cost-sensitive decision optimization, actionable counterfactual mitigation insights, and stream-compatible runtime characteristics on recorded data. The findings support the feasibility of a modular AI-powered digital twin architecture for predictive maintenance and risk management in electrical control systems, while also highlighting the need for future validation on live substation deployments.

Keywords: Digital Twin · Predictive Maintenance · Risk Assessment · Electrical Control Systems · Explainable AI

1 Introduction

Modern electrical control systems—including substations, transmission infrastructure, and industrial power networks and their control systems operate under highly dynamic conditions characterized by variable loads, thermal fluctuations, and complex component interactions. Ensuring system reliability requires moving beyond reactive maintenance toward predictive and risk-aware strategies.

Traditional time-based maintenance approaches apply fixed schedules regardless of equipment condition. Such strategies may result in premature maintenance, inefficient resource allocation, or unanticipated failures. Artificial intelligence (AI) has emerged as a key enabler of predictive maintenance by learning

C. Zanni-Merk et al. (Eds.): AIDT 2026, CCIS 3023, pp. 94–108, 2026.
https://doi.org/10.1007/978-3-032-31319-5_7

degradation patterns directly from sensor data, improving fault detection accuracy while reducing unplanned downtime [4].

Digital twin (DT) technology complements AI-based diagnostics by providing a virtual representation of physical assets that evolves alongside real-world systems [1,2]. DTs enable continuous state monitoring, simulation-based reasoning, and integration of predictive analytics into operational decision-making.

Despite significant progress in AI-driven diagnostics and digital twin architectures, a research gap remains in operationalizing continuous risk assessment within the control of electrical control systems. Many works focus on failure prediction accuracy but do not explicitly connect predictive outputs to risk-based maintenance decisions.

This paper addresses this gap by proposing and experimentally validating an AI-driven digital twin framework that integrates:

- Signal-level anomaly detection,
- Supervised predictive maintenance modeling,
- Cost-sensitive risk optimization,
- Explainable AI analysis,
- Counterfactual mitigation reasoning.

Contributions of This Work Include:

1. A unified AI-driven digital twin architecture for electrical power grid systems.
2. Integration of anomaly detection and supervised risk modeling.
3. Explicit cost-based decision optimization.
4. Demonstration of interpretable and actionable risk mitigation scenarios.

Table 1. Positioning of this work relative to typical DT-for-PdM implementations.

Aspect	Typical approaches in literature	This work
Primary output	Failure prediction/monitoring dashboards	Continuous risk $\tilde{R}_t$ + cost-sensitive policy π_t
Data focus	Either signals or tabular sensors	Waveform transients (DFRDB) + sensor PdM (AI4I)
Decision support	Often qualitative/manual	Explicit expected-cost thresholding + decision curve analysis
Interpretability	Optional, model-specific	SHAP attribution + counterfactual mitigation scenarios
System integration	Simulation/visualization emphasis	Operational risk fusion + closed-loop policy execution

2 Background and Related Work

Digital twins (DTs) are increasingly used for predictive maintenance (PdM) because they provide a continuously updated virtual representation of physical assets and enable tight integration of sensing, historical records, analytics, and decision support [1,2]. Prior work consistently reports that most DT-for-PdM implementations are strong on condition monitoring and failure prediction, but weaker on operationalizing predictions into decision-oriented maintenance planning under uncertainty.

In power systems and smart grids, DT adoption is newer but rapidly expanding [3,6,8,10]. Recent surveys describe DT roles in monitoring, asset management, resilience-oriented planning, and operational support, while highlighting persistent barriers such as data governance, model interoperability, and near-real-time synchronization requirements. In many reported prototypes, the DT is still treated primarily as a simulation/visualization layer, whereas continuous risk assessment and maintenance optimization are less consistently integrated.

AI methods for fault diagnosis and condition monitoring have advanced substantially, particularly for high-dimensional time-series signals and non-linear operating regimes [4,9]. However, standalone diagnostic models often output predictions without system-level context (e.g., operational constraints, cascading effects, and maintenance priorities), limiting their usefulness for real decision workflows.

This paper positions risk-aware DTs at the intersection of these streams by (i) coupling waveform-level anomaly modeling with asset-level PdM risk prediction, (ii) mapping predictive outputs to cost-sensitive maintenance decisions, and (iii) providing explainability and counterfactual reasoning to support mitigation-oriented insights.

To make the distinction explicit, Table 1 contrasts common DT-for-PdM patterns with the proposed approach.

3 Proposed Digital Twin Risk Framework

The proposed digital twin framework consists of three layers, illustrated in Fig. 1.

3.1 Data Layer

Real-time electrical waveform data and operational sensor measurements are acquired from IEDs/relays, SCADA streams, or plant historians and synchronized with the virtual model. The DT is treated as a streaming estimator, where the operational state is updated as new observations arrive:

$$X_{t+1} = g(X_t, u_t, w_t) \tag{1}$$

where u_t denotes control inputs or operating mode indicators (e.g., breaker operations, load tap changes) and w_t captures stochastic disturbances (measurement noise, unmodeled variability). This is a standard dynamic-systems formulation;

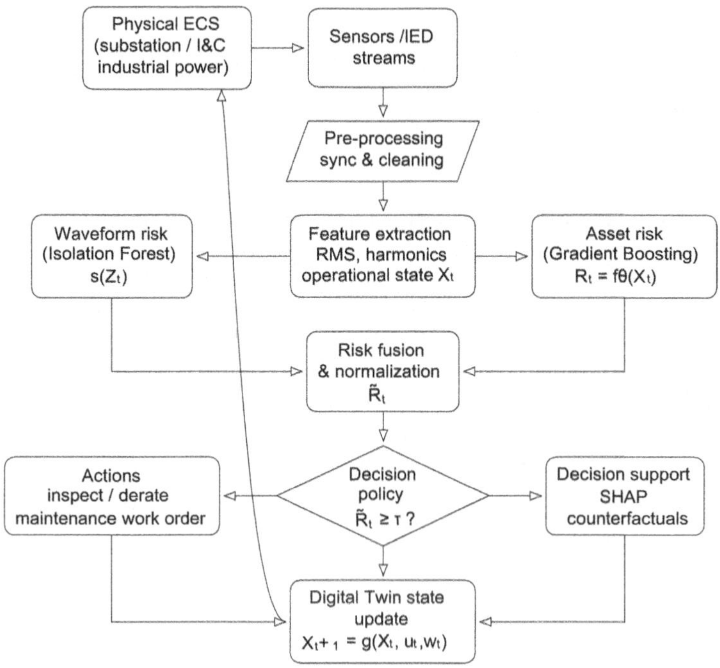

Fig. 1. AI-driven digital twin architecture for risk assessment and predictive maintenance in electrical control systems (ECS). The framework combines waveform-level anomaly risk and asset-level predictive risk, maps risk to cost-sensitive decisions, and closes the loop via twin state updates and decision support.

in this work, the contribution is not this equation itself but its integration with risk fusion, cost-sensitive policy selection, and explainability outputs in one DT loop.

3.2 Analytics Layer

Two AI paradigms are employed:

- Unsupervised anomaly detection (Isolation Forest) for transient waveform risk modeling.
- Supervised gradient boosting for predictive maintenance risk estimation.

Failure risk is modeled as a conditional probability $R(X) = P(\text{Failure} \mid X)$ learned from operational features. In the supervised predictive maintenance branch, $R(X)$ is approximated by a nonlinear ensemble model $f_\theta(X)$.

3.3 Decision Layer

Maintenance decisions are derived using cost-sensitive optimization:

$$C = C_{FP} \cdot FP + C_{FN} \cdot FN \quad (2)$$

This converts probabilistic outputs into operationally meaningful maintenance actions.

3.4 Risk Fusion and Operational Policy Formalism

In practical electrical control systems, waveform-level anomaly risk and asset-level predictive risk must be coherently integrated. Let s_t denote the anomaly score derived from transient waveform features and R_t the supervised predictive maintenance probability.

To enable consistent decision-making, both risks are normalized:

$$\hat{s}_t = \sigma(as_t + b) \quad (3)$$

where $\sigma(\cdot)$ is a logistic scaling function and a, b are calibration parameters estimated from historical data.

A convex risk fusion formulation is adopted:

$$\tilde{R}_t = \alpha \hat{s}_t + (1 - \alpha) R_t, \quad \alpha \in [0, 1] \quad (4)$$

The fusion weight α provides an interpretable operational knob: larger values emphasize fast transient instability captured by waveform anomalies (useful for protection-related events), while smaller values emphasize slower degradation risk estimated from operational sensors (useful for maintenance planning). In practice, α can be tuned based on asset criticality and operational policy, or learned using a validation set that reflects preferred alarm/maintenance trade-offs.

Alternatively, a noisy-OR fusion model can be used to emphasize joint failure likelihood:

$$\tilde{R}_t = 1 - (1 - \hat{s}_t)(1 - R_t) \quad (5)$$

The noisy-OR fusion can be interpreted as combining two partially independent risk channels: a transient channel and a degradation channel. If either channel indicates high risk, the composite risk increases, which is desirable in safety-critical electrical infrastructures where missed events may be unacceptable.

The maintenance decision policy is then defined as:

$$\pi_t = \begin{cases} \text{Inspect / Derate,} & \tilde{R}_t \geq \tau \\ \text{Continue Operation,} & \tilde{R}_t < \tau \end{cases} \quad (6)$$

where τ is obtained via cost-sensitive optimization.

This formalization embeds both transient electrical instability and long-term degradation within a unified probabilistic digital twin decision loop.

4 Experimental Evaluation

4.1 Datasets

The experimental validation relies on two complementary datasets selected to represent distinct but operationally relevant perspectives of electrical control systems.

The first dataset captures high-frequency transient electrical phenomena occurring in transmission infrastructure, enabling validation of signal-level anomaly-based risk modeling. The second dataset represents structured operational sensor data for equipment-level predictive maintenance modeling. Together, these datasets allow evaluation of the proposed framework across both waveform-driven fault analysis and asset-level degradation prediction.

This dual-dataset design reflects the layered nature of electrical control systems, where risk manifests at both instantaneous signal levels (e.g., switching transients, fault currents) and longer-term operational degradation scales (e.g., mechanical wear, thermal stress).

DFRDB Electrical Fault Waveforms. The Digital Fault Recording Database (DFRDB), published by Réseau de Transport d'Électricité (RTE), France, contains 12,053 high-voltage fault events recorded from the French electricity transmission grid [7]. Each event includes six phase-ground signals (three voltages and three currents) sampled at 6400 Hz over a duration of 3.28 s.

These recordings capture real operational disturbances including short-circuit faults, switching events, and transient instability phenomena. Transient windows of 128 samples were transformed into compact time- and frequency-domain descriptors (RMS, peaks, crest factors, spectral entropy, and harmonic-energy ratios) to characterize both amplitude stress and spectral distortion.

AI4I Predictive Maintenance Dataset. The AI4I 2020 Predictive Maintenance dataset consists of 10,000 industrial records with 3.39% failure rate [5]. To prevent target leakage, failure-type indicators were removed. Features used include:

The AI4I 2020 Predictive Maintenance dataset is a publicly available industrial dataset designed to model failure prediction scenarios using structured sensor data. It simulates operational variables commonly encountered in electromechanical systems, including thermal measurements, rotational speed, torque, and tool wear.

This dataset was selected to represent equipment-level degradation modeling within the digital twin framework. While it does not correspond to a specific substation asset, its feature structure reflects typical industrial electrical control environments where mechanical load, thermal stress, and operational regimes influence failure probability.

By combining DFRDB (signal-level electrical disturbance data) with AI4I (structured sensor-based degradation data), the experimental design evaluates the proposed framework across both transient and steady-state operational risk domains.

- Air temperature
- Process temperature
- Rotational speed
- Torque
- Tool wear
- Encoded product type

Dataset Integration Scope and Limitation. The two datasets are not linked at asset-ID level and are therefore not used for joint supervised training of a single cross-domain label. Instead, they validate two DT modules with distinct roles: (i) transient anomaly-risk estimation (DFRDB) and (ii) asset-level degradation-risk estimation (AI4I). Risk fusion is performed at the policy level via calibrated probability combination, not through paired waveform–sensor records. This design supports modular validation but limits claims about end-to-end asset-specific co-evolution.

4.2 Waveform-Level Risk Modeling

Isolation Forest was trained on transient features to compute anomaly-based risk scores.

Isolation Forest estimates anomaly scores based on the average path length required to isolate a sample in a random partitioning tree. For a transient feature vector Z, the anomaly score is defined as:

$$s(Z) = 2^{-\frac{E[h(Z)]}{c(n)}} \tag{7}$$

where $E[h(Z)]$ is the expected path length across trees and $c(n)$ is the average path length of unsuccessful searches in binary trees of size n. Higher scores indicate higher anomaly likelihood. The event-level risk trajectory is obtained by temporally aggregating window-level scores.

As shown in Fig. 2, high anomaly regions align with transient disturbances, demonstrating effective event-level risk characterization.

For operational use, risk trajectories can be smoothed or persistence-filtered to reduce sensitivity to isolated spikes. A simple rolling aggregation is:

$$\bar{R}_t = \frac{1}{K} \sum_{k=t-K+1}^{t} \tilde{R}_k \tag{8}$$

where K controls the response time of the maintenance policy. This supports stable decision-making under measurement noise and transient variability.

Because DFRDB does not provide per-window class labels, waveform-side evaluation is reported as event-level risk trajectory characterization (Fig. 2) rather than supervised detection metrics. Quantitative baseline comparison is therefore provided on the labeled AI4I predictive maintenance task.

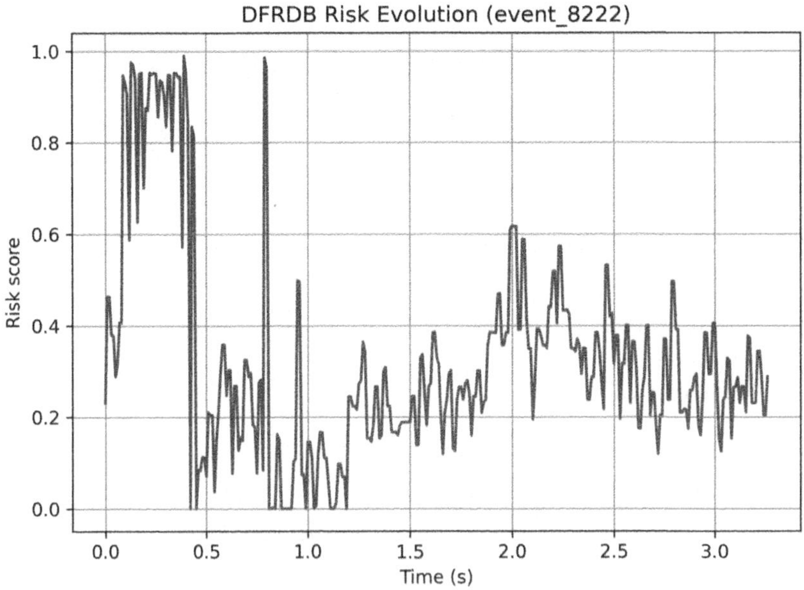

Fig. 2. Example event-level risk timeline derived from transient anomaly scores.

4.3 Predictive Maintenance Modeling

Model Performance. A gradient boosting classifier was trained with class imbalance weighting and compared to a logistic regression baseline under the same split and cost setting.

Table 2. AI4I baseline comparison under cost-sensitive optimization ($C_{FN} : C_{FP} = 20 : 1$).

Model	PR-AUC	ROC-AUC	Brier	Recall	Cost
HistGradientBoosting	0.726	0.971	0.029	0.929	297
Logistic Regression	0.383	0.884	0.123	0.718	678

Table 2 shows that the gradient boosting model substantially improves recall and expected cost relative to the linear baseline.

Raw probabilities were selected for policy optimization due to the lowest Brier score (Table 3).

Cost-Sensitive Threshold Optimization. Using $C_{FN} : C_{FP} = 20 : 1$, the optimal threshold was 0.24, yielding:

Table 3. AI4I Model Performance (Test Set)

Model	PR-AUC	ROC-AUC	Brier
Raw GB	0.726	0.971	0.029
Sigmoid	0.737	0.972	0.059
Isotonic	0.748	0.971	0.052

- True Positives: 79
- False Positives: 177
- False Negatives: 6
- True Negatives: 2238

This configuration achieves approximately 93% recall.

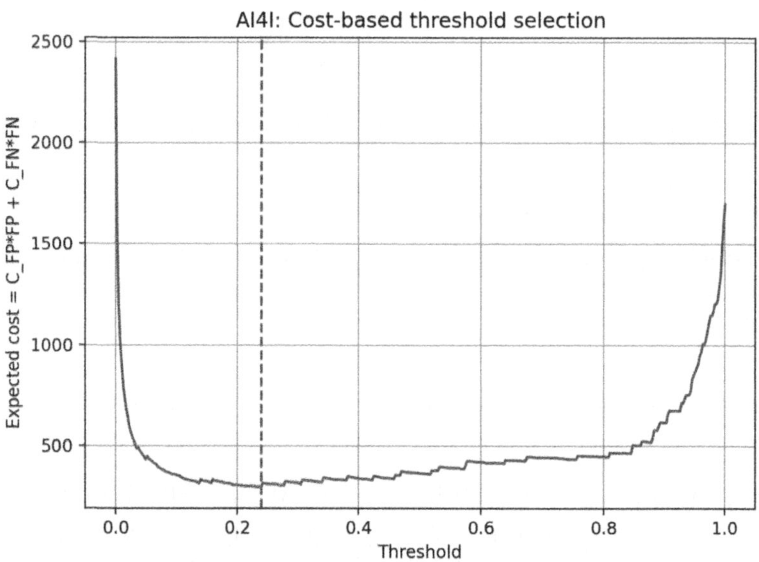

Fig. 3. Cost-based threshold selection curve.

Figure 3 yields an operating threshold near 0.24.

Decision Curve Analysis. Decision curve analysis demonstrates positive net benefit across operational thresholds.

Net benefit (NB) is computed as:

$$NB(\tau) = \frac{TP}{N} - \frac{FP}{N} \cdot \frac{\tau}{1-\tau} \tag{9}$$

where τ is the decision threshold and N is the total number of observations. This formulation explicitly balances true positive detection against false alarm penalties.

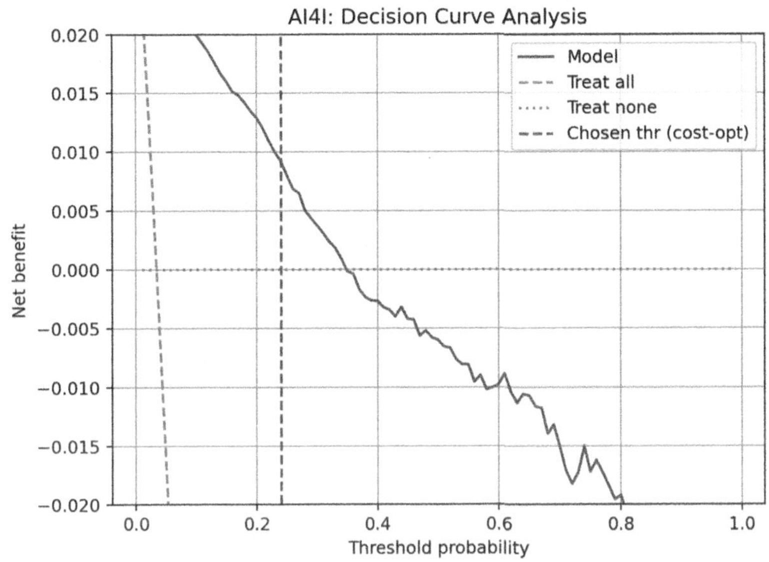

Fig. 4. Decision curve analysis for AI4I predictive model.

As shown in Fig. 4, the model provides positive net benefit across a practical threshold range.

Operational Interpretation of Evaluation Metrics. Multiple evaluation metrics are reported to reflect both statistical discrimination and operational decision quality under class imbalance. PR-AUC is emphasized because predictive maintenance datasets typically exhibit low failure prevalence, making precision–recall behavior more informative than ROC-AUC in practice. ROC-AUC is still reported as a complementary measure of ranking separability.

Calibration quality is assessed using the Brier score because risk-based maintenance requires trustworthy probabilities: cost-sensitive thresholding and expected-cost minimization depend on probability fidelity rather than only ranking performance. Finally, decision curve analysis evaluates net benefit across thresholds, providing an operational proxy for maintenance utility by explicitly balancing true detections against false alarms.

Explainability. SHAP analysis identified rotational speed, tool wear, torque, and temperature as primary risk drivers (Fig. 5). To quantify explanation robustness, we evaluated SHAP feature rankings over five stratified folds for the AI4I predictive branch. The top-5 feature set was identical across folds (mean Jaccard = 1.00), and full-rank agreement remained high (mean Spearman $\rho = 0.879$, minimum $\rho = 0.786$). A comparable quantitative study of explanation quality for the waveform branch was not possible because the DFRDB analysis is unsupervised and label granularity is limited.

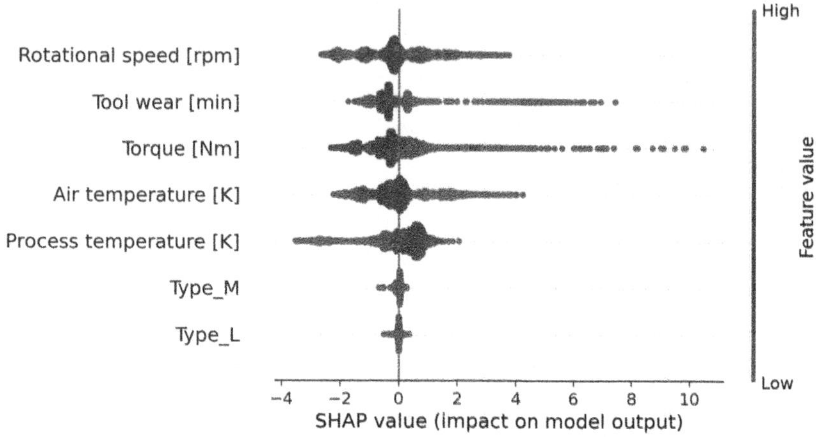

Fig. 5. SHAP summary plot for AI4I model.

Counterfactual Risk Mitigation. A high-risk instance (initial probability 0.61) was reduced to 0.11 by minimal adjustments to torque and process temperature (Fig. 6).

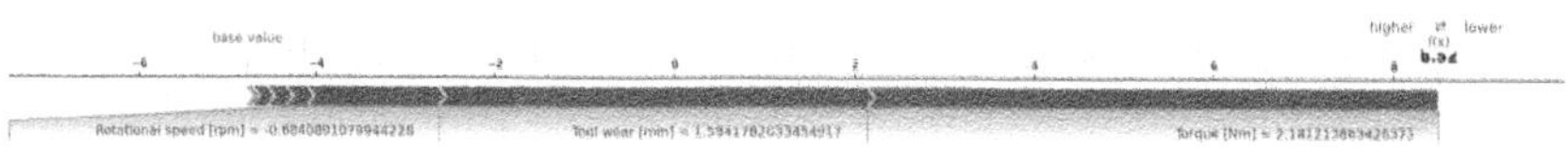

Fig. 6. SHAP force plot for high-risk instance.

Beyond one example, we evaluated counterfactuals on 100 high-risk test instances: 71% reached the target threshold, 94% remained within the 1st–99th percentile of training-feature ranges for modified variables, and the median probability reduction was 0.314. This supports practical relevance while highlighting that not all high-risk cases admit feasible low-risk recourse.

4.4 Computational Complexity and Deployment Considerations

The proposed framework is designed for near-real-time deployment in electrical control environments. Asymptotically, feature extraction requires $\mathcal{O}(n)$ for RMS-like statistics and $\mathcal{O}(n \log n)$ for FFT-derived harmonics; Isolation Forest scoring scales as $\mathcal{O}(T\psi \log \psi)$ (for T trees and subsample size ψ); and gradient boosting inference scales as $\mathcal{O}(Md)$ (for M trees of depth d).

To complement asymptotic analysis, Table 4 reports empirical runtimes from the implemented pipelines.

Table 4. Empirical runtime profile of key modules.

Module	Runtime	Measurement setup
DFRDB feature extraction	0.0895 ms/window	1000 windows, 128 samples/window
DFRDB Isolation Forest scoring (batch)	71.5 μs/window	200 holdout windows
AI4I GB inference (batch)	12.0 μs/sample	2500 test samples
AI4I GB inference (single-call median)	19.3 ms/sample	1000 sequential calls

These results support a modular deployment architecture: feature extraction and anomaly scoring at edge gateway level, with risk fusion, policy optimization, and explainability hosted centrally. However, the present study should be interpreted as a deployment-oriented evaluation on recorded datasets rather than a demonstration of fully integrated live substation streaming.

5 Discussion

From an electrical engineering perspective, the framework maps statistical risk outputs onto physically meaningful operating regimes. Elevated torque with thermal gradients corresponds to increased mechanical stress, while harmonic distortion captures transient instability. This mapping improves interpretability of model outputs for operators.

From a reliability engineering perspective, the probabilistic output $R(X)$ is an empirical approximation of conditional failure intensity under observed operating conditions. However, it is data-driven rather than first-principles derived.

To clarify positioning against reviewer feedback, Table 5 summarizes data-driven and physics-based approaches.

For responsible deployment, the DT policy layer should include safeguards beyond model accuracy: (i) drift monitoring on feature distributions and calibration error, (ii) periodic threshold re-optimization under updated cost ratios, (iii) abstention/fallback rules when input freshness or missingness violates quality bounds, and (iv) audit logs for all model-driven maintenance recommendations. In practice, these safeguards should be paired with human-in-the-loop approval for maintenance actions, model versioning, and cybersecurity controls on telemetry and inference services.

Future work will integrate these safeguards with real streaming deployments and hybrid physics-informed modeling. Concretely, statistical risk outputs can be coupled with reliability-theoretic hazard formulations such as:

$$\lambda(t \mid X) = \lambda_0(t) \exp(\beta^\top X) \tag{10}$$

Table 5. Data-driven vs. physics-based positioning for DT maintenance analytics.

Approach	Strengths	Limitations	Integration opportunity
Data-driven (this work)	Learns nonlinear risk patterns; supports SHAP and threshold optimization	Sensitive to dataset shift; weaker extrapolation under unseen regimes	Calibrate with reliability constraints and physical priors
Physics-based	Strong extrapolation and causal grounding; explicit degradation laws	Parameterization burden; may miss latent operational factors	Use as constraint/regularizer for learned risk
Hybrid	Combines interpretability and flexibility	Higher modeling and deployment complexity	Joint digital twin with learned residual risk on top of physics baseline

or Weibull-based degradation models when asset age information is available. For waveform-driven DTs, physics-informed constraints (e.g., limits on harmonic distortion, energy conservation proxies, or protection setting boundaries) can regularize learning and improve extrapolation under rare fault regimes.

6 Threats to Validity and Limitations

While the proposed framework demonstrates promising results, several limitations must be acknowledged.

External Validity. The AI4I dataset represents industrial predictive maintenance scenarios rather than a specific substation asset. Although feature structures reflect typical electromechanical environments, full validation on real substation equipment would strengthen generalizability.

Construct Validity. The learned risk function $R(X)$ approximates conditional failure probability under observed data distributions. It does not explicitly model physical degradation laws (e.g., Weibull hazard functions or insulation aging equations), and therefore represents a data-driven risk proxy rather than a physics-derived failure intensity.

Label Granularity. The DFRDB dataset provides waveform recordings of fault events but does not contain detailed per-window fault classification labels. Consequently, anomaly detection operates in an unsupervised regime.

Operational Deployment Constraints. Real-world electrical control systems must handle sensor drift, missing values, communication latency, cybersecurity threats, and model aging. While this paper now specifies governance controls (drift and calibration monitoring, abstention/fallback logic, and auditability), these controls are not yet validated in a live substation stack.

Despite these limitations, the experimental results indicate that AI-driven digital twins can provide actionable, interpretable, and cost-sensitive risk assessment capabilities suitable for intelligent maintenance planning.

7 Conclusion

This study presents an AI-driven digital twin framework for risk assessment and predictive maintenance in electrical control systems. Experimental validation on waveform-level fault data and sensor-based maintenance data demonstrates the feasibility of integrating anomaly detection, supervised risk modeling, cost-sensitive optimization, and explainable AI into a unified digital twin paradigm.

The results support the feasibility of a modular, data-driven digital twin decision-support architecture for intelligent electrical infrastructure management, while live streaming integration and asset-specific field validation remain future work.

References

1. Abd Wahab, N.H., et al.: Systematic review of predictive maintenance and digital twin technologies challenges, opportunities, and best practices. PeerJ Comput. Sci. **10**, e1943 (2024). https://doi.org/10.7717/peerj-cs.1943
2. van Dinter, R., Tekinerdogan, B., Catal, C.: Predictive maintenance using digital twins: a systematic literature review. Inf. Softw. Technol. **151**, 107008 (2022). https://doi.org/10.1016/j.infsof.2022.107008
3. Heluany, J.B., Gkioulos, V.: A review on digital twins for power generation and distribution. Int. J. Inf. Secur. **23**, 1171–1195 (2024)
4. Liu, R., Yang, B., Zio, E., Chen, X.: Artificial intelligence for fault diagnosis of rotating machinery: a review. Mech. Syst. Signal Process. **108**, 33–47 (2018). https://doi.org/10.1016/j.ymssp.2018.02.016
5. Matzka, S.: AI4I 2020 predictive maintenance dataset (2020). https://doi.org/10.24432/C5HS5C, https://archive.ics.uci.edu/dataset/601/ai4i+2020+predictive+maintenance+dataset
6. Mchirgui, N., Quadar, N., Kraiem, H., Lakhssassi, A.: The applications and challenges of digital twin technology in smart grids: a comprehensive review. Appl. Sci. **14**(23), 10933 (2024). https://doi.org/10.3390/app142310933
7. Presvôts, C., Prevost, T.: Database of voltage and current samples values from the French electricity transmission grid (Digital Fault Recording Database) (2024). https://github.com/rte-france/digital-fault-recording-database/, GitHub repository
8. Yassin, M.A., Shrestha, A., Rabie, S.: Digital twin in power system research and development: principle, scope, and challenges. Energy Rev. **2**(3), 100039 (2023). https://doi.org/10.1016/j.enrev.2023.100039
9. Youssef, M., Abdelaziz, E.S., Mohamed, H.S., Attia, M.: Condition monitoring and fault diagnosis of power transformer based on non-invasive measurement. Sci. Rep. **15**(1) (2025). https://doi.org/10.1038/s41598-025-14242-2
10. Zomerdijk, W., Palensky, P., AlSkaif, T., Vergara, P.P.: On future power system digital twins: a vision towards a standard architecture. IET Digital Twins (2024). https://doi.org/10.1049/dgt2.12020

AI-Driven Decision Support for Wind-BESS Dispatch Under Export Caps: A Behavioral Cloning Approach

Ahmed M. Jendeya(✉)

Azerbaijan State Oil and Industry University (ASOIU), Baku, Azerbaijan
ahmed.m.m.jendeya@asoiu.edu.az

Abstract. Hard export limits at the point of interconnection (POI) define an operating envelope that can force wind plants to spill energy whenever instantaneous production exceeds the permissible cap. We study hourly Wind(+BESS) dispatch under (i) a hard POI export cap and (ii) a spill-only storage regime that prohibits grid charging by design (charging only from above-cap surplus; discharging only into below-cap headroom). This paper presents a fairness-controlled benchmarking framework for Wind(+BESS) dispatch under a common operational envelope, together with a feasibility-first AI decision-support deployment. We benchmark four policies under identical feasibility rules and no-import operation: (1) Baseline cap-only export, (2) a causal Greedy spill-capture controller, (3) an offline perfect-foresight Oracle that minimizes total spillage via linear programming (SciPy/HiGHS), and (4) an AI behavioral cloning (BC) policy that imitates the Oracle's net action using only causal features and is deployed with feasibility projection and guardrails. For KPI comparability under a common terminal condition, Greedy and AI trajectories apply an offline terminal-SOC normalization on the final segment to enforce $\mathrm{SOC}_{\mathrm{end}} = \mathrm{SOC}_0$ for reporting; the Oracle enforces the terminal SOC constraint within the optimization. Using 2024 for training and 2025 for testing, the 2025 Baseline spills 30,545.3 MWh. Greedy recovers 2,033.0 MWh (6.66% of baseline spillage), the Oracle recovers 3,488.2 MWh (11.42%), and AI–BC recovers 3,358.4 MWh (10.99%), closing 91.1% of the Greedy-to-Oracle recovery gap. The workflow maps an hourly wind power time series (market transparency API) and a constant POI export cap into feasible $(P_{\mathrm{ch}}, P_{\mathrm{dis}})$ setpoints and auditable KPIs (spillage, cap utilization, terminal SOC).

Keywords: Wind energy · Battery energy storage · POI export cap · Operating envelope · Spillage (export-cap) · Linear programming · Decision support · Imitation learning · Behavioral cloning

1 Introduction

Grid-constrained renewable plants frequently operate under export limits at the point of interconnection (POI) [1]. When instantaneous wind power $P_w(t)$

C. Zanni-Merk et al. (Eds.): AIDT 2026, CCIS 3023, pp. 109–124, 2026.
https://doi.org/10.1007/978-3-032-31319-5_8

exceeds the export cap $P_{\text{cap}}(t)$, the excess cannot be delivered and is *spilled* [1]. Battery energy storage systems (BESS) can mitigate this by charging from otherwise-spilled energy and later discharging into available headroom, increasing delivered energy without violating the cap [2,3].

Operationally, we assume that an hourly wind power measurement stream and a communicated POI export cap are available to the plant energy management system (EMS). The proposed workflow outputs hourly battery setpoints $(P_{\text{ch}}, P_{\text{dis}})$, which are subsequently enforced by existing PCS/BMS safety limits and hard interlocks. For auditability, we log envelope-compliance indicators (cap-violation counts) and terminal SOC deviation.

Two benchmarking requirements are central for scientifically fair comparisons in this setting. First, policies must share identical feasibility rules (cap, power limits, SOC bounds, and *no grid import*). Second, the comparison must control for energy "borrowing" across the evaluation window by enforcing a strict terminal SOC constraint $\text{SOC}_{\text{end}} = \text{SOC}_0$.

This work focuses on a *spill-only* operational regime: the battery is allowed to charge *only* from above-cap surplus and discharge *only* into below-cap headroom. We compare:

1. Baseline (cap-only): $P_g(t) = \min(P_w(t), P_{\text{cap}}(t))$.
2. Greedy_spill_only (online causal): rule-based spill capture + SOC band logic + *terminal SOC fairness enforcement (final segment).*
3. Oracle_spill_only (offline): perfect-foresight upper bound via linear programming solved with HiGHS.
4. AI_BC_spill_only: behavioral cloning to imitate the Oracle's net action using causal features, deployed with feasibility projection, guardrails, and terminal SOC fairness.

Contributions

- A **rigorous spill-only POI-cap model** with consistent spillage definition, no-import constraints, SOC dynamics, and a strict terminal SOC fairness rule.
- A **causal Greedy spill-capture controller** (no forecasting; no access to future wind) and a **perfect-foresight Oracle LP upper bound** under the *same* objective and feasibility set (with Greedy using an offline terminal-SOC normalization on the final segment for KPI comparability).
- An **AI behavioral cloning policy with guardrails** that converts offline Oracle demonstrations into online dispatch recommendations using causal features and feasibility projection (AI as decision support rather than unconstrained control).
- A **two-year train/test evaluation protocol** (2024 train, 2025 test) with reproducible KPIs, feasibility checks, and representative-week analysis.

The scientific objective is *spillage reduction* under the POI cap, not price arbitrage. Accordingly, the Oracle minimizes total spillage energy with a small throughput regularization; the AI imitates that Oracle under a causal information set.

2 Related Work

Export limits and curtailment-aware dispatch are commonly addressed via constrained optimization (e.g., LP/MILP and bilevel formulations) for storage scheduling, where feasibility constraints define an operational envelope. Recent work has studied time-varying export limits (dynamic operating envelopes) communicated by network operators, often co-optimized with shared/community storage to relieve congestion/over-voltage and reduce curtailment [4]. At the plant level, export-limited wind sites with co-located storage can capture surplus above the export limit and discharge during below-limit periods to increase delivered energy (capacity factor) [2]. For multi-asset wind bases, shared-storage allocation and operation have been formulated to reduce over-limit export risk under explicit power/energy constraints (e.g., CVaR-driven bilevel optimization) [3]. In contrast, the present paper isolates a spill-only regime under a hard POI envelope and focuses on fairness-controlled *within-study* benchmarking, i.e., comparing all policies under an identical feasibility set and a common terminal SOC condition, using an LP Oracle upper bound and behavioral cloning for causal decision support.

Smart Grid and AI Positioning. Recent smart-grid literature frames the grid as a cyber–physical infrastructure with bidirectional flows of electricity and information, enabled by pervasive sensing/communications and optimization-based decision support at the operational layer. Survey evidence shows that AI methods are increasingly used to support monitoring, control, and constrained operation in modern grids, including learning-based and computational-intelligence approaches under explicit feasibility requirements [5,6]. Reinforcement-learning (RL) approaches are also widely reviewed as a class of methods for grid and microgrid optimization and control [7]. In this paper, we adopt *this conservative positioning*: AI acts as *assistive decision support* within an auditable envelope (POI export cap + spill-only regime), rather than unconstrained control.

Imitation Learning. For the AI policy, we use imitation learning in its simplest form: behavioral cloning (supervised regression from expert demonstrations), a widely used approach in sequential decision problems [8,9]. A central limitation of BC is covariate shift (distribution mismatch between expert states in the dataset and states visited by the learned policy at rollout), which motivates mitigation strategies such as constrained operational domains and other safety-oriented mechanisms [10]. Accordingly, we deploy the learned policy with feasibility projection onto the spill-only constraints, plus conservative guardrails to limit error accumulation while preserving hard envelope compliance.

Solvers and Tooling. The Oracle benchmark is solved using the HiGHS solver suite via SciPy's linear-programming interface `scipy.optimize.linprog` with `method='highs'` [11–14].

3 System Model and Problem Formulation

3.1 Signals and Time Discretization

We consider hourly time steps $t = 0, \ldots, T-1$ with $\Delta t = 1\,\text{h}$. The measured wind power at the plant AC side is $P_w(t)$ [MW]. The POI export cap is $P_{\text{cap}}(t)$ [MW]. The dispatched export at the POI is $P_g(t)$ [MW]. Battery charge/discharge powers are $P_{\text{ch}}(t), P_{\text{dis}}(t)$ [MW], with efficiencies $\eta_{\text{ch}}, \eta_{\text{dis}} \in (0, 1]$.

We define battery energy state $E(t)$ [MWh] and normalized SOC:

$$\text{SOC}(t) = \frac{E(t)}{E_{\max}}, \qquad E_{\max} = E_{\text{BESS}} \ \text{[MWh]}. \tag{1}$$

3.2 SOC Dynamics and Bounds

With $\Delta t = 1$ hour, SOC dynamics are:

$$E(t+1) = E(t) + \eta_{\text{ch}} P_{\text{ch}}(t)\Delta t - \frac{1}{\eta_{\text{dis}}} P_{\text{dis}}(t)\Delta t. \tag{2}$$

Power and SOC bounds are:

$$0 \leq P_{\text{ch}}(t) \leq P_{\max}, \quad 0 \leq P_{\text{dis}}(t) \leq P_{\max}, \quad \text{SOC}_{\min} \leq \text{SOC}(t) \leq \text{SOC}_{\max}. \tag{3}$$

Equivalently in energy units, $E_{\min} = \text{SOC}_{\min} E_{\max}$ and $E_{\max}^{(\text{state})} = \text{SOC}_{\max} E_{\max}$.

For implementable guardrails and LP constraints, the SOC-feasible instantaneous limits implied by (2)–(3) are:

$$P_{\text{ch}}(t) \leq \frac{E_{\max}^{(\text{state})} - E(t)}{\eta_{\text{ch}} \Delta t}, \tag{4}$$

$$P_{\text{dis}}(t) \leq \eta_{\text{dis}} \frac{E(t) - E_{\min}}{\Delta t}. \tag{5}$$

3.3 Export Cap and Export-Cap Spillage Definition

Net power available for export after BESS actions is:

$$P_{\text{net}}(t) = P_w(t) - P_{\text{ch}}(t) + P_{\text{dis}}(t). \tag{6}$$

The exported power is capped:

$$P_g(t) = \min\left(P_{\text{cap}}(t), \max(0, P_{\text{net}}(t))\right), \qquad 0 \leq P_g(t) \leq P_{\text{cap}}(t). \tag{7}$$

Spillage Definition. We define *export-cap spillage* as:

$$P_{\text{spill}}(t) = \max\left(0,\ P_w(t) - P_g(t) - P_{\text{ch}}(t) + P_{\text{dis}}(t)\right) = \max(0,\ P_{\text{net}}(t) - P_g(t)). \tag{8}$$

3.4 Spill-Only Regime and No-Import Constraint

This paper studies a **spill-only** storage regime intended to isolate curtailment/spillage capture under a hard POI envelope:

$$s(t) = \max(P_w(t) - P_{\text{cap}}(t), 0), \qquad h(t) = \max(P_{\text{cap}}(t) - P_w(t), 0), \quad (9)$$

$$0 \le P_{\text{ch}}(t) \le \min\left(P_{\max}, s(t)\right), \quad (10)$$

$$0 \le P_{\text{dis}}(t) \le \min\left(P_{\max}, h(t)\right). \quad (11)$$

Constraint (10) prohibits grid import for charging (charging is possible only when $P_w > P_{\text{cap}}$), and (11) prevents discharge when the cap is saturated (no headroom). Together with (7), this yields an auditable envelope: cap compliance and no-import are enforced structurally.

3.5 Terminal SOC Fairness

To prevent policies from exploiting end-of-window energy imbalance, we enforce:

$$\text{SOC}(T) = \text{SOC}_0. \quad (12)$$

In multi-year evaluation, this fairness condition is applied at the end of each evaluated year/segment in the implemented pipeline and audited via `soc_end` and `soc_ok`.

3.6 Modeling Assumptions and Limitations

We intentionally model only active-power energy balance at hourly resolution to isolate the export-cap spillage mechanism. Reactive power, voltage/frequency constraints, network losses, detailed converter capability limits, and battery degradation/thermal dynamics are not represented (discussed further in Sect. 7).

4 Methods

All methods share the same physical parameters, the same spillage definition (8), the same spill-only constraints (10)–(11), and the same terminal SOC fairness rule (12). The intended differences are the information set (online vs. perfect foresight) and the policy class (rule-based vs. optimization vs. learned imitation).

4.1 Baseline

Baseline applies the cap without storage:

$$P_g^{\text{base}}(t) = \min(P_w(t), P_{\text{cap}}(t)), \qquad P_{\text{spill}}^{\text{base}}(t) = \max(P_w(t) - P_{\text{cap}}(t), 0). \quad (13)$$

Algorithm 1. Greedy_spill_only (online causal) with terminal SOC fairness enforcement

Require: Hourly series $P_w(t), P_{\mathrm{cap}}(t)$; BESS params $(P_{\max}, E_{\max}, \eta_{\mathrm{ch}}, \eta_{\mathrm{dis}}, \mathrm{SOC}_{\min}, \mathrm{SOC}_{\max}, \mathrm{SOC}_0, \mathrm{SOC}_{\mathrm{band}})$

Initialize $\mathrm{SOC}(0) \leftarrow \mathrm{SOC}_0$
for $t = 0$ to $T-1$ **do**
  $s \leftarrow \max(P_w(t) - P_{\mathrm{cap}}(t), 0), \quad h \leftarrow \max(P_{\mathrm{cap}}(t) - P_w(t), 0)$
  Compute SOC-feasible limits from (4)–(5)
  $P_{\mathrm{ch}}(t) \leftarrow \min(P_{\max}, s, P_{\mathrm{ch}}^{\mathrm{soc}})$
  $P_{\mathrm{dis}}(t) \leftarrow 0$
  if $s = 0$ **and** $h > 0$ **and** $\mathrm{SOC}(t) > \mathrm{SOC}_0 + \mathrm{SOC}_{\mathrm{band}}$ **then**
    Limit discharge so $\mathrm{SOC}(t)$ does not drop below $\max(\mathrm{SOC}_0, \mathrm{SOC}_{\min})$
    $P_{\mathrm{dis}}(t) \leftarrow \min(P_{\max}, h, P_{\mathrm{dis}}^{\mathrm{soc}})$
    $P_{\mathrm{ch}}(t) \leftarrow 0$ ▷ single-mode safety
  end if
  Compute $P_g(t)$ via (7) and $P_{\mathrm{spill}}(t)$ via (8)
  Update SOC using (2) and clip to $[\mathrm{SOC}_{\min}, \mathrm{SOC}_{\max}]$
end for
Enforce terminal SOC fairness (12) on the final contiguous valid segment via deterministic terminal-SOC enforcement (72 h lookback, expand if needed; recompute a feasible trajectory).

4.2 Greedy Spill-Capture Policy

The Greedy policy is causal: at each hour t it uses only the current measurements $(P_w(t), P_{\mathrm{cap}}(t), \mathrm{SOC}(t))$, fixed plant parameters, and the remaining time-to-end $r = T - t$ (no forecasts and no access to future P_w). It charges only from instantaneous above-cap surplus (subject to power and SOC headroom) and discharges only into instantaneous export headroom, optionally gated to low-generation periods (e.g., $P_w(t) \leq P_{\mathrm{thr}}$), with mutual exclusion enforced by construction (single-mode operation). To promote terminal SOC fairness, the controller applies a causal endgame mechanism: a time-varying SOC floor $\mathrm{SOC}_{\mathrm{floor}}(r)$ that ramps from $\mathrm{SOC}_{\min}$ toward the reference SOC_0 over the last H_{pre} hours and equals SOC_0 for $r \leq H_{\mathrm{end}}$, together with a last-hour correction when surplus/headroom permits. Because this steering remains subject to actuator limits $(P_{\max}, E_{\mathrm{bess}})$ and export-cap gating, exact satisfaction of $\mathrm{SOC}(T) = \mathrm{SOC}_0$ is not guaranteed by the causal rollout for all sizing points; therefore, in the simulation we compute the terminal deviation $\Delta\mathrm{SOC} \triangleq \mathrm{SOC}(T) - \mathrm{SOC}_0$ and mark runs as fairness-feasible when $|\Delta\mathrm{SOC}| \leq \mathrm{SOC}_{\mathrm{tol}}$. In the implementation used for annual KPI reporting, Greedy rollouts are additionally subjected to an *offline terminal-SOC normalization* applied on the final contiguous valid segment (restricted to the last $H_{\mathrm{lb}} = 72$ h, to enforce $\mathrm{SOC}(T) = \mathrm{SOC}_0$ exactly for comparability across policies. This post-processing does not use future wind information, but it can modify already-issued setpoints and is therefore used only for benchmarking fairness rather than as a deployable online controller.

4.3 Oracle Upper Bound (Offline LP, Perfect Foresight)

The Oracle is a perfect-foresight benchmark on the *same* spill-only feasible set. For each contiguous valid segment, it solves an LP minimizing total spillage energy plus a small throughput regularizer (to reduce degeneracy and unnecessary cycling):

$$\min \sum_{t=0}^{T-1} P_{\text{spill}}(t)\Delta t + \lambda_{\text{thr}} \sum_{t=0}^{T-1} \big(P_{\text{ch}}(t) + P_{\text{dis}}(t)\big)\Delta t. \tag{14}$$

Subject to: (i) power balance $P_g(t) + P_{\text{ch}}(t) - P_{\text{dis}}(t) + P_{\text{spill}}(t) = P_w(t)$, (ii) $0 \leq P_g(t) \leq P_{\text{cap}}(t)$ and $P_{\text{spill}}(t) \geq 0$, (iii) spill-only bounds (10)–(11), (iv) SOC dynamics (2) and bounds (3) (implemented as linear equalities plus bounds on $\text{SOC}(t)$), (v) initial SOC fixed to the segment start value $\text{SOC}(0) = \text{SOC}_{\text{init}}$ (carried over from the previous valid segment; $\text{SOC}_{\text{init}} = \text{SOC}_0$ only at the start of each year when splitting by year), and (vi) terminal SOC enforced only on the final contiguous valid segment: $\text{SOC}(N) = \text{SOC}_0$; for non-final segments, $\text{SOC}(N)$ is free and carried forward.

4.4 Behavioral Cloning Policy (AI) with Guardrails

Expert labels. The Oracle provides expert charge/discharge actions. We compress the expert action into a single scalar:

$$u(t) = P_{\text{dis}}^{\text{oracle}}(t) - P_{\text{ch}}^{\text{oracle}}(t), \tag{15}$$

and learn a regressor $\hat{u}_\theta(t)$ by supervised learning.

Causal Features. The feature vector $x(t)$ uses only information available at time t (or strictly earlier):

- instantaneous: $P_w(t)$, $P_{\text{cap}}(t)$, $\text{SOC}(t)$,
- wind lags: $P_w(t-L)$ for $L \in \{1, 2, 3, 6, 12, 24, 48, 72, 168\}$ restricted to $L \leq L_{\max}$,
- strictly-past rolling statistics (computed from P_w shifted by one hour): mean_6, mean_{24}, std_{24},
- time context: sin / cos of hour-of-day and month-of-year.

On the training year, the SOC feature is taken from the Oracle trajectory; on rollout, it is the policy's own SOC state (a potential distribution shift addressed by guardrails). The present feature set is intentionally restricted to causal plant-level and time- context variables in order to focus the study on the dispatch value of Oracle- guided behavioral cloning under a common operational envelope. Accordingly, other site- and environment-specific factors, such as local meteorological variables, turbulence-related proxies, ambient temperature, and dynamic grid-state measurements, are outside the scope of the present study.

Model. The default model is `HistGradientBoostingRegressor` with (`max_depth` = 3, `max_iter` = 200, `learning_rate` = 0.05) and a median imputer, trained on the 2024 year and evaluated on 2025. (Ridge is supported in the codebase but not used in this run.)

Deployment: Feasibility Projection + Do-No-Harm Guardrails. At each hour the AI predicts $\hat{u}(t)$, but the executed $(P_{\rm ch}, P_{\rm dis})$ is projected onto the spill-only feasible set and augmented with two guardrails:

1. *Spill-capture guardrail:* if spillage exists and SOC headroom permits, always charge greedily from $s(t) = \max(P_w - P_{\rm cap}, 0)$. This prevents under-charging that would trivially worsen spillage.
2. *Discharge guardrail:* discharge is allowed only when not charging, only into headroom $h(t)$, and only above a *causal time-to-end SOC floor* that ramps from $\mathrm{SOC}_{\min}$ toward SOC_0 near the end of the year segment; additionally, discharge is lower-bounded by a Greedy-style trigger so AI cannot be worse than the Greedy discharge decision when discharge is feasible.

Learning vs. Guardrails (Limitation). The deployed AI–BC policy includes feasibility projection and two conservative do-no-harm guardrails. Because these mechanisms can contribute to performance, this study does not fully attribute the recovery gain to learning alone. A formal ablation study (e.g., guardrails-only vs. learned-only under the same projection) is left for future work or as part of the extension work of this paper.

Terminal SOC Normalization (Offline Fairness Step). For the Greedy and AI–BC policies, we first generate a causal rollout (no access to future P_w). To enable annual comparisons under an identical terminal condition, we then apply a deterministic *post-processing* adjustment restricted to the last 72 h of the final contiguous valid segment to enforce $SOC(T) = SOC_0$ exactly for KPI reporting. This normalization does not use future wind information, but it can modify already-issued setpoints and is therefore used *only* for benchmarking fairness, not as a deployable online controller.

5 Dataset and Experimental Setup

5.1 Hourly Wind Dataset, Time Integrity, and Split

Data Source and Acquisition. Hourly wind generation data were obtained from an official market transparency platform via its authenticated web API. We queried the generation injection-quantity endpoint over full calendar years and exported the returned fields (date, hour, wind) to CSV.

Reproducible Query Template. The platform exposes a documented REST API. Our retrieval consists of an HTTP `GET` request to the generation injection-quantity endpoint with query parameters specifying (i) the date range (`start_date`, `end_date`), (ii) the generation asset identifier (`asset_id`), and (iii) the technology/filter fields required by the platform (e.g., wind generation). Access requires an API key issued to registered users of the platform; thus we do not include credentials in the manuscript.

Hourly wind power is provided as two CSV files (one per year) with columns `date`, `hour`, and `wind` [MW]. All timestamps are interpreted in the market-local time convention used by the platform (treated as a naive hourly index for analysis), and we explicitly handle `hour==24` by rolling to the next day at 00:00. Duplicate timestamps are averaged, and the series is reindexed to a complete hourly range while keeping missing hours as NaN (no forward fill). Negative wind values (if any) are clipped to 0 MW.

Table 1. Nominal configuration used for the simulation process

Parameter	Value
Rated power P_{rated}	200 MW
POI cap fraction α_{nom}	0.8
Constant cap P_{cap}	160 MW
BESS power fraction (`bess_pfrac`)	0.1
BESS power limit $P_{\max}$	20 MW
BESS duration (`bess_E_hours`)	2 h
BESS energy $E_{\max}$	40 MWh
Efficiencies $(\eta_{\text{ch}}, \eta_{\text{dis}})$	(0.95, 0.95)
SOC bounds $(\text{SOC}_{\min}, \text{SOC}_{\max})$	(0.10, 0.90)
Initial/terminal SOC SOC_0	0.50
SOC band (`soc_band`)	0.03
Oracle solver	HiGHS via SciPy
Oracle throughput weight λ_{thr}	0.001
Imitator model	HGB (depth = 3, iter = 200, lr = 0.05)
Max lag $L_{\max}$	168 h

The experiment uses a train/test split by calendar year: 2024 for training and 2025 for testing (2024 is a leap year). Missing-hour gaps induce contiguous valid segments; within each year, SOC is initialized to SOC_0 only at the first hour and then carried across segments (SOC held constant over missing hours). For KPI comparability under a common terminal condition, Greedy and AI apply an offline terminal-SOC normalization on the final contiguous valid segment (last 72 h, expanding if needed) to enforce $SOC(T) = SOC_0$ for reporting, while the Oracle enforces the same terminal SOC condition as a hard constraint within the optimization.

5.2 POI Export Cap

For the purpose of this simulation, the cap is assumed constant:

$$P_{\text{cap}}(t) = \alpha_{\text{nom}} P_{\text{rated}}. \tag{16}$$

With $P_{\text{rated}} = 200$ MW and $\alpha_{\text{nom}} = 0.8$, the cap is $P_{\text{cap}} = 160$ MW for all hours.

5.3 BESS Configuration and Solver

Table 1 summarizes the nominal configuration.

Oracle Throughput-Weight Sensitivity. The Oracle objective includes a small linear throughput regularizer $\lambda_{\text{thr}} \sum_t \left(P_{\text{ch}}(t) + P_{\text{dis}}(t)\right)\Delta t$ to break degeneracy among multiple minimum-spillage solutions and to discourage unnecessary cycling. In the nominal case, a brief sensitivity with $\lambda_{\text{thr}} \in \{0, 10^{-4}, 10^{-3}\}$ produces the same Oracle spillage and recovered energy (to the displayed precision) on both 2024 and 2025; thus the spillage-optimal benchmark is effectively unchanged.

5.4 Evaluation Metrics and Integrity Checks

For each year and policy, energies are computed over valid hours:

$$E_w = \sum_t P_w(t)\Delta t, \quad E_{\text{export}} = \sum_t P_g(t)\Delta t, \quad E_{\text{spill}} = \sum_t P_{\text{spill}}(t)\Delta t. \tag{17}$$

Recovered energy vs. Baseline is:

$$E_{\text{rec}} = E_{\text{spill}}^{\text{base}} - E_{\text{spill}}^{(\cdot)}, \tag{18}$$

and the capture rate is $E_{\text{rec}}/E_{\text{spill}}^{\text{base}}$. We audit feasibility and integrity from the exported hourly policy trajectories: (i) cap-violation hours where $P_g > P_{\text{cap}}$, (ii) discharge-over-cap hours where $P_{\text{dis}} > 0$ while $P_w \geq P_{\text{cap}}$, and (iii) mutual-exclusion overlap energy $\sum_t \min(P_{\text{ch}}, P_{\text{dis}})\Delta t$. Terminal SOC fairness is summarized by SOC_{end} and the indicator soc_ok.

For imitation quality, we report regression metrics on $u(t)$ (MAE, RMSE, R^2) per year from the annual evaluation summary.

6 Results

6.1 Annual KPIs (train Year 2024 and Test Year 2025)

Table 2 reports the annual KPI summary (rounded for display). On the test year (2025), Baseline spillage is 30,545.3 MWh. Greedy reduces spillage to 28,512.3 MWh (recovered 2,033.0 MWh, 6.66% capture). The Oracle upper bound reduces spillage to 27,057.0 MWh (recovered 3,488.2 MWh, 11.42% capture). AI–BC achieves 27,186.9 MWh spillage (recovered 3,358.4 MWh, 10.99% capture), close to the Oracle. From these values, the AI–BC policy closes 91.1% of the Greedy-to-Oracle recovery gap.

Table 2. Annual KPIs per policy and year. Energies are reported in MWh; displayed values are rounded. Capture rate is expressed as a percentage of Baseline spillage.

Year	Case	E_w	E_{export}	E_{spill}	E_{rec}	Capture (%)	SOC_{end}/soc_ok
2024	Baseline	612569.3	584819.3	27750.0	0.0	0.00	–
	Greedy_spill_only	612569.3	586762.0	25597.4	2152.6	7.76	0.500 / 1
	AI_BC_spill_only	612569.3	587996.1	24230.0	3520.0	12.68	0.500 / 1
	Oracle_spill_only	612569.3	588071.8	24146.1	3603.9	12.99	0.500 / 1
2025	Baseline	600858.9	570313.7	30545.3	0.0	0.00	–
	Greedy_spill_only	600858.9	572148.4	28512.3	2033.0	6.66	0.500 / 1
	AI_BC_spill_only	600858.9	573344.6	27186.9	3358.4	10.99	0.500 / 1
	Oracle_spill_only	600858.9	573461.8	27057.0	3488.2	11.42	0.500 / 1

6.2 Imitation Quality

Table 3 reports imitation metrics for the net action $u(t)$. Performance degrades from training to test (expected under distribution shift), but remains predictive on the test year.

Table 3. Imitation metrics for $u(t) = P_{\text{dis}}^{\text{oracle}}(t) - P_{\text{ch}}^{\text{oracle}}(t)$

Year	MAE (MW)	RMSE (MW)	R^2	n
2024	0.684	1.958	0.657	8784
2025	0.860	2.261	0.530	8760

6.3 Monthly Spillage Patterns

Figure 1 shows monthly spillage energy for all policies across the full two-year horizon. AI-BC and Oracle remain closely aligned, while Greedy exhibits higher spillage in high-spill months, consistent with the annual recovered-energy gap.

6.4 Representative-Week Behavior: Export and SOC

A representative week is selected by the maximum 7-day rolling sum of baseline spillage (centered on the peak week), yielding the window 2025-02-09 04:00 to 2025-02-16 03:00 in this dataset. This selection is performed *post hoc* for visualization only; it does not influence any controller, hyperparameter choice, or KPI computation (all KPIs are computed over the full-year test window).

Figure 2 shows POI export $P_g(t)$ under the cap alongside wind power and the cap. Storage-equipped policies charge during above-cap hours (reducing spillage) and discharge into headroom during below-cap hours (raising export toward the cap). Figure 3 shows the corresponding SOC evolution.

6.5 Feasibility and Integrity Checks

From the exported hourly time-series outputs, all cases satisfy $P_g(t) \leq P_{\text{cap}}(t)$ for all valid hours (0 cap-violation hours in both years). Discharge never occurs when $P_w(t) \geq P_{\text{cap}}(t)$ (0 discharge-over-cap hours), consistent with (11). Mutual exclusion overlap energy $\sum_t \min(P_{\text{ch}}(t), P_{\text{dis}}(t))\Delta t$ is 0 for all storage-equipped methods.

7 Discussion and Limitations

The study isolates a single service: spill-capture under a fixed POI export cap. In practice, additional services (e.g., ramp support or reserve provision) would

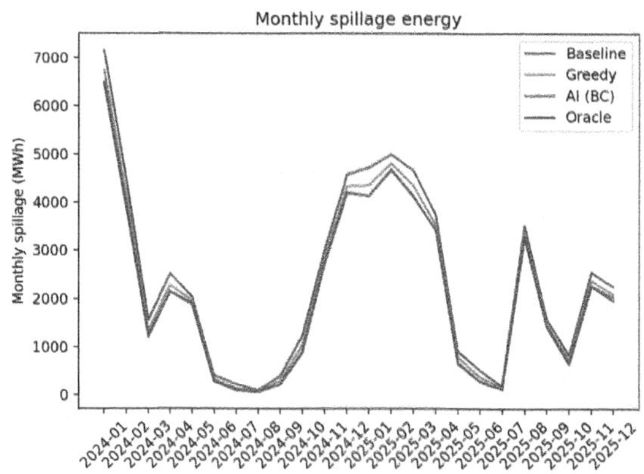

Fig. 1. Monthly spillage energy (MWh). Spillage is defined as $P_{\text{spill}}(t) = \max(0, P_w(t) - P_g(t) - P_{\text{ch}}(t) + P_{\text{dis}}(t))$ (export-cap spillage, not TSO curtailment).

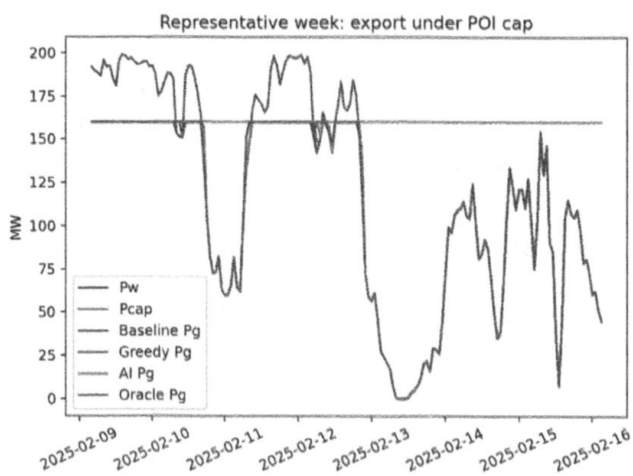

Fig. 2. Representative week: export under POI cap. The cap is constant at 160 MW in this run. Storage policies reshape export by charging from above-cap spill and discharging into headroom.

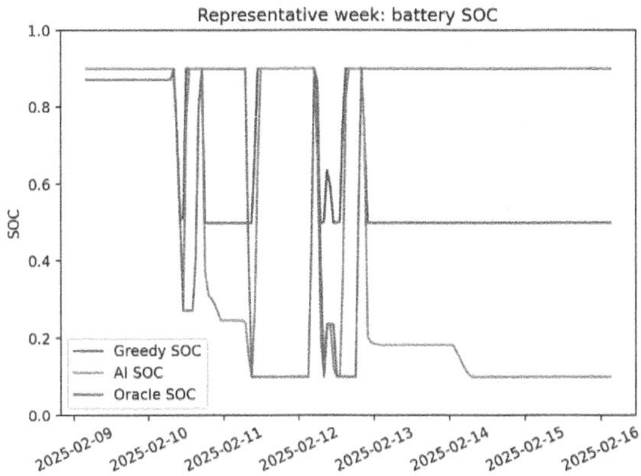

Fig. 3. Representative week: battery SOC for Greedy, AI–BC, and Oracle. All storage policies enforce SOC bounds; KPI reporting applies a common terminal SOC condition via the offline normalization described in Sect. 4.4.

introduce further constraints and/or objectives that are not modeled here; therefore, the $E_{\rm rec}$ should be interpreted only as spill-capture potential under the stated feasibility set.

7.1 Decision Support Under Envelopes

Because the POI export limit is binding, dispatch must be computed in a feasibility-first manner: at each hour we produce setpoints that satisfy $P_g \leq P_{\rm cap}$, SOC bounds, and the no-import rule, then evaluate spillage and terminal SOC compliance. In this sense, the Oracle provides an offline benchmark for what foreknowledge could achieve within the same envelope, while Greedy and AI–BC represent online, deployable policies with auditable guardrails.

7.2 Why AI–BC Approaches the Oracle Under Spill-only Constraints

Under spill-only operation, the most important action is to capture spill whenever it exists. The deployed AI–BC policy includes a do-no-harm guardrail that enforces this behavior deterministically, ensuring it cannot under-charge relative to spill availability. Remaining performance differences are primarily due to when to discharge into headroom, which is where foresight provides advantage.

On the 2025 test year, AI–BC recovers 3,358.4 MWh vs. baseline compared to 3,488.2 MWh for the Oracle, closing 91.1% of the Greedy-to-Oracle recovery gap. This indicates that, within the chosen feature set and under the spill-capture guardrail, a causal imitation policy can approximate a large fraction of the discharge-timing value of foresight.

7.3 Fairness and Feasibility

A strict terminal SOC constraint is essential for fair comparisons across policies. Without $\mathrm{SOC}(T) = \mathrm{SOC}_0$, a policy could increase export by draining energy at the end of the window, artificially inflating recovered energy. Here, all storage-equipped methods satisfy terminal SOC fairness (soc_ok=1 in Table 2); integrity checks confirm no cap violations and no discharge when the cap is saturated, matching the spill-only formulation.

7.4 Responsible-AI Positioning

The proposed AI component is used as operational decision support within the plant dispatch workflow. The learned policy produces charge/discharge recommendations that are subsequently projected onto the spill-only feasible set and remain subject to existing PCS/BMS limits and hard plant interlocks. No human subjects, personally identifiable information, or sensitive personal data are involved in this study; the data consist of plant-level operational time series and simulated dispatch trajectories. From an operational perspective, the primary design principle is feasibility-preserving assistance under explicit physical constraints.

7.5 Limitations

This study deliberately simplifies several aspects to isolate the spill-only export-cap mechanism:

- **Cap model.** The cap is constant in the performed simulation ($P_{\mathrm{cap}} = 160\,\mathrm{MW}$). Time-varying export limits (e.g., dynamic operating envelopes) are not evaluated here.
- **Electrical constraints.** Reactive power, voltage, network losses, and converter limits at the POI are not modeled; the controller operates on active power only.
- **Battery realism.** Degradation, temperature effects, and efficiency variation are omitted; throughput is only lightly regularized in the Oracle objective.
- **Behavioral cloning limitations.** BC may suffer from covariate shift in sequential rollouts. Training uses Oracle SOC as a feature on the training year, while deployment uses the policy's own SOC state; guardrails mitigate but do not eliminate distribution shift.
- **Generalization.** Results are reported for one train/test split (2024→2025) under a constant cap; multi-asset and time-varying-cap evaluation is left for future work.
- **Statistical inference.** The annual and monthly KPI differences are descriptive comparisons under the present simulation setup; formal inferential testing of policy differences is left for future work.

8 Conclusion and Future Work

We presented a fairness-controlled comparison of Wind(+BESS) dispatch policies under a hard POI export cap using an export-cap spillage definition that matches the implementation. Under identical feasibility rules and strict terminal SOC fairness, an offline Oracle LP provides an upper bound on recoverable spillage energy, while an online Greedy policy provides a deployable baseline. An AI behavioral cloning policy trained to imitate Oracle net actions using causal features and deployed with spill-capture and feasibility guardrails closes most of the Greedy-to-Oracle gap on the held-out test year (91.1% gap closure in 2025).

Future work: (i) time-varying caps and additional POI constraints (e.g., ramps, converter capability), (ii) uncertainty-aware dispatch (forecast+scenario MPC) under the same no-import and terminal SOC fairness rules; (iii) broader evaluation across plants/years to quantify generalization and robustness.

Disclosure of Interests. The author has no competing interests to declare.

References

1. Chatzistylianos, E.S., Psarros, G.N., Papathanassiou, S.A.: Export constraints applicable to renewable generation to enhance grid hosting capacity. Energies **17**(11), 2588 (2024). https://doi.org/10.3390/en17112588
2. Hutchinson, A.J., Gladwin, D.T.: Capacity factor enhancement for an export limited wind generation site utilising a novel flywheel energy storage strategy. J. Energy Storage **68**, 107832 (2023). https://doi.org/10.1016/j.est.2023.107832
3. Zhu, W., et al.: Allocating the capacity of shared energy storage for wind farm groups based on the over-limit power export risk. Front. Energy Res. **10**, 1099262 (2023). https://doi.org/10.3389/fenrg.2022.1099262
4. Zabihinia Gerdroodbari, Y., Khorasany, M., Razzaghi, R., Heidari, R.: Management of prosumers using dynamic export limits and shared community energy storage. Appl. Energy **355**, 122222 (2024). https://doi.org/10.1016/j.apenergy.2023.122222
5. Omitaomu, O.A., Niu, H.: Artificial intelligence techniques in smart grid: a survey. Smart Cities **4**(2), 548–568 (2021). https://doi.org/10.3390/smartcities4020029
6. Alam, M.M., Hossain, M.J., Habib, M.A., Arafat, M.Y., Hannan, M.A.: Artificial intelligence integrated grid systems: technologies, potential frameworks, challenges, and research directions. Renew. Sustain. Energy Rev. **211**, 115251 (2025). https://doi.org/10.1016/j.rser.2024.115251
7. Xu, N., Tang, Z., Si, C., Bian, J., Mu, C.: A review of smart grid evolution and reinforcement learning: applications, challenges and future directions. Energies **18**(7), 1837 (2025). https://doi.org/10.3390/en18071837
8. Ly, A.O., Akhloufi, M.: Learning to drive by imitation: an overview of deep behavior cloning methods. IEEE Trans. Intell. Veh. **6**(2), 195–209 (2021). https://doi.org/10.1109/TIV.2020.3002505
9. Wang, Q., et al.: Identifying expert behavior in offline training datasets improves behavioral cloning of robotic manipulation policies. IEEE Robot. Autom. Lett. **9**(2), 1294–1301 (2024). https://doi.org/10.1109/LRA.2023.3342559

10. Mahmoudi, S., Davar, A., Sohrabipour, P., Bist, R.B., Tao, Y., Wang, D.: Leveraging imitation learning in agricultural robotics: a comprehensive survey and comparative analysis. Front. Robot. AI **11** (2024). https://doi.org/10.3389/frobt.2024.1441312
11. Virtanen, P., Gommers, R., Oliphant, T.E., et al.: SciPy 1.0: fundamental algorithms for scientific computing in Python. Nat. Methods **17**, 261–272 (2020). https://doi.org/10.1038/s41592-019-0686-2
12. SciPy documentation: *linprog(method='highs') — SciPy Manual.* Accessed 27 Feb 2026
13. Gommers, R., Virtanen, P., Oliphant, T.E., et al.: scipy/scipy: SciPy 1.9.0. Zenodo (2022). https://doi.org/10.5281/zenodo.6940349
14. Huangfu, Q., Hall, J.A.J.: Parallelizing the dual revised simplex method. Math. Program. Comput. **10**(1), 119–142 (2018). https://doi.org/10.1007/s12532-017-0130-5

Core AI Models and Learning Paradigms

Relational Verification, Feature Sensitivity, and Cross-City Transfer in Heterogeneous Graph Learning for Urban Functional Zone Classification

Etibar Vazirov[1,2(✉)] and Daniele Apiletti[1]

[1] Politecnico di Torino, Turin, Italy
{etibar.vazirov,daniele.apiletti}@polito.it
[2] ADA University, Baku, Azerbaijan

Abstract. Urban functional zone classification is a key component of data-driven urban analytics, supporting applications in planning, mobility, and digital governance. Although point-of-interest (POI) composition is widely used to infer regional functionality, it remains unclear whether predictive performance reflects meaningful relational structure or primarily direct semantic signals derived from the same data source. This study introduces a heterogeneous graph learning framework that models Region–POI, POI–Category, and Region–Region interactions for functional zone classification. Comparable graph representations are constructed for three cities with diverse urban characteristics—Baku, Berlin, and Montevideo—using a uniform grid partition and OpenStreetMap-derived POI data. To enable rigorous evaluation, the analysis incorporates structured relation ablation, cross-city transfer, and a feature sensitivity setting where direct semantic region features are removed.

Experimental results show that conventional models and homogeneous graph baselines experience substantial performance degradation under feature removal, whereas the heterogeneous model maintains strong performance in both in-city and cross-city settings. Relation ablation further indicates that Region–POI interactions contribute most significantly to predictive performance, while spatial adjacency provides comparatively limited explanatory power.

These findings suggest that heterogeneous relational modeling captures meaningful structural dependencies in urban data and provides a more reliable representation of functional organization under varying feature and transfer conditions.

Keywords: Heterogeneous Graph Neural Networks · Urban Functional Zone Classification · Relational Modeling · Cross-City Transfer · Feature Sensitivity

1 Introduction

Urban functional zone classification aims to identify semantic region types such as residential, commercial, industrial, and mixed-use areas from heterogeneous

C. Zanni-Merk et al. (Eds.): AIDT 2026, CCIS 3023, pp. 127–141, 2026.
https://doi.org/10.1007/978-3-032-31319-5_9

urban data. Accurate functional labeling supports data-driven planning, infrastructure optimization, and AI-enabled digital transformation. As urban systems increasingly rely on automated decision-support pipelines, understanding how semantic structure emerges from spatial and relational dependencies becomes a central research challenge. Existing approaches range from rule-based schemes to POI-driven machine learning and multimodal geospatial frameworks [9,13,19]. POI composition is widely adopted due to its availability and interpretability; however, most methods treat regions independently or rely on aggregated statistics, thereby neglecting structured relational dependencies inherent in urban systems. In addition, the reliance on POI-derived features for both labeling and prediction raises concerns about feature–label circularity, potentially overstating model performance. Graph-based modeling provides a principled alternative for capturing spatial interactions. Graph neural networks (GNNs) have shown strong performance in urban forecasting and infrastructure modeling [7,10], while heterogeneous GNNs (HeteroGNNs) explicitly model multiple node and relation types, enabling richer representations of interactions among regions, POIs, categories, and spatial adjacencies [1]. Graph-based functional zoning and multimodal urban representation learning further demonstrate the benefits of relational modeling [3,14]. Despite these advances, three challenges remain. First, the contribution of individual relation types is rarely quantified in a systematic, counterfactual manner; improvements are typically attributed to graph structure without isolating which dependencies drive performance gains. Second, cross-city transferability remains underexplored. Although transfer learning has been studied in traffic and spatiotemporal forecasting [8,11,18], domain shift effects in region-level functional classification remain insufficiently examined. Third, it remains unclear whether the predictive power of existing models reflects genuine relational learning or is dominated by direct semantic features derived from the same POI source. To address these limitations, a heterogeneous graph learning framework is introduced for region-level functional zone classification, modeling three relational components: region–POI containment, POI–category mapping, and region–region spatial adjacency. Using a uniform 750-m grid, comparable heterogeneous urban graphs are constructed for three cities with diverse urban characteristics, namely Baku, Berlin, and Montevideo, based on OpenStreetMap data accessed via OSMnx and Overpass. Beyond predictive evaluation, the analysis incorporates structured relation ablation to quantify the contribution of each relation type, cross-city transfer experiments to assess transferability under domain shift, and a feature sensitivity setting in which direct semantic region features are removed to evaluate reliance on POI-derived attributes.

The contributions of this work are threefold:

1. A unified heterogeneous graph framework for region-level functional zone classification across multiple urban environments;
2. A structured relation ablation strategy for quantifying the contribution of heterogeneous relations;

3. A combined evaluation protocol including cross-city transfer and feature sensitivity analysis for assessing robustness and mitigating feature–label circularity.

2 Related Work

2.1 Urban Functional Zone Modeling

Urban functional zone identification is a central task in smart city analytics and digital urban planning. Early approaches rely on land-use records, remote sensing data, and statistical aggregation of point-of-interest (POI) distributions, while recent surveys highlight a transition toward data-driven and learning-based frameworks [9].

Recent studies increasingly incorporate multimodal geospatial data and graph-based representations to capture spatial dependencies beyond aggregated POI statistics. Multilevel graph neural networks have been applied to functional zone classification using heterogeneous spatial features [19], and multimodal graph fusion approaches integrate POIs, road networks, and auxiliary signals to improve predictive performance [14]. Dynamic and multi-temporal graph models further attempt to capture evolving urban functions [3]. Despite these advances, most approaches focus primarily on predictive accuracy, with limited analysis of how individual relational components contribute to performance.

2.2 Graph Modeling and Heterogeneous Urban Representations

Graph learning has become a key paradigm for modeling complex urban systems, including infrastructure networks and semantic spatial interactions [6,15]. Heterogeneous graph neural networks (HeteroGNNs) extend classical GNNs by modeling multiple node and relation types, making them particularly suitable for urban environments characterized by structural and semantic heterogeneity [1].

Recent developments include spatial graph-based foundation models for large-scale urban representation learning [21] and domain-enhanced multimodal graph models for planning applications [16]. While these methods demonstrate strong representational capacity, they generally do not provide systematic mechanisms for interpreting or verifying the role of different relations within the graph.

2.3 Cross-City Transfer and Domain Shift

Transferability across cities remains a major challenge in urban AI systems. Prior work has extensively studied cross-city transfer in traffic and spatiotemporal prediction [8,11,20], where alignment-based and knowledge graph approaches are used to mitigate structural and distributional differences [18].

However, most existing studies focus on temporal forecasting rather than region-level semantic classification. Systematic cross-city evaluation for functional zone classification remains limited. Recent surveys on graph learning under distribution shift emphasize the importance of robustness and transferability capabilities in graph-based models [17], but empirical evidence at the level of regional functional semantics is still scarce.

2.4 Research Gap

Existing literature demonstrates the effectiveness of graph-based models for urban representation and highlights the importance of transfer learning under domain shift. However, two key limitations remain: (i) the lack of systematic verification of relational dependencies in heterogeneous urban graphs, and (ii) limited empirical evaluation of cross-city transferability for region-level functional classification.

This work addresses these limitations by combining structured relation ablation, cross-city transfer evaluation, and feature sensitivity analysis within a unified heterogeneous graph framework, enabling both relational verification and robust behavior under transfer assessment.

3 Methodology

3.1 Problem Formulation and Graph Construction

Let $\mathcal{C} = \{c_1, \ldots, c_M\}$ denote the set of cities under study. For each city $c \in \mathcal{C}$, the urban area is partitioned into a uniform grid of 750-m spatial regions $\mathcal{R}_c = \{r_1, \ldots, r_N\}$. Each region is associated with a functional label $y_i \in \mathcal{Y}$.

$$\mathcal{Y} = \{\text{Residential}, \text{Commercial}, \text{Industrial}, \text{Leisure}, \text{Mixed}\}. \tag{1}$$

The task is to learn a function

$$f_\theta : \mathcal{R}_c \to \mathcal{Y}, \tag{2}$$

that predicts region-level functional zones from multi-relational urban structure.

For each city, a heterogeneous graph is constructed as

$$\mathcal{G}_c = (\mathcal{V}_c, \mathcal{E}_c), \tag{3}$$

where

$$\mathcal{V}_c = \mathcal{R}_c \cup \mathcal{P}_c \cup \mathcal{K}_c$$

contains region nodes, POI nodes, and semantic category nodes. The edge set consists of three relation types: region–POI containment, POI–category assignment, and region–region spatial adjacency. POIs are obtained from OpenStreetMap data through OSMnx and Overpass, mapped to grid cells using geometric containment, and grouped into semantic super-categories. Region adjacency is defined by shared grid boundaries in projected coordinate space.

This graph design separates spatial proximity from semantic composition, allowing relation-aware learning instead of collapsing all interactions into a homogeneous graph. Heterogeneous graph neural networks are well suited to this setting because they model multiple node and edge types through relation-specific transformations [1]. The resulting schema is shown in Fig. 1.

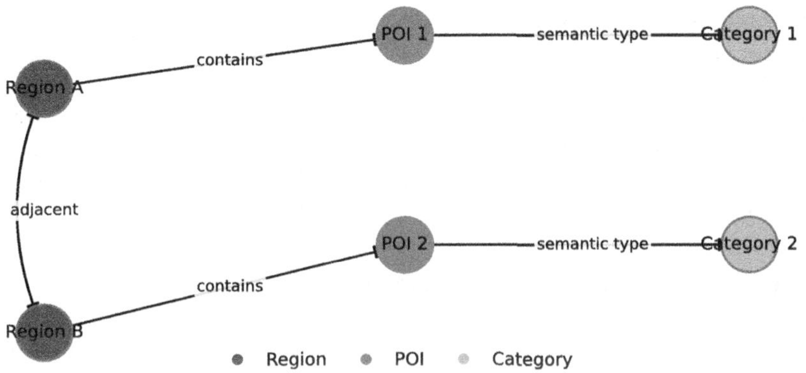

Fig. 1. Schema of the heterogeneous urban graph. Region nodes connect to POIs through containment relations, POIs link to semantic categories, and neighboring regions are connected through spatial adjacency.

3.2 Functional Label Definition and Feature Settings

City-scale authoritative functional labels are often incomplete, unavailable, or inconsistent across urban contexts. To ensure reproducibility and cross-city comparability, functional labels are derived from POI composition within each spatial region.

Let $\mathcal{P}(r_i)$ denote the set of POIs contained in region r_i, and let $\mathcal{P}_k(r_i)$ denote the subset belonging to semantic super-category k. The normalized composition vector of region r_i is defined as

$$\mathbf{c}_i = \left[\frac{|\mathcal{P}_1(r_i)|}{|\mathcal{P}(r_i)|}, \ldots, \frac{|\mathcal{P}_K(r_i)|}{|\mathcal{P}(r_i)|} \right], \tag{4}$$

where K is the number of super-categories.

Functional labels are assigned using a dominance-based rule:

$$y_i = \arg\max_{k \in \mathcal{Y}} \mathbf{c}_{i,k}, \tag{5}$$

subject to a fixed dominance threshold applied consistently across all cities. Regions without a sufficiently dominant category are assigned to the *Mixed* class.

This strategy follows the common use of POI distributions as proxies for regional semantics [9,13], while enabling a controlled and consistent evaluation across cities despite differences in label distribution.

To assess the impact of direct semantic features, two region-feature settings are considered. The *FULL* setting includes direct semantic composition features such as category counts and aggregate diversity indicators. The *NODIRECT* setting removes direct category-count features from region nodes and retains only indirect structural summaries, including total POI count, diversity, active category count, maximum category ratio, second-largest category ratio, and dominance gap.

The heterogeneous graph structure, including POI and category nodes and their relations, is kept unchanged. This setup allows isolating whether predictive performance is driven by explicit semantic attributes or by relational structure encoded in the graph.

3.3 Heterogeneous Message Passing Model

Given $\mathcal{G}_c$, node representations are updated using relation-specific message passing. Let $\mathcal{R}$ denote the set of relation types. At layer ℓ, the embedding of node v is updated as

$$\mathbf{h}_v^{(\ell+1)} = \sigma\left(\sum_{r\in\mathcal{R}}\sum_{u\in\mathcal{N}_r(v)} \frac{1}{|\mathcal{N}_r(v)|}\mathbf{W}_r^{(\ell)}\mathbf{h}_u^{(\ell)}\right), \tag{6}$$

where $\mathcal{N}_r(v)$ is the relation-specific neighborhood of v, $\mathbf{W}_r^{(\ell)}$ is a learnable transformation for relation r, and $\sigma(\cdot)$ is a nonlinear activation function. Mean aggregation is used to reduce sensitivity to node degree.

The model uses stacked heterogeneous convolution layers with relation-wise transformations, ReLU activations, and dropout regularization. All node types are projected into a shared latent space, while only region embeddings are used for classification. After the final layer, the predicted class distribution for region r_i is computed as

$$\hat{\mathbf{y}}_i = \mathrm{softmax}\left(\mathbf{W}_o\mathbf{h}_{r_i}^{(L)}\right), \tag{7}$$

where $\mathbf{W}_o$ is the output projection matrix. The parameters are optimized using cross-entropy loss over labeled training regions:

$$\mathcal{L} = -\sum_{r_i\in\mathcal{R}_c^{\mathrm{train}}}\sum_{k=1}^{|\mathcal{Y}|} y_{ik}\log\hat{y}_{ik}. \tag{8}$$

3.4 Evaluation Protocol

The evaluation consists of three complementary components. First, in-city classification is performed independently for each city using stratified splits, and results are reported as mean and standard deviation over five random seeds. Second, structured relation ablation is used to quantify the importance of each relation type. For a relation r, a modified graph $\mathcal{G}_c^{-r}$ is created by removing all edges of that relation while keeping all other graph components fixed. The relation contribution is measured as

$$\Delta_r = \text{MacroF1}(\mathcal{G}_c) - \text{MacroF1}(\mathcal{G}_c^{-r}). \tag{9}$$

Third, cross-city transfer is used to assess transferability under domain shift. Given a source city c_s and target city c_t, the model is trained on $\mathcal{G}_{c_s}$ and evaluated on $\mathcal{G}_{c_t}$ without fine-tuning:

$$\theta^* = \arg\min_\theta \mathcal{L}_{c_s}(\theta), \quad \text{evaluate on } \mathcal{G}_{c_t}. \tag{10}$$

This setting is important because graph learning models may be sensitive to distribution shifts across domains [17]. Both FULL and NODIRECT settings are evaluated to distinguish ordinary predictive performance from robustness to direct semantic feature removal.

4 Experimental Setup

4.1 Datasets and Graph Construction

Experiments are conducted on three structurally diverse urban environments: Baku (Azerbaijan), Berlin (Germany), and Montevideo (Uruguay). These cities differ in terms of geographic location, scale, and urban organization, providing a controlled yet diverse setting for evaluating transferability under heterogeneous urban distributions. For each city, the urban space is partitioned into a uniform grid of 750-m regions, and inactive cells without POIs are removed.

The resulting heterogeneous graphs contain:

- **Baku**: 1097 regions and 12,566 POIs,
- **Berlin**: 1323 regions and 66,485 POIs,
- **Montevideo**: 579 regions and 9,222 POIs.

Each graph includes three relation types: region–POI containment, POI–category semantic mapping, and region–region spatial adjacency.

To explicitly control potential circularity between POI-based labels and features, two feature configurations are evaluated:

- **FULL**: includes category-level POI composition features,
- **NODIRECT**: excludes direct category features and retains only aggregate statistics (e.g., diversity, dominance, and POI counts).

This design enables systematic evaluation of whether models rely on direct feature-label correspondence or exploit relational structure.

4.2 Baselines and Training Protocol

The proposed heterogeneous GNN is compared against both non-graph and graph-based baselines to provide a controlled evaluation across different modeling paradigms:

- **LGBM**: a feature-based model using region-level statistics,
- **GCN**: a homogeneous graph model using only region adjacency,
- **GIN**: a structure-aware graph baseline with enhanced neighborhood aggregation.

These baselines are selected to isolate the contribution of relational modeling by contrasting feature-only learning, homogeneous graph learning, and more expressive graph-based aggregation. For the heterogeneous model, hyperparameters are selected via grid search based on validation Macro-F1. The search space is summarized in Table 1.

Table 1. Hyperparameter search space for the heterogeneous GNN.

Hyperparameter	Search Values
Hidden dimension (d)	$\{32, 64\}$
Dropout rate	$\{0.3, 0.5\}$
Learning rate	$\{0.005, 0.01\}$
Number of layers (L)	2
Optimizer	Adam

The best configuration is selected on the validation split and re-evaluated across multiple random seeds to ensure stability. The search space is intentionally kept compact to ensure fair comparison and avoid excessive hyperparameter tuning.

4.3 Evaluation Strategy

Performance is evaluated using Accuracy and Macro-F1, with Macro-F1 serving as the primary metric due to class imbalance.

$$\mathrm{MacroF1} = \frac{1}{K}\sum_{k=1}^{K} \mathrm{F1}_k, \tag{11}$$

where $\mathrm{F1}_k$ denotes the F1-score for class k.

To ensure robustness, all experiments are repeated across multiple random seeds and reported as mean ± standard deviation. Beyond standard predictive evaluation, four complementary analysis protocols are employed:

- **Feature decoupling (FULL vs NODIRECT)** to assess dependence on direct POI-derived features,
- **Structured relation ablation** to quantify the contribution of each edge type,
- **Cross-city transfer** to evaluate transferability under domain shift,
- **Deletion-based fidelity** to measure the influence of local POI structures on model predictions.

For cross-city transfer, models are trained on all labeled regions of a source city and directly evaluated on a target city without fine-tuning, providing a strict assessment of domain transferability.

5 Results and Analysis

This section presents a comprehensive evaluation of the proposed heterogeneous graph framework, focusing on relational robustness, structural verification, and transferability under domain shift. All experiments are conducted over five independent random seeds, and results are reported as mean and standard deviation to ensure statistical reliability.

5.1 Robustness to Feature Removal: FULL vs NODIRECT

To assess the dependence of the model on direct region-level features, a controlled experiment is conducted by comparing two configurations: (i) FULL, where all region features are used, and (ii) NODIRECT, where direct region features are removed and only relational information is retained.

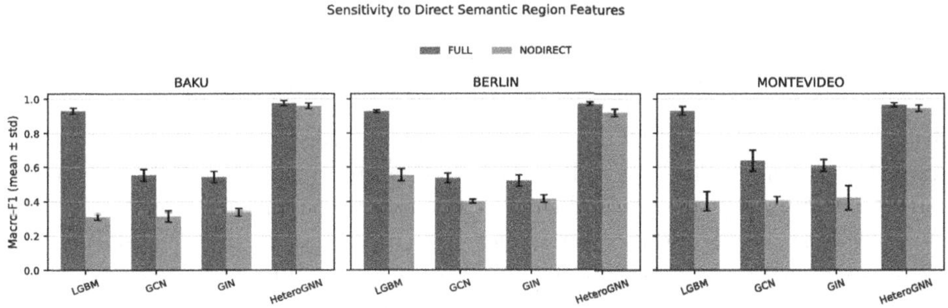

Fig. 2. Performance comparison between FULL and NODIRECT configurations across three cities.

As shown in Fig. 2, the proposed HeteroGNN maintains strong performance even in the absence of direct region features. In Baku and Berlin, the performance degradation remains limited, indicating that relational structure alone provides sufficient information for accurate classification.

In contrast, homogeneous baselines such as GCN and GIN exhibit substantial performance drops under the NODIRECT setting, suggesting that they rely heavily on explicit node features and are less capable of exploiting multi-relational dependencies.

These findings provide empirical evidence that the proposed model does not depend solely on aggregated region-level statistics, but instead leverages structured relational signals, mitigating concerns regarding feature-label circularity.

5.2 Structured Relation Ablation

To quantify the contribution of each relation type, structured relation ablation is performed as defined in Eq. (9). Each relation type is removed independently, and the resulting degradation in macro-F1 is measured.

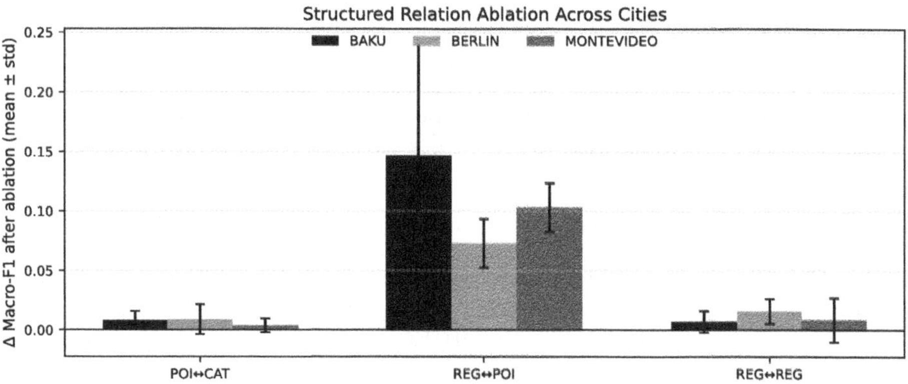

Fig. 3. Relation ablation results across three cities.

Figure 3 shows that removing Region–POI edges consistently leads to the largest performance drop across all cities. This confirms that semantic composition, captured through POI aggregation, is the dominant factor in functional zone classification. The removal of Region–Region adjacency results in comparatively smaller degradation, indicating that spatial proximity provides complementary contextual information rather than acting as a primary signal. The POI–Category relation contributes moderately, improving representation stability but not driving predictions independently. Importantly, the observed patterns remain consistent across all five random seeds and across different urban environments, supporting the robustness of the relational interpretation.

5.3 Cross-City Transfer Under Domain Shift

To evaluate transferability, models are trained on one city and evaluated on another without fine-tuning. The results are presented in Fig. 4 for FULL and NODIRECT settings.

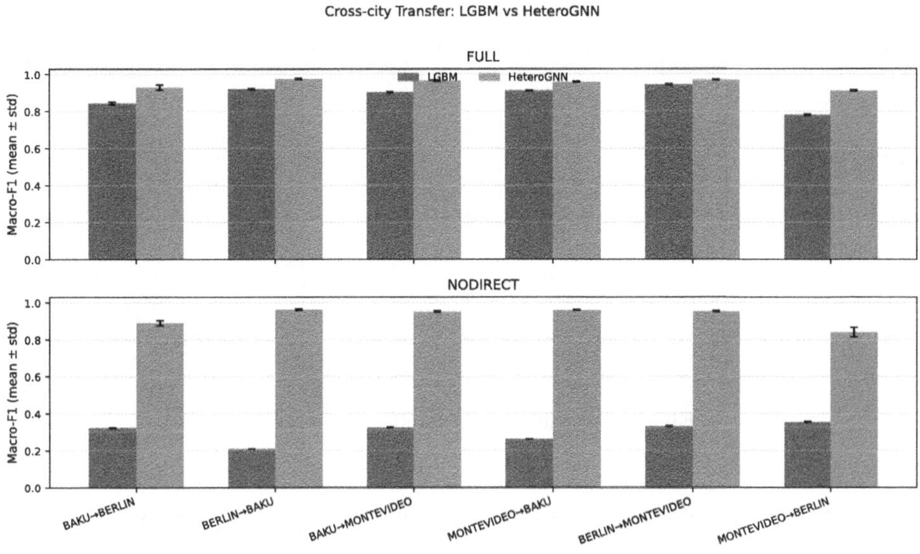

Fig. 4. Cross-city transfer performance under domain shift.

A noticeable performance reduction is observed compared to in-city results, indicating the presence of domain shift across urban environments. However, the proposed HeteroGNN consistently outperforms the LGBM baseline in transfer scenarios, suggesting that relational representations capture partially transferable structural patterns.

The transfer behavior is asymmetric, with certain city pairs yielding stronger results than others, reflecting differences in urban composition and data distribution.

Homogeneous graph baselines (GCN and GIN) are not included in this evaluation due to their significantly lower in-city performance and unstable behavior under domain shift observed in preliminary experiments. As such, they do not provide meaningful reference points for transfer analysis.

It is important to note that these results should be interpreted as evidence of transferability rather than definitive generalization. Given the limited number of cities, the findings indicate robustness under domain shift, while the proposed framework remains extensible to additional urban environments for broader validation.

5.4 Deletion-Based Fidelity Analysis

To further examine model behavior, deletion-based fidelity analysis is conducted by removing the top-K most influential POIs for each region based on gradient-based importance scores.

As shown in Fig. 5, prediction confidence decreases as increasingly important POIs are removed, indicating that the model relies on semantically meaningful

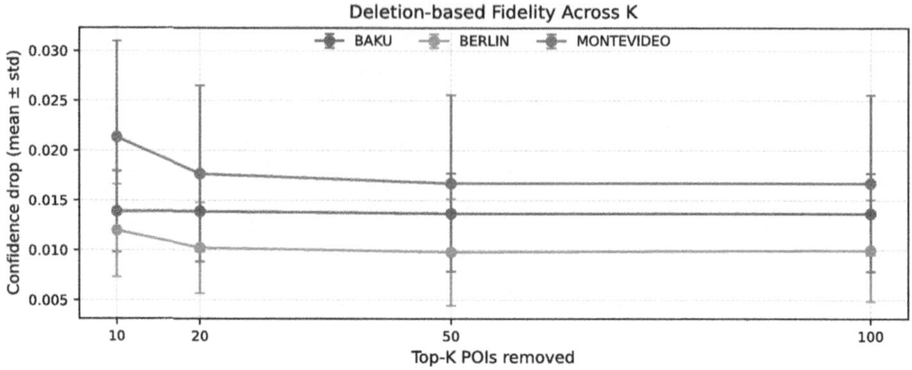

Fig. 5. Deletion-based fidelity across cities.

relational signals. The effect is more pronounced in larger and denser cities such as Berlin, while more moderate variations are observed in Montevideo.

These results provide additional evidence that the model captures meaningful dependencies rather than relying on spurious correlations or noise in the data.

6 Discussion

The results highlight several important aspects of heterogeneous relational modeling for urban functional zone classification.

First, semantic containment between regions and POIs emerges as the dominant signal. Structured ablation consistently shows that removing Region–POI relations leads to the largest performance degradation, while spatial adjacency contributes comparatively less. This indicates that functional zoning is primarily driven by semantic composition, with spatial proximity providing complementary context rather than a primary signal.

Second, cross-city experiments reveal clear distributional differences between urban environments. Although performance decreases under transfer, heterogeneous representations remain more stable than simpler baselines when partial structural alignment exists. These findings suggest that the proposed framework captures transferable relational patterns to a limited extent, while still reflecting city-specific characteristics.

Third, deletion-based fidelity analysis confirms that predictions are driven by semantically meaningful signals. The consistent confidence drop observed after removing influential POIs indicates that the model relies on structured semantic information rather than spurious correlations. Moreover, the stability of this behavior across cities supports the robustness of the learned representations.

From a practical perspective, the framework enables data-driven functional zoning using publicly available POI data, making it applicable in settings where official annotations are incomplete or unavailable. At the same time, the results should be interpreted with appropriate caution, as the evaluation is limited to a small number of cities and relies on POI-based proxy labels.

Finally, the use of publicly available POI data introduces considerations related to data quality, potential biases, and responsible use. While no personal or sensitive individual-level information is involved, POI datasets may reflect uneven spatial coverage or socio-economic biases inherent in volunteered geographic information. These factors can influence model behavior and should be considered when interpreting results or deploying such systems in real-world applications.

7 Conclusion and Future Work

This paper presented a heterogeneous graph neural network framework for region-level urban functional zone classification, explicitly modeling Region–POI, POI–Category, and Region–Region relations within a unified structure. By preserving relational dependencies during message passing, the approach captures semantic patterns that are not accessible through purely spatial or tabular representations.

Experimental results across multiple cities show that heterogeneous modeling consistently outperforms homogeneous graph baselines and remains competitive with strong tabular methods. Structured relation ablation indicates that Region–POI containment is the most influential component, while spatial adjacency provides complementary contextual information. Cross-city transfer experiments reveal measurable performance degradation under distribution shift, suggesting that learned representations are partially transferable but still influenced by city-specific characteristics.

These findings highlight the importance of explicitly modeling semantic composition in urban functional analysis, while also emphasizing the challenges associated with transferring models across heterogeneous urban environments. The reported results should be interpreted as evidence of transferability rather than full generalization, given the limited number of cities considered.

Future work will investigate domain adaptation and cross-city alignment strategies to improve robustness under distribution shift. Extending the framework to a broader set of cities and incorporating additional data sources, such as temporal and multimodal urban signals, may further enhance scalability and applicability.

References

1. Bing, R., Yuan, G., Zhu, M., Meng, F., Ma, H., Qiao, S.: Heterogeneous graph neural networks analysis: a survey of techniques, evaluations and applications. Artif. Intell. Rev. **56**(8), 8003–8042 (2023)
2. Chen, Y., Yang, J., Yang, R., Xiao, X., Xia, J.C.: Contribution of urban functional zones to the spatial distribution of urban thermal environment. Build. Environ. **216**, 109000 (2022)

3. Ding, C., Tang, J., Zhang, M., Wu, H., Wang, B.: Identifying dynamic functions of urban areas with a multi-temporal collaborative graph neural network model. Int. J. Geogr. Inf. Sci. **40**(1), 78–111 (2026)
4. Fan, Y., et al.: Towards predicting urban land use changes: a dynamic graph alignment perspective. ACM Trans. Intell. Syst. Technol. **16**(2), 1–24 (2025)
5. Geng, X.: Multi-source cross-city transfer learning for region-based spatial distribution prediction on ride-hailing demand. Hong Kong University of Science and Technology (Hong Kong) (2024)
6. Javaherian Pour, E., Atazadeh, B., Rajabifard, A., Sabri, S.: Developing a CityGML-based graph data model for utility infrastructure in smart cities. ISPRS Ann. Photogrammetry Remote Sens. Spat. Inf. Sci., 405–412 (2025)
7. Jin, G., et al.: Spatio-temporal graph neural networks for predictive learning in urban computing: a survey. IEEE Trans. Knowl. Data Eng. **36**(10), 5388–5408 (2023)
8. Jin, Y., Chen, K., Yang, Q.: Selective cross-city transfer learning for traffic prediction via source city region re-weighting. In: Proceedings of the 28th ACM SIGKDD Conference on Knowledge Discovery and Data Mining, pp. 731–741 (2022)
9. Liu, B., Deng, Y., Li, M., Yang, J., Liu, T.: Classification schemes and identification methods for urban functional zone: a review of recent papers. Appl. Sci. **11**(21), 9968 (2021)
10. Liu, T., Meidani, H.: End-to-end heterogeneous graph neural networks for traffic assignment. Transp. Res. Part C: Emerg. Technol. **165**, 104695 (2024)
11. Lone, S.N., Abouelela, M., Yu, T., Kim, J., Antoniou, C.: Transfer learning in spatiotemporal graph neural networks for traffic forecasting: taxonomy, analysis, and future directions. Anal. Future Dir. (2025)
12. Lu, K., Zhao, X., Li, M., Zhou, Y., Zhang, B.: An urban change detection method based on multimodal data and knowledge graph technology. Int. J. Digit. Earth **18**(2), 2564902 (2025)
13. Luo, G., Ye, J., Wang, J., Wei, Y.: Urban functional zone classification based on poi data and machine learning. Sustainability **15**(5), 4631 (2023)
14. Tao, Y., et al.: A graph-based multimodal data fusion framework for identifying urban functional zone. Int. J. Appl. Earth Obs. Geoinf. **136**, 104353 (2025)
15. Trudelle, C., Claramunt, C.: A graph-based modelling approach for the representation and analysis of urban conflicts. Comput. Environ. Urban Syst. **114**, 102201 (2024)
16. Wang, C., Wang, Y., Zhao, W.: Domain-graph enhanced multi-task multimodal large models for urban planning. Indoor Built Environ. **34**(9), 1636–1652 (2025)
17. Wu, M., et al.: Graph learning under distribution shifts: a comprehensive survey on domain adaptation, out-of-distribution, and continual learning (2024). arXiv preprint arXiv:2402.16374
18. Zhang, S., et al.: Deep transfer learning for city-scale cellular traffic generation through urban knowledge graph. In: Proceedings of the 29th ACM SIGKDD Conference on Knowledge Discovery and Data Mining, pp. 4842–4851 (2023)
19. Zhang, T., et al.: Classifying urban functional zones via a multilevel graph neural network and multimodal geospatial data. IEEE J. Sel. Top. Appl. Earth Obs. Remote Sens. (2025)

20. Zhang, Y., Wang, X., Yu, X., Sun, Z., Wang, K., Wang, Y.: Drawing informative gradients from sources: a one-stage transfer learning framework for cross-city spatiotemporal forecasting. In: Proceedings of the AAAI Conference on Artificial Intelligence, vol. 39, pp. 1147–1155 (2025)
21. Zhang, Z., Xie, M., Balsebre, P., Huang, W., Luo, S., Cong, G.: UrbanMFM: spatial graph-based multiscale foundation models for learning generalized urban representation. IEEE Trans. Knowl. Data Eng. (2026)

Bridging Prompting and Semantic Grounding: A Unified Evaluation Framework for Document-Level Relation Extraction

Gabriel Henrique Alencar Medeiros[1](✉) and Franco Giustozzi[2]

[1] INSA Rouen Normandie, Normandie Univ, LITIS, UR 4108, Rouen, France
gabriel.alencar_medeiros@insa-rouen.fr
[2] INSA Strasbourg, University of Strasbourg, ICube Laboratory, CNRS (UMR 7357), 67000 Strasbourg, France
franco.giustozzi@insa-strasbourg.fr
https://www.litislab.fr/accueil

Abstract. Document-level relation extraction (DocRE) aims to identify semantic relations between entity pairs by aggregating evidence that may be dispersed across multiple sentences. Large language models (LLMs) offer a flexible alternative to supervised DocRE architectures through zero- and few-shot prompting, yet they remain prone to hallucinated relations, incomplete evidence aggregation, and unstable multi-hop reasoning. This paper presents a structured empirical evaluation of LLM-based DocRE strategies, spanning direct prompting, ontology-guided retrieval, knowledge graph–extended RAG, and planner-driven agentic reasoning, using a large-scale knowledge-rich benchmark as a representative use case and focusing on the development split for reproducible analysis. We compare prompting baselines with progressively grounded approaches that inject structured semantic constraints and external knowledge during inference, while maintaining a shared preprocessing pipeline, output contract, and micro-averaged evaluation protocol. All methods are implemented within a unified open-source framework that standardizes prompts, retrieval modules, and scoring. The comparison establishes strong prompting baselines and demonstrates that planner-driven semantic grounding can yield measurable gains, particularly through improved recall under controlled and auditable tool use in document-level relation extraction.

Keywords: Document-level relation extraction · Retrieval-augmented generation · Knowledge graphs · Ontologies · Agentic reasoning

1 Introduction

Natural language processing (NLP) aims to transform unstructured textual data into structured, machine-interpretable representations. A central task in this

C. Zanni-Merk et al. (Eds.): AIDT 2026, CCIS 3023, pp. 142–156, 2026.
https://doi.org/10.1007/978-3-032-31319-5_10

context is relation extraction (RE), which identifies semantic relations between entities mentioned in text. Extracted relations are commonly represented as relational triples connecting a head entity, a relation, and a tail entity, forming the backbone of knowledge graphs (KGs). KGs enable structured reasoning, analytics, and decision-support across domains [6–9]. While early RE research focused on sentence-level patterns, many real-world relations are expressed implicitly and require aggregating evidence across multiple sentences. This document-level relation extraction (DocRE) setting introduces challenges such as coreference resolution, entity aliasing, and multi-hop reasoning [1].

Traditional DocRE systems rely on supervised neural architectures tailored to fixed relation inventories and annotated datasets [2–4]. Although effective, these approaches depend on task-specific training and extensive labeled data. Large language models (LLMs) offer an alternative paradigm: through prompting, they can perform DocRE without parameter updates, enabling rapid adaptation across domains and schemas [10,13]. However, LLM-based extraction introduces concerns regarding hallucination, semantic drift, and lack of grounding in the input document [14,15]. Addressing these reliability issues while preserving the flexibility of prompting-based approaches motivates the exploration of structured grounding and retrieval mechanisms for DocRE.

Although recent work has explored prompting strategies, retrieval-augmented generation, and structured reasoning for relation extraction, these approaches are typically evaluated under different experimental settings, datasets, and implementation details. This heterogeneity makes it difficult to isolate the effect of semantic grounding and reasoning control mechanisms in document-level relation extraction. In particular, there remains limited systematic analysis of how ontology guidance, knowledge graph retrieval, and agentic tool use influence performance when integrated into a unified and reproducible pipeline.

In this work, we conduct a structured empirical comparison of LLM-based document-level relation extraction strategies using DocRED as a representative knowledge-rich benchmark. We establish strong prompting baselines and progressively introduce semantic grounding and reasoning control mechanisms, enabling controlled analysis of their impact under a consistent evaluation protocol. The novelty of this work lies in a controlled and reproducible comparative evaluation rather than in proposing a new architecture. The main contributions of this paper are threefold:

- We provide a unified and reproducible evaluation of prompting-based DocRE strategies with large language models, including zero-shot, in-context learning, and chain-of-thought prompting.
- We compare semantic grounding paradigms, including ontology-guided retrieval, KG-based retrieval, and agentic retrieval with tool orchestration, within a single experimental framework.
- We release an open-source evaluation pipeline that standardizes preprocessing, prompting, retrieval, and scoring to support transparent and extensible benchmarking of LLM-based DocRE methods.

This paper is organized as follows. Section 2 presents the background and related work, introducing document-level relation extraction, the evaluation setting adopted in this study, and prior approaches based on supervised modeling, prompting, semantic grounding, and agentic reasoning. Section 3 then describes the extraction strategies compared in this work. Section 4 reports the experimental protocol and empirical results under a reproducible evaluation setting, followed by an in-depth analysis in Sect. 5. Finally, Sect. 6 concludes the paper.

2 Related Work

LLMs introduced a prompting-based alternative to DocRE that avoids task-specific training. Zero-shot prompting, in-context learning (ICL), and chain-of-thought (CoT) reasoning enable relation extraction through natural language instructions alone [10–13]. While prompting improves flexibility and reduces engineering overhead, it also raises reliability concerns, including hallucinated relations and unstable outputs under small prompt variations [14,15]. These limitations motivate grounding mechanisms that constrain generation and encourage predictions that remain consistent with explicit semantics and evidence.

Retrieval-augmented generation (RAG) improves factual consistency by conditioning generation on retrieved context [17,19]. In DocRE, retrieval typically does not aim to discover missing facts since the primary evidence is in the document, but to provide semantic constraints, structured priors, or verification signals that guide prediction; this is particularly relevant in knowledge-aligned benchmarks such as DocRED [1]. In our setting, ontologies primarily contribute at the *TBox* level (schema constraints such as relation semantics and type compatibility), and ontology-guided RAG retrieves relation definitions and domain/range constraints to reduce label ambiguity; Schmidt et al. propose GrowlRAG, which integrates ontology fragments directly into prompting for DocRE and improves semantic consistency without finetuning [5].

Knowledge graph-based integration has also emerged as a major direction for grounding LLMs, as surveyed in the roadmap of Pan et al. [25]. In DocRE, KG-RAG emphasizes the *ABox* level by retrieving instance triples as structured evidence to support multi-hop reasoning [20]. BYOKG-RAG extends this paradigm by combining multiple graph retrieval strategies and robust entity linking within a unified LLM-driven framework [22]. Although introduced for KGQA, its triple-centric retrieval and multi-strategy grounding design transfers naturally to DocRE, where entity pairs can be treated as relational queries over structured knowledge, motivating our adoption of a BYOKG-inspired triple retrieval mechanism as the representative KG-grounded variant.

More recently, agentic RAG frameworks incorporate explicit planning and tool orchestration, allowing a model to iteratively retrieve, verify, and refine decisions under a bounded control loop [23,24]. Unlike fixed retrieval pipelines, agentic systems dynamically decide when additional structured knowledge or external evidence is required. For DocRE, where some entity pairs are trivially decidable and others require multi-hop aggregation or constraint validation, this adaptive control is particularly appealing. We therefore include a single-agent configuration that enforces ontology and KG retrieval as mandatory steps and selectively escalates to Web or Wikidata under controlled policies, isolating the effect of tool-based reasoning without introducing multi-agent complexity.

Taken together, existing work suggests three increasingly structured forms of semantic grounding beyond pure prompting: ontology-guided constraint retrieval, knowledge graph–based triple retrieval, and agentic tool orchestration. These paradigms form the basis of the comparative framework evaluated in this study.

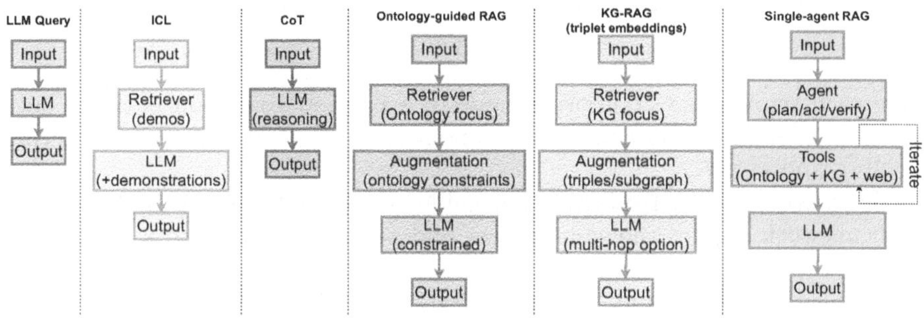

Fig. 1. Overview of the six document-level relation extraction strategies compared in this work. From left to right: single-pass prompting, in-context learning (ICL), chain-of-thought prompting (CoT), ontology-guided RAG (GrowlRAG), KG-RAG (BYOKG-RAG-style triple retrieval), and a single-agent RAG controller with tool use. Each column highlights how retrieval, structured knowledge, and iterative control are incorporated before producing relation triples.

3 Proposed Comparison Framework for Extraction Strategies

This section presents the six extraction strategies compared in this study and the unified framework used to ensure fair comparison. All methods consume the same document-level inputs (document text, entity set, relation inventory) and produce relation triples under a shared structured output contract (Subsect. 3.2). The pipelines are implemented as interchangeable components in our framework, so performance differences are attributable to retrieval, semantic grounding, and reasoning control rather than to preprocessing or evaluation artifacts.

Figure 1 summarizes the six strategies, ordered by increasing semantic grounding and control.

3.1 Problem Inputs and Notation

Given a document d with raw text x and a set of entities $E = \{e_1, \ldots, e_n\}$, each entity e_i is associated with a canonical string name and a set of mentions (spans) in x [1]. We denote the relation inventory by $\mathcal{R}$ (excluding NA). For each ordered entity pair (e_i, e_j) with $i \neq j$, the model predicts either a label $r \in \mathcal{R}$ or NA. Multiple relations per pair are possible in principle; in our implementation we allow multi-label outputs but enforce that each predicted triple (h, r, t) is unique in the final set. Unless otherwise stated, we evaluate all ordered pairs (e_i, e_j) for $i \neq j$ (no pruning), to avoid method-specific biases.

3.2 Unified Output Contract and Postprocessing

All strategies produce a set of predicted relation instances per document:

$$\hat{Y}(d) = \{(h_i, r_i, t_i)\}_{i=1}^{m},$$

where each triple (h_i, r_i, t_i) corresponds to a predicted relation between entity identifiers in the input document (or canonical entity strings mapped deterministically), and $r_i \in \mathcal{R}$.

We enforce a shared machine-parseable schema (JSON-like list of triples). Our framework applies the same parser and postprocessing across methods: (i) canonicalization of entity identifiers, (ii) label normalization (`null`/NA removed), (iii) duplicate removal, and (iv) optional directionality checks when the label is known to be asymmetric. If parsing fails, the system falls back to a conservative empty prediction for the affected document to avoid inflating false positives.

All methods are ultimately compared using micro-averaged precision, recall, and F1 computed over predicted triples (reported in Sect. 4).

3.3 Prompting Baselines

We begin with three prompting-only baselines that differ in the amount of scaffolding provided to the LLM: single-pass extraction, in-context learning (ICL), and chain-of-thought prompting (CoT).

Single-Pass LLM Extraction. The single-pass baseline measures how far a single inference call can go without additional retrieval or iterative refinement. The prompt includes: (i) document text (subject to a fixed token budget), (ii) an entity table (IDs and canonical names; optionally mentions), (iii) the relation inventory $\mathcal{R}$ (with short descriptions when available), and (iv) a strict output schema.

> ***Task:*** *extract DocRE relations using* only *the provided entity IDs.*
> ***Constraints:*** *use* only *labels in $\mathcal{R}$; abstain (NA) if unsupported by the document.*
> ***Output:*** *JSON list of unique triples (h, r, t).*

In-Context Learning (ICL). ICL conditions the model using k demonstrations sampled from the training portion of the dataset [10,13]. Each demonstration includes a short document excerpt, its entity list, and the corresponding gold triples. We consider two selection policies: (i) **Random-k**, which uniformly samples k training documents, and (ii) **Similarity-k**, which retrieves k demonstrations by embedding similarity between the current document (or entity-pair context) and the training pool.

> ***Add:*** *k labeled demonstrations (document excerpt + entities + gold triples).*
> ***Same constraints:*** *only $\mathcal{R}$ labels; only provided IDs; abstain if unsupported.*
> ***Output:*** *JSON triples (h, r, t).*

Chain-of-Thought Prompting (CoT). CoT prompting encourages intermediate reasoning that can help combine evidence distributed across sentences before committing to final relation triples [11,12]. In our setting, CoT is implemented as an *iterative self-refinement* procedure: the model is allowed to call the LLM multiple times (bounded by a step budget B), producing intermediate drafts (e.g., tentative triples and brief evidence cues) that are fed back into the next step to revise and improve predictions; only the final step output is parsed and evaluated under the shared JSON output contract.

> ***Iterate:*** *up to B LLM calls; keep intermediate drafts (tentative triples + notes).*
> ***Refine:*** *revise uncertain pairs and resolve cross-sentence evidence before commit.*
> ***Output:*** *only the final step returns the JSON triple list for evaluation.*

3.4 AOntology-Guided RAG Inspired by GrowlRAG

We implement an ontology-guided retrieval module tailored to DocRE, inspired by GrowlRAG [5], whose objective is to inject semantic constraints (relation definitions and type expectations) so that the model can better disambiguate labels and filter invalid relations.

The ontology can be provided as an external artifact or constructed in a dataset-agnostic manner from the source text by inducing relation/type signals and linking them to canonical identifiers in Wikidata [7]. For instance, in the DocRED use case, this alignment is facilitated by the fact that relation labels are grounded in Wikidata properties [1], allowing us to reuse property metadata (labels, descriptions, and type expectations) as a structured constraint store.

At inference time, we (i) retrieve a bounded ontology fragment relevant to the current document or entity pair, (ii) package it as compact prompt context (glosses and lightweight constraints), and (iii) request constrained generation under the shared output contract.

> ***Retrieve:*** *ontology fragment (relation glosses + type/constraint signals).*
> ***Use:*** *prune incompatible labels; reduce ambiguity among semantically close relations.*
> ***Output:*** *JSON triples consistent with retrieved constraints.*

3.5 KG-RAG with Triplet Embeddings Inspired by BYOKG-RAG

Inspired by KG-augmented retrieval strategies [20] and BYOKG-RAG [22], we build a retrieval index over verbalized triples that can be injected as structured context during relation classification. The KG can be reused if available, or built from the same sources used for ontology construction by exporting aligned triples into a canonical (s, p, o) store.

Each triple is verbalized into a canonical textual form (e.g., `"s [p] o"`) to support dense embedding and retrieval. At inference time, we embed the query from (i) the entity pair (e_i, e_j), (ii) local document context around mentions, and (iii) optional label hints, then retrieve the top-K triples and inject them into the prompt. We optionally enable a bounded second hop by expanding the query with intermediate entities from the top-K set.

Retrieve: *top-K relevant KG triples (optionally bounded 2-hop).*
Goal: *provide structured bridges when document evidence is sparse or implicit.*
Output: *JSON triples under the shared schema.*

3.6 Single-Agent RAG with Ontology, KG, Web, and Wikidata Tools

Following the agentic paradigm [23,24], we implement a single-agent controller with tool access. Unlike fixed pipelines, the agent dynamically decides when additional retrieval is needed, but it always starts from the same structured resources as the previous two methods.

The agent first queries structured tools (ontology and KG retrieval), which are mandatory and can be either provided or built as described above. It can then escalate to external sources (Web and Wikidata) under a controlled policy. The controller is implemented using LangGraph for orchestration, but the experimental focus is the tool policy and the resulting extraction behavior rather than on the framework itself. The agent uses a planner that produces a variable-length sequence of steps (bounded by a budget B), and follows: plan → retrieve → verify → decide → stop. Stopping occurs when (i) constraints uniquely determine a label, (ii) additional retrieval yields no new information, (iii) confidence exceeds a threshold, or (iv) the budget is exhausted. Because the dataset is Wikipedia-derived, Web/Wikidata retrieval is restricted (allowlist + caching) and is not intended to fetch the exact page corresponding to the current document.

Policy: *ontology+KG first (mandatory); Web/Wikidata only if constraints are insufficient. Budget B tool calls.*
Stop: *when label is determined or budget exhausted.*
Output: *always return final triples in the shared JSON schema.*

Table 1. Comparison of extraction strategies by retrieval, structured grounding, and control characteristics.

Method	Retrieval	Structured knowledge	Iterative control	Multi-hop option
Single-pass LLM	none	none	no	limited (implicit)
ICL	demonstration retrieval (train)	none	no	limited (implicit)
CoT	none	none	no	yes (prompted)
Ontology-guided RAG (GrowlRAG)	ontology fragments	relation definitions + constraints	no	limited (via constraints)
KG-RAG (BYOKG-RAG-style)	triple retrieval	KG triples	optional (bounded 2-hop)	yes
Single-agent RAG	ontology + KG + Web/Wikidata	ontology + KG + external evidence	yes	yes

3.7 What Differs Among Methods

Table 1 summarizes the main axes of variation. In all cases, the predicted triple set is evaluated against gold triples using micro precision, recall, and F1 in our experimental evaluation.

4 Experimental Evaluation and Results

This section describes the experimental protocol and reports the empirical results obtained for the six extraction strategies introduced in Sect. 3. All methods are evaluated under the same input/output contract and scoring procedure so that observed differences are attributable to retrieval, grounding, and reasoning control rather than to preprocessing or evaluation artifacts. All experiments are executed within our library RAGTree, which standardizes document I/O, prompt execution, parsing, and scoring across strategies to enable auditable and reproducible comparisons.[1]

4.1 Dataset and Evaluation Split

We evaluate on DocRED using the publicly available dev split, following the official entity and relation annotations [1]. We focus on dev for reproducibility; the hidden test labels are not publicly available. We exclude `NA/null` labels from scoring and compute micro-averaged precision, recall, and F1 over predicted relation triples.

[1] https://github.com/Gabriel382/RAGTree.

4.2 Systems, Decoding, and Structured Outputs

All strategies are executed through the same LLM backend interface implemented in RAGTree. Unless explicitly stated otherwise, we keep decoding parameters fixed across methods (e.g., temperature, maximum output length) and enforce a common structured output schema (JSON-like list of triples) to enable deterministic parsing and comparable postprocessing. Concretely, each input artifact is a document JSON file (e.g., doc1.json) containing document_id, raw text, and entity annotations; each method produces an output JSON (e.g., doc1.<method>.<model>.json) that preserves the original structure and appends the predictions under pred_relations. This uniform contract is shared by all prompting, RAG-based, and agentic variants.

4.3 Experimental Controls and Evaluation

All strategies are executed with the same backbone model, gpt-oss-20b, and share identical decoding parameters and structured output constraints, ensuring that differences are attributable to the extraction strategy rather than to model or generation settings. As described in Sect. 3, retrieval components (ontology, KG, and optional Web/Wikidata tools) are bounded by fixed budgets and cached to guarantee deterministic reruns. Because DocRED is Wikipedia-derived, external retrieval in the agentic setting is additionally restricted by policy (domain allowlist, caching, and explicit avoidance of directly fetching the originating Wikipedia page of the current document unless justified). For ontology-guided retrieval, we also leverage publicly released ontology resources when available, including the ontology artifact accompanying GrowlRAG.[2]

Evaluation follows the document-level protocol defined earlier: for each document, predicted triples (h, r, t) are compared against gold triples, and global TP/FP/FN counts are aggregated over the dev. We report micro-averaged precision, recall, and F1, excluding NA/null relations.

4.4 Results on DocRED dev

Table 2 reports micro-averaged precision, recall, and F1 for the six methods, as well as average runtime per document. All experiments use gpt-oss-20b on a workstation with an Intel Xeon w5-2465X CPU, 256 GiB RAM, and an NVIDIA RTX 6000 Ada GPU. Among prompting-only baselines, ICL yields the strongest F1, improving both precision and recall relative to single-pass prompting. In contrast, CoT substantially reduces precision without improving recall, suggesting that unconstrained intermediate reasoning can increase unsupported relation generation in this setting.

To illustrate how document-level extraction is realized in practice, Fig. 2 shows a representative example from our pipeline. The left panel corresponds

[2] https://github.com/boschresearch/growlrag_semantics2025/blob/main/data/Re-DocRED/ontology.ttl.

to the input artifact (`doc1.json`), while the right panel reflects the method output artifact that augments the same structure by adding `pred_relations`. This deterministic input/output interface is central to RAGTree: it enables method interchangeability and reproducible scoring while preserving all original document annotations.

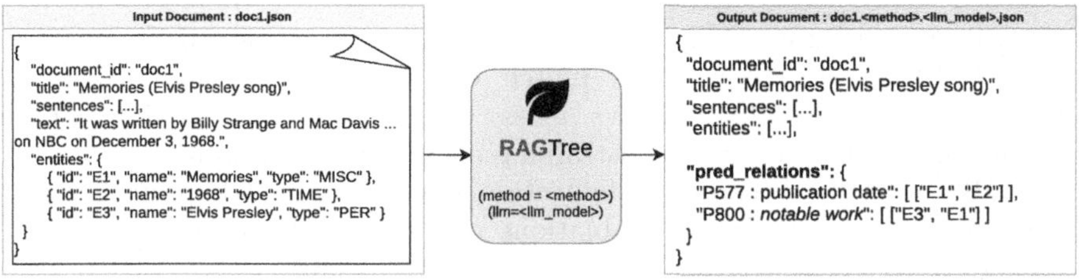

Fig. 2. Example of document-level processing in RAGTree. The input JSON document is preserved and extended with predicted relation triples under the `pred_relations` field in a method-specific output JSON.

Table 2. Micro-averaged relation extraction results on DocRED dev with average runtime (s/doc).

Method	Precision	Recall	F1	Time (s/doc)
Single-pass LLM	0.2139	0.1118	0.1468	0.19
ICL (k-shot)	0.2466	0.1263	**0.1670**	0.14
CoT prompting	0.1171	0.1159	0.1165	0.41
GrowlRAG	0.2484	0.1040	0.1466	26.05
BYOKG-RAG	**0.9717**	0.0896	0.1641	17.58
Single-agent RAG	0.2510	**0.1799**	**0.2096**	0.34

Across semantic strategies, distinct operating regimes emerge. GrowlRAG increases precision relative to the single-pass baseline but reduces recall, yielding a similar F1, consistent with constraints acting as a pruning mechanism when linking or constraint coverage is imperfect. BYOKG-RAG exhibits extremely high precision (0.9717) but very low recall (0.0896), indicating a conservative acceptance policy.

Finally, the single-agent strategy (mandatory ontology+KG retrieval with planner-driven escalation to Web/Wikidata under a fixed budget) achieves the strongest overall performance, obtaining the highest recall (0.1799) and the best micro-F1 (0.2096), while still improving precision relative to all prompting baselines. This indicates that iterative, tool-based reasoning can recover supported cross-sentence relations that fixed pipelines miss, without sacrificing semantic control. These trends motivate the detailed analysis in Sect. 5.

4.5 Ablation Plan and Diagnostic Extensions

The results indicate that the main challenge for fixed semantic pipelines lies in balancing precision gains with recall preservation, whereas the single-agent strategy already improves both recall and overall F1. We therefore design ablations to disentangle whether performance differences stem from retrieval coverage, constraint strictness, planner configuration, or decision calibration.

- **Ontology guidance (GrowlRAG):** ontology context on/off; fragment size budget; linking strategy (exact match vs. embedding-assisted); constraint packaging variants (gloss-only vs. gloss+domain/range) to analyze whether recall loss is driven by over-constraining or incomplete linking.
- **BYOKG-RAG:** retrieval configuration and budgets; KG source variants (reused KG vs. rebuilt KG); ablations over retrieval strategies and multi-hop depth to assess whether low recall is due to insufficient coverage or overly conservative acceptance.
- **Single-agent:** ontology+KG-only vs. ontology+KG+Web/Wikidata; planner step budget; tool-call statistics (calls/document and calls/pair); stop-criterion sensitivity to determine how iterative control contributes to recall gains and whether additional steps translate into meaningful improvements.

In addition, we extend the analysis by reporting (i) average number of predicted triples per document, (ii) distribution of predicted relation types, and (iii) detailed tool-usage traces for the agent. These diagnostics connect qualitative failure modes and computational cost with the observed precision–recall profiles and clarify how semantic grounding and iterative reasoning affect extraction behavior.

5 Discussion and Limitations

This section interprets the trends in Table 2 and highlights limitations that affect generalization. Prompting baselines provide non-trivial performance but remain sensitive to calibration and evidence aggregation, while semantic grounding changes the operating point in different ways. Fixed constraint/retrieval pipelines (GrowlRAG, BYOKG-RAG) tend to favor precision but can under-recover relations, whereas the planner-driven single-agent strategy achieves the strongest overall F1 (0.2096) by substantially increasing recall (0.1799) while also improving precision relative to prompting baselines.

5.1 Prompting Baselines: Performance Characteristics

Among prompting-only strategies, ICL improves both precision (0.2466 vs. 0.2139) and recall (0.1263 vs. 0.1118), yielding the strongest prompting-only F1 (0.1670), consistent with demonstrations providing practical calibration of label semantics [10,13]. CoT yields lower precision (0.1171) without improving recall (0.1159 vs. 0.1118), resulting in the weakest prompting-only F1 (0.1165), suggesting that iterative reasoning does not automatically translate into better evidence aggregation for DocRE under the current setup.

5.2 Semantic Grounding and Tool Use: Conservative Fixed Pipelines vs Agentic Gains

GrowlRAG improves precision (0.2484 vs. 0.2139) but reduces recall (0.1040), leading to an F1 comparable to single-pass (0.1466 vs. 0.1468), consistent with constraints acting primarily as pruning when linking or constraint coverage is imperfect. BYOKG-RAG reaches extremely high precision (0.9717) with low recall (0.0896), reflecting a conservative acceptance regime that may suit high-precision KG population but is less suitable for balanced DocRE unless coverage and acceptance criteria are relaxed. In contrast, the single-agent strategy attains the highest recall (0.1799) and best overall F1 (0.2096), indicating that planner-driven tool use (mandatory ontology+KG grounding and selective Web/Wikidata escalation) can recover supported relations that fixed pipelines miss, motivating ablations on planner budgets, escalation policy, and verification prompts.

5.3 Limitations and Threats to Validity

First, results are reported on DocRED dev for fully reproducible evaluation and should not be interpreted as hidden-test generalization. Second, LLM extraction is prompt- and formatting-sensitive; although we standardize the output schema and parsing, complex pipelines remain exposed to strict parsing failures. Third, DocRE evaluation depends on exact entity matching and correct directionality, and canonicalization heuristics can introduce systematic errors under aliasing. Fourth, semantic methods depend on ontology/KG alignment and coverage; incomplete mappings can cause abstention or conservative filtering. Finally, external retrieval introduces leakage and temporal-drift risks; our controlled policy (allowlist, caching, and avoidance of the originating Wikipedia page) mitigates these issues and should be paired with transparent reporting of tool usage and cache versions.

Overall, the results highlight that semantic grounding is most beneficial when paired with calibrated decision control: fixed constraints improve precision but may reduce coverage, while agentic tool orchestration can improve recall and overall F1 under bounded, auditable retrieval.

6 Conclusion

We presented a structured survey and unified empirical comparison framework for LLM-based document-level relation extraction, instantiated on the DocRED dev benchmark under a single, reproducible evaluation protocol, using the same `gpt-oss-20b` backend across methods. Across six strategies spanning prompting-only baselines and progressively stronger semantic grounding, our results show that few-shot prompting via in-context learning provides the strongest prompting-only baseline, while the planner-driven single-agent strategy achieves the best overall performance by substantially improving recall (highest

0.1799) and micro-F1 (0.2096), while still improving precision over the prompting baselines. In contrast, fixed semantic pipelines tend to shift the operating point toward higher precision at the cost of reduced recall: ontology-guided prompting (GrowlRAG) improves semantic consistency through explicit constraints but can increase abstentions when linking or constraint coverage is imperfect, and BYOKG-RAG can yield highly selective, high-precision outputs under conservative acceptance policies. These findings highlight that semantic grounding and reasoning control are not uniformly beneficial by default; their impact depends critically on coverage, alignment quality, and calibration of abstention, escalation, and stopping criteria.

Future work will extend this study in two complementary directions. First, we will deepen the empirical analysis within the current benchmark by conducting systematic ablations of (i) prompt templates and structured output constraints, (ii) demonstration selection policies and k in ICL, (iii) ontology linking accuracy and fragment budgeting for ontology-guided RAG, (iv) BYOKG-RAG retrieval strategies, KG reuse vs. rebuilding, and retrieval budgets (including controlled multi-hop variants), and (v) planner design, tool ordering (mandatory ontology/KG, then selective Web/Wikidata), and budgeted stopping rules for the agentic strategy. These ablations will be accompanied by cost and behavior reporting, including tool-call statistics, average predicted triples per document, label distributions, and error typologies that distinguish hallucinated relations, directionality errors, and missed long-range evidence.

Second, we will continue and broaden the survey aspect of this work by benchmarking additional state-of-the-art instantiations of semantic grounding and agentic extraction. Concretely, we plan to compare multiple ontology resources and ontology-guided prompting schemes (e.g., alternative constraint packaging and type systems), multiple KG sources and KG-RAG variants (e.g., alternative graph-retrieval backends and reasoning traces), and a wider range of agentic controllers (single-agent vs. multi-agent, reactive vs. deliberative planners) under shared evaluation conditions. To strengthen external validity, we will extend the comparison to additional document-level relation extraction datasets and relation inventories, enabling a larger-scale survey that characterizes which combinations of prompting, retrieval, and control generalize beyond a single benchmark setting. All configurations and evaluation scripts will be released as part of RAGTree to support reproducibility and future extensions.

References

1. Yao, Y., et al.: DocRED: a large-scale document-level relation extraction dataset. In: Proceedings of ACL (2019)
2. Christopoulou, F., Miwa, M., Ananiadou, S.: Connecting the dots: document-level neural relation extraction with edge-oriented graphs. In: Proceedings of EMNLP-IJCNLP (2019)
3. Nan, G., Zeng, J., Zeng, J., Liu, Z.: Reasoning with latent structure refinement for document-level relation extraction. In: Proceedings of ACL (2020)

4. Zhou, W., Chen, M., Han, X.: Document-level relation extraction with adaptive focal loss and evidence reasoning. In: Proceedings of NAACL (2021)
5. Schmidt, W.J.: Document-level relation extraction with ontology-guided RAG. In: Studies on the Semantic Web. IOS Press (2025). https://doi.org/10.3233/SSW250006
6. Nickel, M., Murphy, K., Tresp, V., Gabrilovich, E.: A review of relational machine learning for knowledge graphs. Proc. IEEE **104**(1), 11–33 (2016). https://doi.org/10.1109/JPROC.2015.2483592
7. Hogan, A., Blomqvist, E., Cochez, M., et al.: Knowledge graphs. ACM Comput. Surv. **54**(4), 1–37 (2021). https://doi.org/10.1145/3447772. Article 71
8. Ehrlinger, L., Wöß, W.: Towards a definition of knowledge graphs. In: Joint Proceedings of the Posters and Demos Track of SEMANTiCS 2016 and SuCCESS'16. CEUR Workshop Proceedings, vol. 1695, pp. 13–16 (2016)
9. Paulheim, H.: Knowledge graph refinement: a survey of approaches and evaluation methods. Semantic Web **8**(3), 489–508 (2017). https://doi.org/10.3233/SW-160218
10. Brown, T.B.: Language models are few-shot learners. In: Advances in Neural Information Processing Systems (2020)
11. Wei, J.: Chain-of-thought prompting elicits reasoning in large language models. In: Advances in Neural Information Processing Systems (2022)
12. Kojima, T.: Large language models are zero-shot reasoners. In: Advances in Neural Information Processing Systems (2022)
13. Min, S.: Rethinking the role of demonstrations: what makes in-context learning work? In: Proceedings of EMNLP (2022)
14. Ji, Z., et al.: Survey of hallucination in natural language generation. ACM Comput. Surv. **55**(12), 1–38 (2023). https://doi.org/10.1145/3571730. Article 248
15. Lin, S.: TruthfulQA: measuring how models mimic human falsehoods. In: Proceedings of ACL (2022)
16. Yeh, C.-K.: Uncertainty-aware reasoning for large language models. arXiv preprint (2023)
17. Lewis, P.: Retrieval-augmented generation for knowledge-intensive NLP tasks. In: Advances in Neural Information Processing Systems (2020)
18. Izacard, G., Grave, E.: Leveraging passage retrieval with generative models for open domain question answering. In: Proceedings of EACL (2021)
19. Thayaparan, M.: Knowledge-enhanced large language models: a survey. arXiv preprint (2024)
20. Linders, J., Tomczak, J.M.: Knowledge graph-extended retrieval augmented generation for question answering. arXiv preprint (2025)
21. Kühn, R.: Enhancing rhetorical figure annotation: an ontology-based web application with RAG integration. In: Proceedings of COLING (2025)
22. Mavromatis, C., et al.: BYOKG-RAG: multi-strategy graph retrieval for knowledge graph question answering (2025). arXiv preprint arXiv:2507.04127
23. Yao, S.: ReAct: synergizing reasoning and acting in language models. In: Proceedings of ICLR (2023)
24. Singh, A.: Agentic retrieval-augmented generation: a survey on agentic RAG. arXiv preprint (2025)
25. Pan, S., Luo, L., Wang, Y., Chen, C., Wang, J., Wu, X.: Unifying large language models and knowledge graphs: a roadmap. IEEE Trans. Knowl. Data Eng. **36**(7), 3580–3599 (2024)

A Divide-and-Conquer Framework for Query-Efficient Causal Discovery with Large Language Models

Louis Hernandez[1,2](✉), Matthieu Boussard[1], and Alessandro Leite[2]

[1] Craft AI, Paris, France
louis.hernandez@example.com
[2] INSA Rouen Normandy, University of Rouen Normandy, LITIS UR 4108, Rouen, France

Abstract. Causaldiscovery from observational data in high-dimensional spaces is computationally challenging. Large Language Models (LLMs) can provide semantic priors, but eliciting them with pairwise prompting requires $\mathcal{O}(N^2)$, where N is the number of variables or queries. We propose a divide-and-conquer framework that clusters variables using text embeddings, elicits directed relations between clusters with an LLM, and integrates these relations as soft penalties in DAGMA. On semantically structured synthetic linear SEM (Structural Equation Model) benchmarks ($N = 20$), our method improves unconstrained DAGMA (SHD (Structural Hamming Distance) 32.667 → 25.889) and direct LLM baselines (SHD $\geq$ 36), while classical PC remains strongest on this setting (SHD 15.222). We further report scaling stress tests up to $N = 200$, including timeout-constrained classical baselines.

Keywords: Causal Discovery · Large Language Models · Continuous Optimization · DAGMA · Divide-and-Conquer

1 Introduction

Inferring underlying causal mechanisms from observational data is a fundamental challenge in fields ranging from social sciences to biology. Addressing causal questions requires explicit causal modeling [18,19]. Constructing a causal graph from data has traditionally relied on statistical and computational methods, including constraint-based discovery (PC) [24] and score-based search (GES) [8]. However, these approaches face scalability limits in high-dimensional settings because the graph search space grows rapidly with the number of variables. At the same time, these classical methods can be very strong when their assumptions are well matched to the data-generating process (e.g., linear synthetic SEMs with sufficient samples). Given these limitations, graphs by human experts are still massively preferred over those generated by these methods.

Recent work has explored the potential of large language models (LLMs) for causal tasks, including causal discovery [2,10,17]. They often emulate human

C. Zanni-Merk et al. (Eds.): AIDT 2026, CCIS 3023, pp. 157–172, 2026.
https://doi.org/10.1007/978-3-032-31319-5_11

experts; similarly, they generate graphs based on "common sense" and without seeing the data, potentially enabling better scaling. Despite this promise, existing methods often rely on pairwise variable queries, which incur $\mathcal{O}(N^2)$ query complexity and become expensive at scale. Breadth-first prompting strategies reduce nominal query complexity to $\mathcal{O}(N)$ [15], but robustness remains limited when context windows are stressed. In addition, multiple studies report that LLMs can confuse correlation with causation or produce plausible but incorrect causal claims [12–14,28].

To integrate prior knowledge into query-efficient data-driven discovery, priors can be translated into soft constraints in continuous optimization solvers [3,4, 9,29]. Some other hybrid works take the reverse approach: the causal graph is fully generated by a traditional data-based method, then refined by an LLM [16], which does not really solve the issues, as we still need a query for each edge, which, in the worst case, is still $\mathcal{O}(N^2)$. To address both complexity and instability, we first partition variables using semantic embeddings, then reason over cluster-level relations.

We propose a two-step approach to scale LLM reasoning. In the first step, we employ a sentence-transformer to produce dense vector embeddings for each variable. We then apply agglomerative hierarchical clustering to group semantically related variables. In the second step, we query the LLM to infer causal relations among clusters, not individual variables. Each high-level hypothesis is translated into low-level structural constraints. Finally, these constraints are integrated into a continuous causal discovery algorithm (DAGMA) as penalty terms in the optimization objective.

The contributions of this paper are:

- **LLM Query Efficiency:** The framework reduces LLM elicitation from variable-level pairwise prompting ($\mathcal{O}(N^2)$) to cluster-level querying ($\mathcal{O}(K^2)$, $K \ll N$), with K the number of clusters, with bounded prompt size through hierarchical merge/split refinement.
- **Reproducibility:** The method uses deterministic clustering and schema-constrained prompting, designed to improve run-to-run reproducibility.
- **Empirical Performance:** We show that cluster-level semantic priors improve structure recovery over direct LLM querying and unconstrained DAGMA. On these synthetic linear benchmarks, classical constraint-based methods remain strong; within LLM-assisted continuous methods, our framework gives the best aggregate recovery.

2 Related Work

Causal discovery aims to infer structural relationships from observational data, balancing statistical identifiability with computational efficiency [6]. This section reviews the progression from traditional statistical methods to hybrid approaches that incorporate Large Language Models (LLMs) as structural priors.

2.1 Statistical Foundations and Continuous Optimization

Traditional causal discovery relies primarily on constraint-based and score-based algorithms. Constraint-based methods, such as the PC algorithm [24], use conditional independence tests to identify causal skeletons and orient edges. Score-based methods, like Greedy Equivalence Search (GES) [8], optimize a penalized likelihood score over Markov Equivalence Classes (MECs). While theoretically sound, these methods scale poorly in high-dimensional settings because the search space of Directed Acyclic Graphs (DAGs) grow super-exponentially with the number of variables (N).

To address these scalability limits, recent work has reformulated causal discovery as a continuous optimization problem. The NOTEARS algorithm [29] replaced the discrete acyclicity constraint with a differentiable trace-exponential function, enabling the use of standard gradient-based optimizers. Because the matrix exponential can lead to numerical instabilities in larger graphs, DAGMA [3] introduced a log-determinant characterization based on M-matrices to improve gradient stability. These differentiable frameworks are particularly useful because they allow prior knowledge to be integrated directly into the loss function as soft regularization terms.

2.2 LLMs as Causal Priors

Alongside continuous optimization, researchers have explored using LLMs to extract causal priors from semantic metadata [2,7,17]. Early methods prompted LLMs to determine causal directions between variable pairs. While effective on small datasets, this pairwise approach requires $\mathcal{O}(N^2)$ queries, making it computationally expensive for large systems. Additionally, evaluating pairs in isolation removes global context, which can lead to transitive errors. Multiple studies have observed that LLMs may over-index on semantic association rather than interventional causal logic [13,14,26,28], which can produce plausible but incorrect edges [12].

2.3 Scalable Discovery and Graph Partitioning

To improve scalability, subsequent research has investigated divide-and-conquer and graph traversal architectures. Jiralerspong et al. [15] proposed a Breadth-First Search (BFS) method that queries an LLM for all potential descendants of a node simultaneously, reducing query complexity to $\mathcal{O}(N)$. However, this method requires accurate identification of root nodes and can exceed the context window of smaller models.

For high-dimensional datasets, Sokolov et al. [23] utilized semantic clustering to group variables into functional domains, while Shah et al. [22] provided theoretical guarantees that valid causal graph partitioning preserves the true Markov Equivalence Class. Other approaches, such as Causal-LLM [21], use one-shot generation to output adjacency matrices, although they face a similar context length limitations on larger graphs.

2.4 Hybrid Architectures and Soft Constraints

Recent frameworks commonly adopt hybrid architectures, using LLMs to generate priors that guide continuous optimization algorithms [10]. Methods such as CauScientist [20] incorporate statistical scoring (e.g., BIC) to verify LLM-proposed edges, while CARE [11] uses instruction fine-tuning to improve alignment with numerical data. Conversely, approaches like ALCM [16] invert this dynamic by executing data-driven algorithms first to generate an initial graph, which an LLM then evaluates and refines downstream.

In differentiable discovery, priors are commonly integrated as soft constraints [1,4,5,9]. This is achieved by adding a regularization penalty scaled by prior confidence. Soft constraints bias optimization toward the semantic prior but still allow the data-driven loss to override incorrect hypotheses when statistical evidence is strong. Our framework follows this hybrid principle, combining semantic clustering with soft constraint integration to improve both scalability and structure recovery.

3 Methodology

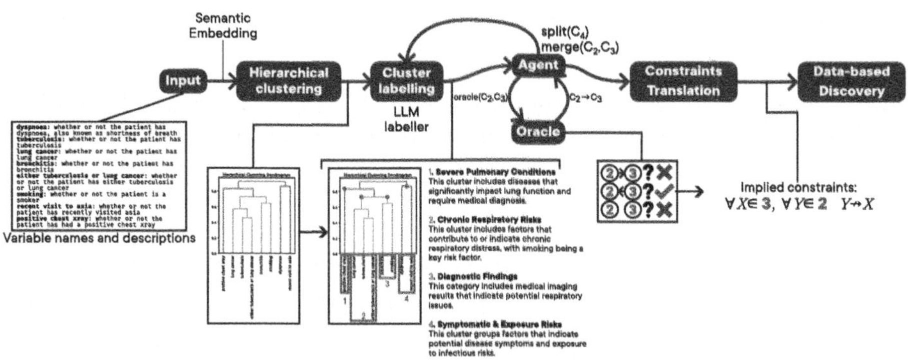

Fig. 1. The architecture of our proposed framework. In the diagram, the "oracle" block denotes an LLM relation-query module, not access to ground-truth causal labels.

Let $\mathbf{X} \in \mathbb{R}^{n \times N}$ be observational data over N variables, and let $\mathbf{W} \in \mathbb{R}^{N \times N}$ denote the weighted adjacency matrix of the causal graph to be estimated. Our framework decomposes causal discovery into three stages: (i) semantic partitioning of variables, (ii) cluster-level causal elicitation with an LLM, and (iii) constrained continuous optimization with DAGMA (Fig. 1).

3.1 Divide: Semantic Partitioning

Each variable is represented by its textual descriptor (name + description). We encode descriptors with a SentenceTransformer embedding model and cluster the embeddings using agglomerative clustering (Ward linkage). In all reported experiments, we use `sentence-transformers/all-MiniLM-L6-v2` (mean pooling). Embeddings are used directly (no additional L2 normalization), so Ward clustering operates in Euclidean space. This yields a partition $\pi : \{1, \ldots, N\} \rightarrow \{1, \ldots, K\}$ with $K \ll N$.

The initial cluster count is set by the pipeline default $K_0 = \min(5, N)$, then the agent can refine this partition through valid merge/split operations constrained by the hierarchical tree. This conservative initialization is used to bound prompt size before refinement.

Compared to pairwise variable querying ($\mathcal{O}(N^2)$ prompts), cluster-level querying reduces the search space to $\mathcal{O}(K^2)$. This is the main mechanism by which we reduce LLM elicitation cost.

3.2 Conquer: Cluster-Level Causal Elicitation

We use a tool-augmented LLM agent that operates on clusters rather than raw variables. At each turn, the agent receives: current cluster compositions, already recorded relations, mergeable sibling pairs from the hierarchical tree, splittable clusters, and progress statistics. The agent selects exactly one action per turn from a fixed tool set. One run outputs a set of directed cluster relations and associated rationales, as detailed in Algorithm 1.

Prompt Structure and Constrained Output. To make the LLM step reproducible, we use fixed prompt templates and constrained tool-calling. The system prompt enforces cluster-level causal reasoning and valid tool arguments only. The user prompt includes cluster names/descriptions, variables per cluster, existing constraints, mergeable sibling pairs, splittable clusters, and progress indicators.

The model output is constrained to one tool call per turn: `label(...)` (3 args: source cluster, target cluster, explanation), `merge(...)` (2 args: cluster 1, cluster 2), `split(...)` (1 arg: cluster id), or `finish()`. The concrete argument names in implementation are `cluster_a`, `cluster_b`, `cluster_1`, `cluster_2`, and `cluster_id`.

In this setup, confidence is not emitted per relation. Instead, confidence is estimated downstream from frequency across repeated runs when building the soft-constraint matrix.

3.3 Constraint Translation to Variable Level

Cluster-level relations are converted into variable-level forbidden directions. For a directed relation $C_a \rightarrow C_b$, we forbid reverse edges from variables in C_b to variables in C_a, defined as $\mathcal{F}_{a,b} = \{(j, i) \mid \pi(i) = C_a, \pi(j) = C_b\}$. Thus, the LLM prior is used as a *directional anti-prior*: it penalizes orientations that contradict the semantic hypothesis, while still allowing data-driven discovery elsewhere.

Algorithm 1: Constraint Generation with Tool-Calling Agent

Input: Initial state S_0, number of runs T_{llm}, recursion limit $R_{\max}$
Output: Constraint results $\{\mathcal{R}^{(t)}\}_{t=1}^{T_{\text{llm}}}$
for $t \leftarrow 1$ **to** T_{llm} **do**
  initialize state $S \leftarrow S_0$ and $\mathcal{R}^{(t)} \leftarrow \emptyset$;
  for $r \leftarrow 1$ **to** $R_{\max}$ **do**
    construct prompt from current clusters, existing relations, mergeable sibling pairs, and splittable clusters;
    call LLM with required tool choice in {`label_causal_relationship`, `merge_clusters`, `split_cluster`, `finish`};
    if *tool is* `finish` **then**
      break;
    end
    if *tool is* `label_causal_relationship`(a, b, e) **then**
      validate (a, b) exist and pair not already labeled under current cluster snapshot;
      append directed relation $a \to b$ with explanation e to $\mathcal{R}^{(t)}$;
    end
    if *tool is* `merge_clusters`(c_1, c_2) **then**
      execute only if c_1, c_2 are direct siblings in the clustering tree;
      update assignments/tree and relabel merged cluster;
    end
    if *tool is* `split_cluster`(c) **then**
      execute only if $|c| > 1$; split via tree children (fallback: half split), then relabel both parts;
    end
    apply safety guards (retry limit, repeated invalid operations, repeated loops); force finish on persistent errors;
  end
end
return $\{\mathcal{R}^{(t)}\}_{t=1}^{T_{\text{llm}}}$;

Because LLM outputs are stochastic, we repeat this process over T_{llm} independent runs and aggregate constraints into a confidence matrix $\mathbf{C}$:

$$C_{ij} = 1 - \exp\left(-\gamma \sum_{t=1}^{T_{\text{llm}}} \mathbb{I}\{(i,j) \in \mathcal{F}^{(t)}\}\right), \tag{1}$$

where $\gamma > 0$ is a scaling hyperparameter controlling the sharpness of the confidence penalty, T_{llm} is the total number of independent LLM query runs (e.g., $T_{\text{llm}} = 3$ for the main $N = 20$ benchmark and $T_{\text{llm}} = 1$ for scaling stress tests), and $\mathcal{F}^{(t)}$ is the set of forbidden directional edges generated during the t-th independent LLM run, obtained as the union of the relation-level sets $\mathcal{F}_{a,b}$ defined above over all predicted cluster relations $C_a \to C_b$ in run t.

3.4 Background: DAGMA and Proposed Soft Constraints

DAGMA replaces strict acyclicity constraints with a differentiable characterization $h(\mathbf{W})$ and solves a non-constrained objective:

$$\min_{\mathbf{W}} \mu(\mathcal{L}(\mathbf{W}) + \lambda_1 \|\mathbf{W}\|_1) + h(\mathbf{W}), \tag{2}$$

where $\mathcal{L}$ is the data-fit term, λ_1 controls sparsity, and $\mu > 0$ is the continuation weight balancing data-fit/regularization against the acyclicity term.

We then introduce our LLM prior through a confidence-weighted penalty:

$$\min_{\mathbf{W}} \mu \left(\mathcal{L}(\mathbf{W}) + \lambda_1 \|\mathbf{W}\|_1 + \alpha \sum_{i,j} C_{ij} |W_{ij}| \right) + h(\mathbf{W}), \tag{3}$$

where α controls prior strength and $h(\mathbf{W})$ is the acyclicity characterization from DAGMA [3]. This formulation biases the solution away from LLM-inconsistent directions without hard-masking the graph space.

Geometrically, the added term acts as a targeted, weighted L_1 regularization. While $\lambda_1 \|\mathbf{W}\|_1$ applies uniform sparsity pressure across all edges, the α term applies an asymmetric, prior-driven penalty. Edges that contradict the LLM semantic consensus ($C_{ij} \to 1$) face a steeper cost. Because this term is linear in $|W_{ij}|$, it preserves the local optimization structure of DAGMA and does not introduce additional non-convexity beyond the base formulation.

4 Experimental Setup

4.1 LLM-Architected Synthetic Benchmark

We evaluate on semantically structured synthetic datasets generated by an LLM-architected pipeline. For each (domain, N, r) configuration, we generate one dataset with a deterministic seed. Each dataset is a $5000 \times N$ table (5000 observations, N variables).

The generation process has four steps:

1. **LLM graph proposal.** The LLM outputs a JSON object containing semantic layers and directed edges. The prompt enforces exactly N variables and a bounded edge budget.
2. **Normalization and sanitization.** We normalize variable identifiers, remove invalid/self-loop edges, and enforce exact variable count (trim/pad if needed). We then sanitize the graph to enforce acyclicity, weak connectivity, and bounded sparsity (target average degree in $[1.5, 3.0]$). These rigorous programmatic checks ensure that the LLM-proposed graphs strictly respect the conditions specified in the experimental setup before any data is simulated.
3. **Semantic-violation injection.** For violation ratio $r > 0$, we rewire a fraction of edges to non-adjacent layers (skip links), while preserving DAG structure and at least one parent for non-root nodes.

4. **SEM simulation.** We sample linear SEM data in topological order:

$$x_j = \epsilon_j + \frac{1}{\sqrt{|\mathrm{Pa}(j)|}} \sum_{i \in \mathrm{Pa}(j)} w_{ij} x_i, \tag{4}$$

with $\epsilon_j \sim \mathcal{N}(0, 1)$ and signed edge weights $w_{ij} \in [0.5, 1.5]$. We generate 5000 samples per dataset and apply per-variable z-score normalization.

Concretely, semantic violations are injected by replacing a proportion r of valid adjacent-layer edges with skip links that jump one or more semantic layers. This creates controlled mismatches between semantic proximity and causal depth while retaining acyclicity. The construction is useful for testing whether methods can recover structure when semantic priors are imperfect. In this paper, we use r as a dataset-generation parameter to diversify graph structure and semantic-causal alignment, not as a standalone robustness claim in the main results.

The parent-count scaling factor $1/\sqrt{|\mathrm{Pa}(j)|}$ and post-hoc z-score normalization are used to reduce variance inflation across graph depths and in-degree values. Without these two steps, downstream nodes can exhibit unstable scale, which can dominate discovery metrics and make cross-dataset comparisons harder to interpret.

For reproducibility, each dataset is saved with: adjacency matrix, sampled data, variable descriptions/codebook, and generation metadata (seed, layer structure, and edge breakdown).

We generate data for three operational domains: E-Commerce Supply Chain, SaaS Cloud Infrastructure, and Manufacturing Plant Operations. Using multiple domains is intended to reduce topic-specific bias and provide a more representative view of overall method performance across heterogeneous application contexts.

For each domain and each size $N \in \{20, 50, 100, 200\}$, we use three semantic-violation levels $r \in \{0, 0.10, 0.15\}$. This yields $3 \times 4 \times 3 = 36$ datasets in total: 9 datasets for the main $N = 20$ comparison and 27 datasets for scaling ($N \in \{50, 100, 200\}$).

4.2 Compared Methods

We compare:

- **Ours (with clustering)**: semantic clustering + LLM constraints + DAGMA.
- **DAGMA pure**: unconstrained continuous optimization baseline.
- **PC** and **GES**: standard score/constraint-based baselines [8,24].
- **Direct Pairwise LLM**: one JSON-constrained LLM decision per variable pair [2,10,17].
- **Direct BFS LLM**: root-and-expansion JSON prompting baseline [15].

All LLM baselines use the same served model and endpoint to ensure fair comparison. For metric comparability, outputs of all methods are evaluated under the same directed-adjacency protocol (detailed below), including methods that naturally return partially directed graphs.

4.3 Implementation Details

Experiments run on Jean Zay (1 node, 4 A100 GPUs) with a local vLLM server hosting `Qwen/Qwen3-30B-A3B-Thinking-2507-FP8` [27]. For FP8 compatibility on this setup, tensor parallelism is set to 2. Constraint generation and dataset synthesis use OpenAI-compatible JSON-constrained decoding through the same local endpoint.

For DAGMA, we use the linear solver with $\lambda_1 = 0.02$, $T_{\text{dagma}} = 4$, warm iterations $= 3 \times 10^4$, max iterations $= 6 \times 10^4$, and learning rate $= 3 \times 10^{-4}$. Soft-prior strength is tuned on $\alpha \in \{0.5, 1.0, 2.0\}$; confidence scaling uses $\gamma = 2.0$. For the main $N = 20$ LLM benchmark, each α configuration is repeated over 3 runs. For scaling stress tests and timeout sweeps, we use one run per method/dataset configuration.

4.4 Evaluation Protocol

We report SHD, FDR, TPR, FPR, and F1 using recovered binary adjacency against ground truth. The main structure-recovery comparison is evaluated on the 9 $N = 20$ datasets (3 domains $\times$ 3 violation levels), averaging per-dataset metrics. Scaling completion is evaluated on 9 datasets per size for $N \in \{50, 100, 200\}$ (27 datasets total). Generation uses deterministic per-dataset seeds (domain- and size-dependent hashing), and the pipeline supports resume/checkpoint behavior to avoid recomputing completed datasets.

Metrics Formulation. Let $\hat{G} = (V, \hat{E})$ be the estimated directed graph and $G^\star = (V, E^\star)$ the ground truth. For each ordered pair (i, j) with $i \neq j$, we compute TP, FP, FN, and TN from edge presence/absence in $\hat{E}$ versus $E^\star$, then derive FDR, TPR, FPR, and F1 using their standard definitions. Structural Hamming Distance (SHD) is the minimum number of directed edge additions, deletions, and reversals needed to transform $\hat{G}$ into $G^\star$.

Directed Comparability for PC/GES. PC and GES (causal-learn implementations) return graph endpoint matrices with symbolic edge codes. To make all methods directly comparable under the same directed metrics, we apply one shared deterministic post-processing pipeline: (i) map any non-zero endpoint code to candidate directed adjacency ($|A_{ij}| > 10^{-3}$), and (ii) if cycles remain, project to a single DAG by deterministic topological pruning (iteratively select nodes with no incoming edges in the remaining subgraph; if none exist, break ties by minimum incoming weight, then keep only edges consistent with the resulting order). Metrics are computed on this projected DAG for all methods. This yields one deterministic representative DAG per run for SHD/FDR/TPR/FPR comparison.

Execution Protocol. All methods are executed under the same evaluation pipeline and compared on the completed $N = 20$ suite (9 datasets). For direct LLM baselines, individual API requests use bounded retries (default: 3 attempts with short backoff). Scaling runs are executed in dataset-sharded batches: one clustering-only LLM run family for our method and separate timeout-constrained classical run families for PC and GES.

Confidence Intervals. For method-level summaries on the $N = 20$ suite, we compute 95% percentile bootstrap confidence intervals from dataset-level scores (10,000 resamples over the 9 datasets). In tables/text, we report these as mean $\pm$ half-width of the 95% interval. For pairwise method comparisons, we use paired bootstrap intervals of mean differences over datasets and treat differences as statistically significant when the 95% interval excludes zero.

5 Results

5.1 Structure Recovery

Table 1 reports structure-recovery results on the $N = 20$ suite (9 datasets). PC obtains the strongest overall recovery (SHD 15.222 ± 3.167, F1 0.670 ± 0.062). Compared with GES, our method has lower SHD (25.889 ± 4.167 vs 28.222 ± 9.612) but lower F1 (0.449 ± 0.087 vs 0.611 ± 0.089). Using paired bootstrap differences over datasets, PC has significantly lower SHD than both our method and GES, while the SHD and F1 differences between our method and GES are not statistically significant at the 95% level.

Relative to unconstrained DAGMA (the solver component), divide-and-conquer reduces SHD from 32.667 to 25.889 and increases F1 from 0.255 to 0.449; both improvements are statistically significant. Relative to direct LLM baselines (the prior component without solver integration), divide-and-conquer also improves substantially in SHD, with statistically significant gains over both direct pairwise and direct BFS.

Table 1. Structure recovery metrics on graphs with $N = 20$ variables (averaged over 9 datasets). SHD and F1 are reported as mean $\pm$ half-width of the 95% bootstrap interval (10,000 resamples over datasets). FDR, TPR, and FPR are reported as means. Lower SHD, FDR, FPR, and higher TPR, F1 indicate better performance.

Method	SHD ↓	F1 ↑	FDR ↓	TPR ↑	FPR ↓
PC Algorithm	**15.222** $\pm$ 3.167	**0.670** $\pm$ 0.062	0.290	0.636	0.061
GES	**28.222** $\pm$ 9.612	**0.611** $\pm$ 0.089	0.466	0.741	0.180
DAGMA (No Constraints)	**32.667** $\pm$ 3.334	**0.255** $\pm$ 0.060	0.711	0.232	0.133
Direct Pairwise LLM	**36.778** $\pm$ 2.445	**0.062** $\pm$ 0.042	0.538	0.028	0.010
Direct BFS LLM	**37.778** $\pm$ 2.834	**0.066** $\pm$ 0.054	0.430	0.033	0.015
Divide-and-Conquer (**Ours**)	**25.889** $\pm$ 4.167	**0.449** $\pm$ 0.087	0.464	0.390	0.081

The metric pattern is consistent across methods. Direct LLM baselines are conservative: they have low FPR (0.010–0.015), but very low TPR (0.028–0.033), which yields low F1. Our method has significantly higher TPR than direct LLM baselines, but also significantly higher FPR than those baselines. This increase in FPR is likely due to the LLM hallucinating plausible but incorrect causal links between semantically related clusters; when these false priors are integrated as soft constraints, they can bias the continuous optimizer into retaining incorrect edges that a purely conservative direct LLM approach would omit. Compared with GES, our method has significantly lower FPR and significantly lower TPR.

PC is expected to be strong on this benchmark because the data are generated from linear SEMs with large sample size. Under these conditions, independence tests are well matched to the data-generating process. The main empirical contribution here is therefore not global superiority over classical methods, but improved use of LLM priors within a continuous optimization pipeline.

5.2 Scalability

We first analyze LLM-query complexity and then report a budgeted scaling experiment. Direct pairwise prompting requires $\mathcal{O}(N^2)$ LLM calls, whereas the proposed method performs iterative cluster-level querying with $\mathcal{O}(MK^2)$ calls ($K \ll N$), where M is the maximum number of agent refinement rounds. In our implementation, M is fixed and treated as a small constant.

This reduction is independent of solver choice and provides the core motivation for using semantic partitioning before optimization.

We also report a scaling stress test on $N \in \{50, 100, 200\}$, with 9 datasets per size (3 domains $\times$ 3 violation levels). Table 2 presents both the completion rates (number of successfully completed datasets out of 9, under a 900 s per-dataset timeout budget) and the total execution times in seconds (mean $\pm$ standard deviation).

Crucially, the execution time reported for our method encompasses the entire pipeline, fully including the time taken for the LLM API calls (semantic clustering and constraint generation) as well as the downstream DAGMA continuous optimization. Because our framework queries the LLM at the cluster level rather than the variable level, this temporal overhead remains bounded and scales gracefully. It does not become a bottleneck, allowing the pipeline to avoid the severe timeout failures observed with classical constraint-based and score-based methods (PC and GES) at $N = 100$ and $N = 200$.

6 Discussion and Future Work

6.1 The Role of Classical Baselines vs. Semantic Priors

Our results show that PC achieves the lowest SHD on the $N = 20$ synthetic datasets, with statistically significant SHD gains over both our method and GES. This is consistent with the benchmark construction: data are generated from linear SEMs under settings favorable to conditional-independence testing.

Classical constraint-based methods rely on the Causal Markov and Faithfulness assumptions. In practice, finite-sample noise, weak effects, or near-canceling

Table 2. Scaling comparison with completion rates and completion times. Time is reported as mean ± SD (seconds) on completed runs only. PC/GES scaling sweeps use a 900 s per-dataset timeout budget.

Method	Metric	$N = 20$	$N = 50$	$N = 100$	$N = 200$
Ours (DAGMA+LLM)	Completion	9/9	9/9	9/9	9/9
	Time (s)	407.6 ± 219.4	508.8 ± 339.2	233.3 ± 231.5	653.4 ± 289.6
PC	Completion	9/9	9/9	0/9 (timeout)	0/9 (timeout)
	Time (s)	21.4 ± 1.4	23.1 ± 4.3	–	–
GES	Completion	9/9	3/9	0/9 (timeout)	0/9 (timeout)
	Time (s)	63.4 ± 20.7	810.8 ± 50.2	–	–

paths can make conditional-independence outcomes unstable, effectively violating faithfulness from the estimator perspective. Because PC orientation rules are sequential, one early test error can propagate and alter multiple downstream edge directions.

At the same time, these results do not invalidate the utility of semantic priors. Recent frameworks such as ALCM [16] address statistical limitations by using an LLM to post-process and refine the algorithmically generated graphs, directly modifying or deleting edges. However, because LLMs are prone to hallucinating causal relationships, relying on them as a downstream adjudicator risks permanently overwriting valid statistical signals with semantic errors. In contrast, our proposed approach uses a global continuous objective with soft penalties. Consequently, local prior errors do not force hard graph decisions and can instead be safely corrected by the data-fit term when statistical evidence is strong. Furthermore, because purely statistical approaches recover structures only up to a Markov Equivalence Class, integrating semantic priors provides essential directional signals to break these ties without overriding the underlying data.

6.2 Interpretation

Classical methods are strongest on the current clean linear synthetic setup. Within the LLM-assisted setting, the central empirical finding is that cluster-level semantic priors provide an effective interface between language-derived knowledge and continuous optimization. In this comparison family, divide-and-conquer significantly improves over direct pairwise/BFS querying on SHD and over unconstrained DAGMA on both SHD and F1. Crucially, while our method introduces an execution time overhead compared to unconstrained DAGMA due to the LLM querying phase, this cost remains practically bounded and scales gracefully. It constitutes a worthwhile trade-off given the significant improvements in structure recovery. Against GES, SHD and F1 differences are not statistically significant at 95%, while GES trades higher recall for higher false-positive rate and suffers from severe scalability limits at $N \geq 100$.

6.3 Threats to Validity

Several threats remain. First, the experiments are conducted solely on synthetic datasets. While this provides a controlled environment, real-world descriptor quality and latent confounding may differ significantly. Future evaluations must involve real-world data, which necessitates rigorously handling ethical concerns, particularly regarding potential biases embedded in LLM priors when applied to sensitive domains. Second, LLM-generated priors and LLM-based baselines currently use a single model family (Qwen). Future work should expand the list of LLM backbones—varying by architecture type, parameter volume, and pre-training data—to systematically validate how different LLM backbones affect the performance and shape of the generated constraints. Third, constrained decoding behavior is implementation-dependent (model/version/server settings), which can affect failure rates and runtime. **Additionally, the high variance observed in execution times for our method is largely driven by the autoregressive nature of the LLM and server-side queuing, meaning real-world runtime depends heavily on the deployment infrastructure.** Fourth, directed comparability for PC/GES requires projection from partially directed outputs to a DAG representation; although the projection is deterministic and shared across methods, it may still influence absolute classical scores. Fifth, scaling completion rates for PC/GES depend on the selected timeout budget (here 900 s per dataset); different budgets can change absolute timeout rates while preserving the same protocol-level comparison logic.

6.4 Practical Implications

The method is most relevant when domain descriptions are available and one wants to combine semantic priors with data-driven fitting. In this setting, soft constraints provide a controlled mechanism to bias the solution while still allowing the optimization objective to override incorrect priors.

6.5 Future Work

Priority follow-ups are: (i) adaptive confidence calibration for noisy semantic priors, (ii) support for cross-layer exceptions during constraint translation, and (iii) validation on non-synthetic datasets with expert-written variable descriptions and heterogeneous noise models.

Looking ahead, our framework can serve as the starting point for fully automated causal analysis. As the field develops new autonomous agents, these tools will need a reliable initial graph to work from. Once our pipeline efficiently generates a causal graph backed by statistical data, it can be handed off to other specialized systems. For example, the graph could be sent to refinement tools like ALCM for closer, edge-by-edge inspection. Alternatively, it could be passed to causal inference agents like CAIS [25] to calculate the actual numerical effects of specific interventions. Connecting our efficient graph discovery with these downstream calculation tools is a clear next step for the field.

7 Conclusion

This work presents a divide-and-conquer framework for integrating LLM semantic knowledge into continuous causal discovery. The method reduces query complexity by shifting from variable-level to cluster-level reasoning and converts semantic hypotheses into soft constraints for DAGMA. On the reported benchmarks, classical PC remains strongest on synthetic linear settings. Among the evaluated LLM-assisted continuous optimization methods, our framework significantly improves over direct LLM prompting on SHD and over unconstrained DAGMA on SHD and F1. Differences versus GES on SHD/F1 at $N = 20$ are not statistically significant, while our method maintains higher completion rates at larger N under fixed timeout budgets, demonstrating that the LLM querying overhead is a manageable trade-off for improved scaling and accuracy over unconstrained approaches.

References

1. Amirkhani, H., Rahmati, M., Lucas, P.J., Hommersom, A.: Exploiting experts' knowledge for structure learning of Bayesian networks. IEEE Trans. Pattern Anal. Mach. Intell. **39**(11), 2154–2170 (2016)
2. Ban, T., Chen, L., Wang, X., Chen, H.: From query tools to causal architects: harnessing large language models for advanced causal discovery from data. arXiv:2306.16902 (2023)
3. Bello, K., Aragam, B., Ravikumar, P.: DAGMA: learning DAGs via m-matrices and a log-determinant acyclicity characterization. In: Advances in Neural Information Processing Systems, vol. 35, pp. 8226–8239 (2022)
4. Borboudakis, G., Tsamardinos, I.: Incorporating causal prior knowledge as path-constraints in Bayesian networks and maximal ancestral graphs. In: 29th International Conference on Machine Learning, pp. 427–434 (2012)
5. Borboudakis, G., Tsamardinos, I.: Scoring and searching over Bayesian networks with causal and associative priors. In: 29th Conference on Uncertainty in Artificial Intelligence, pp. 102–111 (2013)
6. Camps-Valls, G., et al.: Discovering causal relations and equations from data. Phys. Rep. **1044**, 1–68 (2023)
7. Chen, L., Ban, T., Wang, X., Lyu, D., Chen, H.: Mitigating prior errors in causal structure learning: towards LLM driven prior knowledge. arXiv:2306.07032 (2023)
8. Chickering, D.M.: Optimal structure identification with greedy search. J. Mach. Learn. Res. **3**, 507–554 (2002)
9. Constantinou, A.C., Guo, Z., Kitson, N.K.: The impact of prior knowledge on causal structure learning. Knowl. Inf. Syst. **65**(8), 3385–3434 (2023). https://doi.org/10.1007/s10115-023-01858-x
10. Darvariu, V.A., Hailes, S., Musolesi, M.: Large language models are effective priors for causal graph discovery (2024). arXiv:2405.13551
11. Dong, J., Liu, Y., Aloui, A., Tarokh, V., Carlson, D.E.: CARE: turning LLMs into causal reasoning expert (2025). arXiv:2511.16016
12. Ji, Z.: Survey of hallucination in large language models. ACM Comput. Surv. (2023)

13. Jin, Z., et al.: CLadder: assessing causal reasoning in language models. In: 37th Conference on Neural Information Processing Systems (2023)
14. Jin, Z., et al.: Can large language models infer causation from correlation? In: The 12th International Conference on Learning Representations (2024)
15. Jiralerspong, T., Chen, X., More, Y., Shah, V., Bengio, Y.: Efficient causal graph discovery using large language models (2024). arXiv:2402.01207
16. Khatibi, E., Abbasian, M., Yang, Z., Azimi, I., Rahmani, A.M.: ALCM: autonomous LLM-augmented causal discovery framework (2024). arXiv:2405.01744
17. Kıcıman, E., Ness, R., Sharma, A., Tan, C.: Causal reasoning and large language models: opening a new frontier for causality (2023). arXiv:2305.00050
18. Pearl, J.: Models, Reasoning and Inference. Cambridge University Press (2000)
19. Pearl, J.: Causal Diagrams for Empirical Research. Biometrika (1995)
20. Peng, B., Chen, S., Xu, L., Lu, C.: CauScientist: teaching LLMs to respect data for causal discovery (2026). arXiv:2601.13614
21. Roy, A., Devharish, N., Ganguly, S., Ghosh, K.: Causal-LLM: a unified one-shot framework for prompt- and data-driven causal graph discovery. In: Findings of the Association for Computational Linguistics, pp. 8259–8279 (2025)
22. Shah, A., DePavia, A.F., Hudson, N.C., Foster, I., Stevens, R.: Causal discovery over high-dimensional structured hypothesis spaces with causal graph partitioning. Trans. Mach. Learn. Res. (2025)
23. Sokolov, A., Sabelli, F., Azadie Faraz, B., Li, W., Seco, L.A.: Towards automating causal discovery in financial markets and beyond (2024). SSRN 4679414
24. Spirtes, P., Glymour, C., Scheines, R.: Causation, Prediction, and Search. MIT Press (2000)
25. Verma, V., et al.: Causal AI scientist: facilitating causal data science with large language models. In: NeurIPS 2025 Workshop on CauScien: Uncovering Causality in Science (2025)
26. Wan, G.: Large language models for causal discovery: current landscape and future directions (2024). arXiv:2402.11068
27. Yang, A., Team, Q.: Qwen3 technical report (2025). arXiv:2505.09388
28. Zečević, M., Willig, M., Dhami, D.S., Kersting, K.: Causal parrots: large language models may talk causality but are not causal. Trans. Mach. Learn. Res. (2023)
29. Zheng, X., Aragam, B., Ravikumar, P., Xing, E.P.: DAGs with no tears: continuous optimization for structure learning. In: Advances in Neural Information Processing Systems (2018)

Towards a Non-monotonic Neuro-Symbolic Framework: Application to Outlier Mitigation During Training

Nathan Olejniczak(✉) and Anne Jeannin-Girardon

ICube UMR 7357, University of Strasbourg, Strasbourg, France
nathan.olejniczak2@etu.unistra.fr

Abstract. Artificial intelligence relies primarily on logical reasoning systems to analyze information and make decisions. However, these systems often operate in dynamic environments where knowledge is partial or subject to change. In this context, monotonic reasoning, which assumes that any new information only reinforces or complements existing knowledge, has its limitations. To overcome these shortcomings, non-monotonic reasoning has been developed to enable AI systems to adapt their conclusions as data evolves.

This article proposes a method that integrates this non-monotonic aspect by coupling a logical program, composed of rules and facts, with a neural network in a classification framework. Unlike traditional approaches, this work aims to initiate a correction of neural predictions through logical revision, treating ASP (Answer Set Programming) as an active reasoning layer capable of flagging and rectifying inconsistent outputs.

To this end, a new loss function that uses feedback from the logic program to guide the neural network has been introduced. The results obtained demonstrate the effectiveness of this method, in particular its increased ability to handle outliers without disrupting the classification of standard data, thus validating the value of such an approach.

1 Introduction

Neuro-symbolic AI has gained significant prominence for its capacity to reconcile the statistical learning of neural networks with the formal rigor of symbolic logic, offering a path toward more interpretable and data-efficient models. While numerous approaches have been proposed to address the semantic grounding of knowledge, to guide the learning of sub-symbolic models, or to design "deep reasoners", a research gap persists regarding the exploitation of non-monotonic frameworks within these architectures.

Placing this work in the context of guiding sub-symbolic model learning via domain knowledge, we consider such knowledge as defeasible, meaning it holds true unless contradicted by specific circumstances. This approach must recognize exceptions to rules while actively avoiding the pitfalls of rigidity and maintenance

C. Zanni-Merk et al. (Eds.): AIDT 2026, CCIS 3023, pp. 173–188, 2026.
https://doi.org/10.1007/978-3-032-31319-5_12

difficulty that characterized early expert systems [11]. In this context, Answer Set Programming (ASP) [5] establishes itself as a logic programming language specifically designed to structure this non-monotonic reasoning in a formal and practical way.

The work presented here constitutes a first step in this direction by designing a neural-symbolic model capable of identifying outliers during training through a default logic [10] mechanism. Specifically, this article studies the possibility for a classification model to handle anomalies that can severely distort the learning process, particularly on datasets with few instances. It is assumed that by identifying these outliers and reducing their impact using ASP rules, it is possible to improve the model's accuracy and robustness, i.e., its ability to correctly classify disturbances without compromising generalization to statistically normal data. To enable the network to reconsider the initial classification of an outlier and adjust its weights accordingly, the loss function of our model, here binary cross-entropy is extended by incorporating terms derived from ASP feedback, coupled with a semantic loss.

2 Literature

The primary interest of non-monotonic logic lies in its ability to mirror "common sense", where conclusions are drawn based on incomplete information and retracted when new evidence arises. Unlike classical logic, which is monotonic, non-monotonic reasoning allows "defeasible" inferences: conclusions that are valid by default can be invalidated by specific exceptions introduced later. Answer Set Programming (ASP) has emerged as a main programming language for this purpose, as its semantics are specifically designed to handle such defeasible rules through explicit negation, which allows the system to differentiate what is stated false and a lack of information. This section reviews how recent neuro-symbolic architectures leverage ASP, before examining the foundations of non-monotonic reasoning.

2.1 Coupling ASP and Neural Networks

Recent advancements in neuro-symbolic AI have successfully demonstrated the benefits of coupling neural network with symbolic reasoning through ASP. Two prominent approaches, NeurASP and ASPer, illustrate the potential of this integration.

NeurASP [13] offers a simple but effective extension of ASP programs to help neural networks perform certain tasks more quickly. The main idea is to treat the output of a neural network as a probability distribution, rather than a binary decision, over atomic facts in an ASP program. Specifically, NeurASP introduces the concept of "neural atoms" that link an input tensor, processed by the neural network, to symbolic predicates. The article discusses the case of Sudoku solving, where the neural network is solely responsible for perception (identifying the numbers in an image), while the ASP solver takes care of solving the grid using logical reasoning. This method allows the network to learn effectively with very little data. For example, with only 17 training examples, NeurASP achieved a perception accuracy of 62%, compared to 31% for a standard neural network. Although effective, the approach suffers from scalability issues, as gradient calculation requires the enumeration of stable models, which is a computationally expensive process.

ASPer [6] proposes a more direct method by modifying the loss function of the neural network to include logical constraints. Unlike the usual complex architectures, ASPer uses a shallow neural network coupled with an ASP solver. The methodology is based on a combined loss function that aggregates three components:

- Standard loss (cross-entropy)
- The loss of constraints, calculated from the outputs of the ASP solver, which penalizes solutions that violate the rules of the game.
- The loss of expertise, which incorporates specific business knowledge (for example, the expected sum of numbers in a row).

ASPer demonstrates outstanding results on very small datasets. On a training set of only 12 Sudoku grids, the model achieves perfect accuracy (1.00). However, limitations appear as the dataset grows or the puzzles become more complex, suggesting that constraints alone are not always sufficient to guide the network in difficult search spaces.

Although these studies (NeurASP and ASPer) successfully combine logic and learning to improve the performance of neural networks, they focus primarily on compliance with static rules (Sudoku rules do not change) rather than non-monotonic behavior. It is precisely this shortcoming in revising conclusions in the face of exceptions or contradictory information that is addressed in this article.

2.2 Non-monotonic Foundations: Default Logic and Circumscription

To move beyond fixed constraints and represent a form of "common sense" capable of handling exceptions, we must look to the founding works of non-monotonic reasoning. These concepts are essential because they provide the theoretical basis for "defeasibility", allowing a system to maintain validity without manual enumeration of every possible exceptions unless a contradiction arises.

Default Logic. Reiter introduced Default Logic [10] to formalize reasoning in the absence of complete information. This approach is based on a specific rule: "in the absence of information to the contrary, assume that...". This principle can be illustrated by the following classic example: "if x is a bird, and there is no evidence to the contrary, then x can fly". This is formalized by the following rule, known as *the default rule*:

$$\frac{\text{BIRD}(x) : MFLY(x)}{\text{FLY}(x)}.$$

In this formula, the operator M introduces a consistency condition: "if it is consistent to assume that x flies, then x can fly". This rule allows us to conclude that the bird flies, unless this conclusion contradicts other established knowledge, such as the following exceptions:

$$(x).\,\mathrm{PENGUIN}(x) \supset \neg\mathrm{FLY}(x)$$

$$(x).\,\mathrm{OSTRICH}(x) \supset \neg\mathrm{FLY}(x)$$

$$\ldots$$

This formalization allows a program to produce valid inferences even when knowledge is partial. However, as soon as an exception is identified, for example, the bird turns out to be a penguin, the inference process is updated and blocks the default conclusion. This mechanism, which is the main element for non-monotonic reasoning, provides a theoretical starting point for writing ASP code that will integrate a revision capability within a neural system.

Circumscription. In a more theoretical approach, McCarthy introduced Circumscription [7] as a means of formalizing common sense reasoning without having to enumerate all possible exceptions, a concept also known as the "qualification problem" [8]. This concept refers to the theoretical impossibility of exhaustively listing all conditions required for an action to succeed or an inference to be valid. Circumscription allows a program to jump to certain conclusions: in the absence of evidence to the contrary, it is assumed that normal conditions apply. For instance, if no explicit rule prohibits boats to cross a river, such as "lack of oars", the system will deduce that boats can cross the river.

Rather than defining circumscription as a separate logic, McCarthy presents it as an extension of first-order logic aimed at minimizing the extension of predicates. Specifically, suppose that a formula A contains a predicate $P(\bar{x})$. Circumscription amounts to replacing P with a predicate expression Φ, imposing a minimality constraint:

- $A(\Phi)$ remains valid despite the substitution.
- The extension P is minimal, i.e., $P(\bar{x})$ is considered false for any tuple $\bar{x}$, unless it must necessarily be true to satisfy formula A.

This formalization can be expressed as follows:

$$A(\Phi) \wedge \forall \bar{x}.(\Phi(\bar{x}) \supset P(\bar{x})) \supset \forall \bar{x}.(P(\bar{x}) \supset \Phi(\bar{x})).$$

This formalization is essential in this case: by treating non-monotonic reasoning as an extension of first-order logic, it can be implemented into logical programming languages such as ASP, which is based on this logic.

2.3 ASP

Answer Set Programming (ASP) is a logical programming language specifically designed to model non-monotonic reasoning. Unlike other languages such as Prolog, which rely primarily on "negation as failure", which assumes that anything that cannot be proven true is false, ASP allows for the expression of "explicit negation" [2]. This type of negation allows the system to differentiate between the absence of information ("I don't know if X is true") and what is declared false ("I know that X is false"). This nuance makes it possible to revise or cancel previously established conclusions when a contradiction emerges from new data.

A classic example, known as the *flying penguin problem*, illustrates this mechanism:

```
bird(pigeon).
bird(alfred).
flies(X) :- bird(X), not penguin(X).
penguin(alfred).
#show flies/1.
```

Although this example does not cover all the capabilities of ASP, it isolates the mechanism necessary for the issue addressed in this article. This program deduces the entities that fly according to the rule *flies(X)*, so here the solver will only return *flies(pigeon)*. Indeed, for Alfred, although he is defined as a bird, the addition of the fact *penguin(alfred)* activates the negation of the rule *not penguin(X)*, which blocks the deduction that he flies. On a larger scale, ASP is therefore an essential tool for introducing non-monotonic behavior, allowing the dynamic injection of new rules or exceptions into a system coupled with a neural network, which will enable new inferences over time.

The integration of neural networks and symbolic logic has been proven effective for structured tasks such as perception and constraint satisfaction with NeurASP and ASPer. However, the inherent flexibility of non-monotonic formalisms remains difficult to operationalize within current neuro-symbolic architectures. While Default Logic and Circumscription establish the formal necessity of defeasible reasoning to manage exceptions, these principles are rarely the focus of frameworks designed around static logical constraints. Utilizing the specific properties of ASP to represent two principles facilitates a more adaptive learning process for tasks such as the identification and management of outliers. The subsequent sections detail the formalization of these concepts and their integration into a program capable of handling exceptions through logical backpropagation.

3 Methods

3.1 Formalization of Non-monotonicity in ASP

Although formalized differently, Circumscription and Default Logic are based on a common approach. This common basis allows us to extract the central principles in order to transpose them into an ASP program.

The approach is initially based on the operator M in Reiter's formula, meaning "in the absence of contrary information": if a list of exceptions can be easily represented in ASP, the predicate M cannot be expressed literally and requires adaptation via failure negation. McCarthy's Circumscription formula, although more theoretical, reveals a structure conducive to algorithmic adaptation. By combining these two ideas, an initial classic implementation, based on a list of explicit exceptions, can be illustrated as follows:

```
bird(Alfred).
bird(Ben).
bird(Charlie)
penguin(Alfred).
ostrich(Charlie).
flies(X) :- bird(X), not penguin(X), not ostrich(X).
```

This example reproduces the central idea of Reiter and McCarthy's default logic. However, introducing new exceptions at the individual instance level proves complex when data sets contain anonymous or generic entities (the name of the ostrich is not necessarily known). A generalization of the code at the class level is therefore required to handle the dynamic addition of new species:

```
bird(pigeon).
bird(penguin).
bird(ostrich).
flies(X) :- bird(X), not bird(penguin), not bird(ostrich).
```

This program is more concise and makes adding exceptions easier: simply add a new *bird(X)* fact for the species concerned and update the main rule to exclude it.

3.2 Handling Outliers

The integration of non-monotonic reasoning within a neural network requires the selection of a specific application framework. One possible choice is the treatment of outliers, as these exceptions challenge classical monotonic reasoning. Indeed, particularly on small datasets, these anomalies often cause model instability and reduced accuracy. Studying these atypical cases will not only highlight the limitations of standard learning models, but also explore how non-monotonic logic can be leveraged to offer a model with better performance and less sensitivity to these values.

The case of the penguin among birds perfectly illustrates this problem. Although the general rule associates birds with the ability to fly, the penguin is an exception to this expectation. During learning, the presence of such instances may bias the classification of other animals in the test set: entities that share similar characteristics with the penguin but follow the general rule (and are therefore capable of flight) could be incorrectly classified due to this anomaly.

3.3 Data

To evaluate the proposed approach, this study focuses on a binary classification task using the Zoo Animal Extended dataset [1]. This dataset is divided into two tables that have been concatenated, totaling 113 instances divided into 7 taxonomic classes (Mammal, Bird, Reptile, etc.). Each animal is described by a vector of 16 characteristics (mostly boolean, such as hair, eggs, milk, airborne, and numerical for legs). The classification task here focuses on predicting the binary attribute "feathers".

The primary objective is to ensure that a neural network can correctly identify outliers similar to those encountered during training, without disrupting the classification of standard data. More specifically, the objective is to minimize two types of errors on the target attribute "feathers": the erroneous prediction of feathers on an animal that does not have them (false positive) and the opposite (false negative).

Oversampling. As this dataset is limited, *oversampling* was applied to increase the number of instances in order to obtain more stable performance. One of the problems with outliers is the fluctuation in the model's predictions, which could, for some iterations, correctly classify certain outliers and, for another iteration, incorrectly classify these outliers but correctly classify others. In response to this problem, a larger dataset would allow us to see a clear, albeit variable, improvement between a baseline model without ASP coupling and a model implementing such coupling. Two different approaches were considered: SMOTE [3] and data duplication by sampling with replacement.

SMOTE, for Synthetic Minority Over-Sampling Technique, is a technique for generating examples for a designated minority class. It generates new examples in the feature space by selecting a point from this minority class and creating a new similar example based on an algorithm based on the k-nearest neighbors. Sampling with replacement method is a more naive approach that clones a randomly selected instance from the dataset. Sampling with replacement was preferred to SMOTE in order to preserve taxonomic categories, avoiding generating instances whose feature combination would not correspond to any known animal from the dataset. Such a procedure was considered more appropriate for creating outliers deliberately close to the current data to confuse the model, which must be able to classify them correctly. In this case, the data volume was multiplied by 10, resulting in 1,130 instances, while maintaining the initial ratio between feathered and featherless classes. Multiple multiplication values were tested to find a balance between model accuracy and the risk of bias inherent to high data redundancy.

Outliers Generation. The selected dataset representing animals does not contain any outliers, so it is necessary to create synthetic outliers in order to evaluate the performance of our approach. These outliers were generated based on arbitrary criteria[1]:

- With the "feathers" attribute set to zero, the challenge for the model is to correctly predict the absence of feathers in the model while correctly predicting the presence of feathers in the nearest bird.
- The generated outlier must differ from at least one bird by exactly one characteristic, the aim being to ensure that the generated outlier is close to at least one of the birds in order to have a point of confusion to study.
- The other birds must differ by at least one characteristic.
- Animals without feathers (i.e., mainly land animals) must differ by at least three characteristics, the aim being to ensure that the model does not use animals without feathers to classify these isolated cases.

Several methods can be used to generate these outliers, including the Monte Carlo approach [9] and the Best-First Search method [4]:

- For the first method, an animal is generated with random characteristics (the "feathers" attribute is manually set to zero). If the animal does not meet the specified constraints, it is rejected; otherwise, it is added.
- In order to perform a best-first search, an animal with the attribute `feathers = 1` is randomly selected and, based on this animal's characteristic vector, all possible modifications are made until a case is found that meets the constraints. However, the model is constrained so as not to randomly select the same bird twice, in order to avoid producing two similar outliers.

The first method is the simplest and most naive. Although easy to implement, its theoretical execution time can be infinite in the worst case. The second approach guarantees results if the solution exists and is therefore preferred for generating outliers.

Borderline Instances vs. Outliers. In order to evaluate the model's ability to sufficiently penalize the generated outliers, the outliers are introduced into the training set and borderline instances are generated and stored in the test set: these borderline instances are based on the previously generated outliers with a randomly selected deviation characteristic. This mutated characteristic cannot be the same as the one that was initially mutated to create the synthetic animal, so as not to revert to the original bird, and the variant cannot be the same as an existing animal either. If the variant were to have the same characteristics as another animal except for the feathers characteristic that is to be predicted, the model would have no way of differentiating between the two animals. Table 1 illustrates an outlier generated from a feathered animal and its alternative.

The number of outliers was defined to respect a percentage of the dataset, and this threshold was set at 10% by default. Although a 10% proportion of

[1] This article serves as a preliminary study; in a real-world application, these criteria would be defined by expert knowledge.

outliers can, in some areas, mean that they are a population in their own, this high value was necessary in order to observe their impact on a small dataset. This value could then be reduced with a larger dataset.

This initial configuration allows us to evaluate the network's generalization ability: it must be able to correctly classify these never-seen-before borderline cases close to outliers, while maintaining its performance on real animals, e.g., it must be able to classify the alternative version of an outlier generated from the canine macaw as "featherless" by learning from the outlier present in the training set, even though the canine macaw is also present in the dataset.

Table 1. Illustrative table containing characteristics of a feathered animal, the outlier created from it, and the alternative version of the outlier

Animal name	Feathers	Eggs	Air.	Aquatic	Predator.	Venom	Tail	Domestic	...
Canine macaw	1	1	1	0	0	0	1	1	...
Fake animal	0	1	0	0	0	0	1	1	...
Borderline instance	0	1	0	0	0	1	1	1	...

In this specific case, even if the outlier is one characteristic away from the borderline instance and the original animal, the model needs to find similarities between the original animal and the other feathered animals in the dataset to correctly classify it despite its proximity to the outlier. At the same time, the borderline instance that mutated the "venom" attribute, which is an extremely rare characteristic in feathered animals, is probably easier to isolate.

3.4 Logical Backpropagation

Penalization of Aberrant Classifications. The neural network undergoes training via backpropagation, minimizing the binary cross-entropy loss. However, to incorporate outlier management directly into the learning process, this loss function is modified: in order to mitigate the influence of outliers, the penalty associated with their misclassification is increased. This strategy aims to prevent these exceptions from disrupting the convergence of the model on regular instances. This is modified by adding corrective terms based on feedback from the ASP program.

These additional terms apply only when the ASP program explicitly knows the animal, thus signaling a known exception (e.g., a bird without feathers). In all other cases, according to Reiter's definition, the absence of contradictory information implies the application of the default rule (the animal has feathers). This mechanism is intentional: only identified exceptions should trigger a logical correction, and for standard cases, the model follows the general distribution without intervention from the symbolic module, only interfering when a rule is violated is the core mechanism of default logic. This method requires the implementation of an ASP flag that cancels the contribution of additional penalty terms when the instance is not a recognized exception.

Semantic Loss. It is essential to define a penalty method adapted to prediction errors made by the neural network on outliers. Applying a uniform penalty for all errors is ineffective because it does not take into account the output probability distribution, which may be close to the correct answer despite an incorrect outcome.

To illustrate this nuance, let's consider the case of a prediction assigning a 51% probability to the absence of feathers and a 49% probability to their presence: although technically incorrect, this decision reflects an uncertainty that does not justify the same severity as a 100% to 0% error. Therefore, the penalty is based on the confidence of the model: the more confident an incorrect prediction, the greater the penalty. Semantic Loss [12] formalizes this approach by penalizing the network proportionally to the probability it assigns to a violated logical constraint:

$$L_s(\alpha,\ \mathrm{p}) = -\log\left(\sum_{x\models\alpha}\left(\prod_{i:x\models X_i} p_i\right) \times \left(\prod_{i:x\models\neg X_i} (1-p_i)\right)\right).$$

In this formula, α represents the propositional logic constraint that the system must respect, and p denotes the vector of probabilities predicted by the neural network. The term x corresponds to the assignment of truth values to variables and the notation $x \models \alpha$ indicates that the sum only applies to worlds that satisfy the constraint. Specifically, this function calculates the total probability that the network generates a state that complies with the logical rule, then takes the negative logarithm to maximize this probability.

In the specific context of outlier treatment, the constraint α corresponds to the negation of the target attribute $\alpha = \neg X$, imposed by the ASP program when an exception is detected, for instance: "this animal must not have feathers". This constraint restricts the space of possible worlds to those where variable X is false. Consequently, any term of the sum involving $X = 1$ is excluded. The first product of the equation, relating to true variables, becomes 0 for this attribute, while the second product is reduced to the probability of the attribute's absence. The expression is then simplified as follows:

$$-\log(1 \times (1 - p_i)) = -\log(1 - p_i).$$

In this study, Semantic Loss is integrated as an additional regularization term to the standard loss function in order to specifically target anomalies identified by the logic part. Binary cross-entropy (BCE) is the main component, dedicated to binary classification. It is then supplemented by the addition of semantic loss, whose activation is conditioned by a logical indicator from the ASP program.

This mechanism acts as a filter: the multiplier coefficient derived from the ASP cancels out the contribution of semantic loss when the instance being processed does not correspond to a known exception. Thus, the logical correction only applies to necessary cases, i.e. outliers known by the ASP program. Unidentified outliers are treated as standard data and are exempt from the semantic component. To distinguish these instances, we introduce a binary flag a_i, set to 0 for unknown data and 1 for known data. A known outlier is given a penalty proportional to the deviation from the target probability multiplied by an arbitrarily defined weight λ, allowing the network to learn standard patterns without interference. The combination of these elements results in the following expression of the overall loss function:

$$\ell_i = \underbrace{-\left[y_i \log(p_i) + (1 - y_i)\log(1 - p_i)\right]}_{\text{BCE}_i} + \lambda_{\text{sem}} \times a_i \underbrace{\left(-\log(1 - p_i)\right)}_{\text{SemLoss}_i}.$$

4 Results and Discussion

In order to evaluate the effectiveness of the proposed approach, the performance of the neural network was compared in three different scenarios. These scenarios make it possible to isolate the impact of outliers on standard learning, then measure the contribution of the semantic loss function coupled with the ASP program

Table 2 shows the average performance metrics obtained (Loss, Accuracy, Precision, Recall, and F1-Score) for each experimental configuration over 10 training and testing phases. For the model equipped with the semantic loss component, several λ values were tested.

Table 2. Performance comparison for different experimental conditions: control (*data + BCE* and *augmented data + BCE*) and ASP-Neural network coupling (*augmented data + semantic loss*)

Scenario	λ	Accuracy	Precision	Recall	F1-Score
Classic Data + BCE	-	1.00	1.00	1.00	1.00
Augmented Data + BCE	-	0.97	0.90	0.91	0.91
Augmented Data + SemLoss	0.5	0.98	0.91	1.00	0.95
	1.25	**0.99**	**0.95**	**0.99**	**0.97**
	3	0.98	0.93	0.96	0.95

Since outliers are minority instances in the dataset (10%), the metrics presented in Table 2 are not sufficiently representative of the model's behavior, as improvements... on such a small subset are easily overshadowed by the majority class. To better understand the influence of λ, Table 3 shows the confusions matrix obtained for $\lambda = 0.5$, $\lambda = 1.25$ and $\lambda = 3$. While a high λ increases the false negative rate, a low λ reduces these errors to zero as the cost of an increased number of false positives. Selecting an intermediate λ maintained a near-zero false negative rate while simultaneously keeping the frequency of false positives low.

Table 3. Confusion matrices representing the combined classifications of models using the semantic loss function with different weights over 10 iterationss

	$\lambda = 0.5$			$\lambda = 1.25$			$\lambda = 3$	
	Pred. 0	Pred. 1		Pred. 0	Pred. 1		Pred. 0	Pred. 1
Act. 0	1821	39	Act. 0	1839	21	Act. 0	1834	26
Act. 1	0	400	Act. 1	1	399	Act. 1	17	383

4.1 Discussion

In the first scenario (*Classic Data + BCE*), using the original Zoo dataset with 10x oversampling without adding synthetic outlier birds, the model achieves optimal performance with an accuracy of 1.00. This result serves as a baseline and confirms that the network architecture is sufficient for predictions on a noise-free dataset.

The introduction of outlier animals and their borderline instances in the second scenario (*Augmented Data + BCE*) reveals the limitations of the classic loss function. Although the accuracy remains high, a slight drop in recall and precision can be observed. This deterioration results in the misclassification of feathered animals in the original dataset as well as the generation of outliers. This suggests that the presence of strong exceptions, such as a featherless bird, during training disrupts the convergence of the model, compelling it to generalize incorrectly in an attempt to accommodate these anomalies, which disrupts the decision boundaries for standard data.

The third scenario (*Augmented Data + SemLoss*) implements the method presented in this article by replacing the standard loss function with the new loss function incorporating the logical feedback from the ASP, which allows the model to recover near-perfect stability for the optimal value of the weight λ as described in Sect. 4.2. Each metric shows a higher score than the original model, most notably recall. This improvement suggests that penalizing outliers refines the decision boundary, given that recall specifically measures the capacity to capture rare instances like outliers. Accuracy is the lowest score for this third scenario, which can be explained by the complex task of having only a Hamming distance of one between the original animal and the outlier. Even if the mutated characteristics from the outlier may be rarer for this type of animal (e.g., venom), if the mutated characteristic is common (e.g., eggs), then even the model with the semantic loss function will experience some difficulty in predicting the correct value for this case, which will be close to the characteristics of an existing feathered bird.

This choice of minimal difference between real animals and generated animals was forced by the perfect score of the baseline on the original dataset, which required imposing strong cases of confusion to lower its performance and attempt to correct the prediction errors thanks to our semantic-aware loss function. This distance value is comparable to training under extreme conditions: in a massive dataset that naturally contains outliers, the Hamming distance between outliers and statistically normal instances could naturally be higher, making the outliers easier to isolate and resulting in better results. In addition, oversampling due to the small size of the dataset also required the duplication of normal instances, allowing the base model to achieve better results than on a dataset composed of unique instances. Indeed, as the control model was naturally better at predicting the target characteristic of already known animals, it was easier to correctly identify unique outliers. Several sizes of this same dataset were tested by modifying the oversampling multiplier. The value 10 yielded the most stable results. Higher values led to a degree of duplication that became too much of a bias for the results. Consequently, it would be interesting to characterize the degree of oversampling, more specifically the point at which the data ceases to be a valid representation for training.

An ideal dataset for repeating this protocol should naturally be larger, contain only unique instances without outliers (so that they can be added manually), and obtain imperfect scores with a basic model. In this case, the minimum Hamming distance between the randomly selected animal and the generated synthetic animal could be greater, which would prevent these confusions dependent on the mutated characteristic from creating the alternative synthetic animal.

4.2 Weight Sensitivity

The effectiveness of logical correction using the semantic loss function also depends on the calibration of the weight λ in the loss function. A compromise must be found regarding the value of this penalty:

- If λ is too low, the correction is insufficient to counteract the importance of outliers via cross-entropy.
- If λ is too high, predictions on the original animals will also be impacted.

Indeed, a λ value that is too low will reduce the contribution of the semantic part, which, in extreme cases, will be nullified. With the semantic contribution being zero, the remaining part is the cross-entropy function alone, returning to the basic model. For a value of $\lambda = 0.5$, the slight contribution made it possible to restore the recall of the original model, but the accuracy, based on the classification of alternative outliers, remains close to the result of the original model, thus failing in its main purpose.

Conversely, an excessively high penalty caused by an elevated λ value creates a gradient imbalance that hinders overall learning. Mathematically, this phenomenon stems from gradient backpropagation: when the attribute a_i is active, the total gradient for instance i is defined as $\nabla l_i = \nabla \text{BCE}_i + \lambda_{sem} \nabla \text{SemLoss}_i$. If λ_{sem} is disproportionately high, the semantic term dominates weight adjustments, forcing the network to blindly obey logical rules. This extends the influence of the constraint to distant instances, leading to a global adherence that stifles features learned from normal data. In the presence of outliers, while the BCE gradient pushes toward neighboring instances, the semantic gradient pulls strongly toward logical constraints; an excessive λ_{sem} saturates this competition.

Consequently, in extreme cases, the model may assign a value of zero to all other animals. For $\lambda = 3$, performance begins to decline compared to $\lambda = 1.25$, as the model fails to correctly classify original animals and recall drops. Due to weight sharing, the network prioritizes canceling the semantic penalty over minimizing classification error, driving all predictions toward zero.

However, this new loss function has two objectives: it must correct the known exception while allowing the network to recognize similar data. By correcting an exception identified by the ASP program, it must enable the network to adjust its decision boundaries locally. Semantic loss then acts as a regularization term that guides the model toward consistent generalization thanks to expert knowledge, and an optimal calibration of the weight λ makes it possible to achieve this regulation.

Experiments have shown that a value of $\lambda = 1.25$ yields the best results. The near-perfect recall shows that the ideal compromise has been found: only one real bird was incorrectly classified. A higher value would have lowered this result, while a lower value would not have improved it. Although recall can often be artificially inflated by over-classification, the consistently high precision score demonstrates that this effect was successfully avoided. The accuracy obtained for this weight is the highest among the three trials; a higher value would have converged towards an accuracy equal to one due to the penalty imposed on the entire model. Here, the value $\lambda = 3$ is not sufficient to represent such a case and even shows a decrease in accuracy.

With λ close to one, the penalty is almost applied as its original magnitude without further scaling, making the inclusion of this weighting factor practically unnecessary for this specific case. However, the possibility of adjusting λ is important in anticipation of different datasets with different distributions of outliers. For example, the value of this weight should be increased slightly if the minimum values of the Hamming distances between animals, outliers, and alternative outliers were also increased. A version of the models with a Hamming distance of two between the generated animals and the real animals was also tested: for this experiment, the value λ producing the most conclusive results was $\lambda = 3$. Consequently, it may be beneficial to dynamically derive the value of λ based on the Hamming distances calculated within the dataset.

5 Conclusion

In this article, the integration of non-monotonic reasoning via Answer Set Programming (ASP) within a neural network was explored, with the aim of integrating exception handling. Drawing inspiration from seminal work on default logic and circumscription, we have shown that ASP provides a natural framework for encoding default rules, going beyond the simple principle of "in the absence of contrary information, assume that...".

Unlike current systems such as NeurASP or ASPer, which use symbolism mainly to improve overall performance, the approach presented here stands out by placing non-monotonicity at the heart of the model. Our results demonstrate that including a semantic term to the loss function of the neural model, conditionally activated by the logical return of ASP, limits the impact of outliers without disrupting the classification of standard data or generalization to similar instances.

Although this implementation focuses on simple logical rules, it paves the way for extensions to more complex temporal or conditional logics. The next step will be to validate this model on larger datasets with real outliers in order to test the robustness of this loss function on a large scale.

References

1. Agajorte: Zoo animals extended dataset. Kaggle (2020). https://www.kaggle.com/datasets/agajorte/zoo-animals-extended-dataset. Accessed 26 June 2025
2. Aguado, F., Cabalar, P., Fandinno, J., Pearce, D., Pérez, G., Vidal, C.: Revisiting explicit negation in answer set programming. CoRR abs/1907.11467 (2019). http://arxiv.org/abs/1907.11467
3. Chawla, N.V., Bowyer, K.W., Hall, L.O., Kegelmeyer, W.P.: SMOTE: synthetic minority over-sampling technique. J. Artif. Intell. Res. **16**, 321–357 (2002). https://doi.org/10.1613/jair.953. arXiv:1106.1813 cs
4. Hart, P.E., Nilsson, N.J., Raphael, B.: A formal basis for the heuristic determination of minimum cost paths. IEEE Trans. Syst. Sci. Cybern. **4**(2), 100–107 (1968). https://doi.org/10.1109/TSSC.1968.300136
5. Liu, M., Roijers, R.: Answer Set Solving in Practice. Synthesis Lectures on Artificial Intelligence and Machine Learning, 1st edn. Springer International Publishing, Cham (2013). https://doi.org/10.1007/978-3-031-01561-8
6. Machot, F.A.: Bridging logic and learning: a neural-symbolic approach for enhanced reasoning in neural models (ASPER) (2023). https://doi.org/10.48550/ARXIV.2312.11651
7. McCarthy, J.: Circumscription–a form of non-monotonic reasoning. Artif. Intell. **13**(1–2), 27–39 (1980). https://doi.org/10.1016/0004-3702(80)90011-9
8. McCarthy, J.: Epistemological problems of artificial intelligence. In: International Joint Conference on Artificial Intelligence (1987). https://api.semanticscholar.org/CorpusID:11102028
9. Metropolis, N., Ulam, S.: The Monte Carlo method. J. Am. Stat. Assoc. **44**(247), 335–341 (1949). https://doi.org/10.1080/01621459.1949.10483310
10. Reiter, R.: A logic for default reasoning. Artif. Intell. **13**(1–2), 81–132 (1980). https://doi.org/10.1016/0004-3702(80)90014-4
11. Russell, S., Norvig, P.: Artificial Intelligence: A Modern Approach, 3 edn. Prentice Hall (2010)
12. Xu, J., Zhang, Z., Friedman, T., Liang, Y., Broeck, G.V.D.: A semantic loss function for deep learning with symbolic knowledge (2017). https://doi.org/10.48550/ARXIV.1711.11157
13. Yang, Z., Ishay, A., Lee, J.: NeurASP: embracing neural networks into answer set programming (2023). https://doi.org/10.48550/ARXIV.2307.07700

Lexical Ambiguity in French: Linguistic Issues and Interpretation by Artificial Intelligence Models

Vafa Quliyeva[1,2(✉)] and Vazeh Asgarov[1,2,3]

[1] Azerbaijan University of Languages, Baku, Azerbaijan
{vafa.quliyeva,vazeh.asgarov}@ufaz.az
[2] French-Azerbaijani University, Baku, Azerbaijan
[3] Azerbaijan State Oil and Industry University, Baku, Azerbaijan

Abstract. Lexical ambiguity is a natural language feature that plays an important part in how people communicate. A vast number of lexemes in French can have many different meanings and have various unique senses, the interpretation of which depends on by the linguistic, discursive, and pragmatic contextual factors. While human speakers intuitively handle these confusions, artificial intelligence systems continue to face significant challenges in interpreting them. This article provides a comprehensive look of lexical ambiguity in French, combining a theoretical linguistic approach with a critical review of current language models. Using concrete examples, we investigate lexical disambiguation techniques, AI system performance, and the natural limitations of distributional approaches to meaning.

Keywords: lexical ambiguity · polysemy · homonymy · lexical semantics · natural language processing · artificial intelligence

1 Introduction

Human language is rich, adaptable, and complex. It also stands out by its ability to create potentially infinite numbers of utterances from a finite set of units. This creation, however, includes one key feature: uncertainty. Rather than being a rare phenomenon, uncertainty enhances the diversity and adaptability of natural languages. It arises at three levels: phonological, syntactic, and semantic; however, lexical ambiguity is central to linguistic research.

Since the beginning of lexical semantics, academics have stressed the complexities of the link between language forms and the meanings they communicate. Many lexemes might have several meanings, requiring speakers to continuously interpret them depending on context and world knowledge. Lexical ambiguity is concerned with the link between language form and meaning. It arises when a single term may be connected with numerous different meanings. This problem is very frequent in French, affecting both daily and specialist vocabulary. It

C. Zanni-Merk et al. (Eds.): AIDT 2026, CCIS 3023, pp. 189–204, 2026.
https://doi.org/10.1007/978-3-032-31319-5_13

manifests across several categories: polysemy, homonymy, metaphor, and adjective positioning each presenting distinct interpretive challenges that vary in their degree of contextual dependency.

In most cases, ambiguity does not interfere with comprehension in human communication. Speakers automatically rely on discourse context, inferential ability, encyclopedic knowledge, and cognitive and pragmatic skills to appropriately comprehend statements. However, linguistic ambiguity provides substantial difficulty with artificial intelligence systems. Unlike human interpreters, AI models cannot draw on communicative intent or world knowledge intuitively, making certain categories of ambiguity, particularly figurative and homonymous uses especially resistant to automatic resolution.

With the advent of natural language models such as BERT, GPT, DeepL, and other natural language processing (NLP) systems capable of producing sophisticated and grammatically acceptable texts, the subject of machine comprehension has emerged as a key theoretical concern. These theories are based on a distributional approach to meaning, which states that "a word is defined by the contexts in which it appears" (Firth 1957). While beneficial in many situations, this strategy is basically correlational. Lexical ambiguity thus provides an ideal setting for investigating the linguistic underpinnings of such models and comparing them to classical language theories.

According to Bender and Koller (2020), these models alter language forms that are tied to usage rather than meaning in the cognitive sense. Difficulties emerge, particularly in settings that are not explicit, entail uncommon senses, or rely on figurative usage that necessitates world knowledge or pragmatic inference. This paper takes a critical linguistic approach and seeks to examine how artificial intelligence reads polysemous French words. The purpose is to investigate if automatic disambiguation mechanisms are similar to or different from human interpretive processes, as well as to highlight the conceptual constraints of present models.

To this end, we will empirically test three small, resource-efficient language models: CamemBERT-base, mBERT-base, and Qwen2.5-0.5B-Instruct on a purpose-built dataset of 150 French ambiguous sentences spanning polysemy, homonymy, metaphor, and adjective position. While large SOTA LLMs can effectively resolve contextual ambiguities, we deliberately prioritize lightweight models, reflecting the broader industry shift toward scalable and resource-conscious NLP deployment, while critically examining where their limitations lie.

2 Lexical Ambiguity and Linguistic Semantics

Lexical ambiguity is the persistence of several meanings linked with a single lexical term. It is separated from syntactic ambiguity, which results from numerous grammatical structures, and phonological ambiguity, which is associated with homophony. In lexical ambiguity, the lexeme itself can be interpreted in numerous ways. In lexical ambiguity, the lexeme itself carries several possible interpretations. According to Cruse (1986), lexical ambiguity is a normal property of the lexicon and should not be viewed as a linguistic malfunction. It reflects the ability of languages to reuse existing forms to express new or abstract realities.

Since the seminal work of Ferdinand de Saussure, linguists have generally agreed that a word also referred to as a "linguistic sign" is the result of a conventional association between a signifier and a signified. The signifier corresponds to the material form of the word, whether phonological or orthographic, while the signified refers to the semantic content evoked. Some linguists interested in the problems of lexical semantics, such as G. Kleiber, highlight the referential dimension of meaning. According to this approach, the meaning of a word is not limited to an abstract definition or a set of semantic features but also involves the relationship that this word maintains with the entities in the real world to which it refers. Reference thus plays a central role in the understanding and interpretation of lexical units in discourse.

From this perspective, a word is said to be ambiguous when a single signifier is capable of conveying several signified. According to Asgarov (2016) en français, the present participle can be considered as a verbal form and as a pure adjective that creates ambiguity as for example *adhérant* - present participle and *adhérant* - verbal adjective.

However, such ambiguity does not always originate from the same linguistic mechanism. Sometimes, several unrelated words share the same form: this is homonymy. For example, avocat may denote a fruit or a legal professional; *livre* may refer to a book or a unit of weight. In other cases, a single word has multiple related senses derived metaphorically or via semantic extension: this is polysemy. For instance, *branche* can refer to part of a tree, a professional specialty, or a division within a company. Similarly, *cour* may denote a courtyard, royal entourage, or a court of law.

Ambiguity does not just affect language or words, but it may also occur at the sentence level. This indicates that certain lexical ambiguities are intimately related to syntactic structure in the generative framework. Depending on its properties, a word might trigger multiple syntactic projections, resulting in diverse meanings. This method distinguishes lexical and structural ambiguity while also describing their relationships (Chomsky, 1965). For example:

- *Elle a rencontré la sœur de Paul en voyage* - may mean either that the meeting took place during a trip or that Paul's sister was travelling.
- *Il regarde l'homme avec le télescope* - may mean either that he uses a telescope to look or that the observed man has a telescope.
- *À midi, elle est partie* - is polysemous

Thus, the statement "She met Paul's sister while traveling" can be interpreted in two ways: either the meeting took place during a trip, or Paul's sister was traveling. Similarly, the sentence "He is watching the man with the telescope" can be interpreted in two ways: either he is using a telescope to watch, or the man being observed owns a telescope. On the other hand, a sentence like "At noon, she left" is polysemous, as it can mean either a specific departure at noon, or the fact that she was no longer present after noon.

As can be seen, syntactic construction plays an important role in defining lexical meaning. This connection between semantics and syntax is most visible with verbs, whose interpretation is heavily influenced by the argument structure inside of which they are contained. Indeed, the nature, quantity, and type of complements, either direct, indirect, or prepositional, help to activate different semantic meanings of the same verb.

Thus, the verb *prendre* (to take) does not refer to the same semantic content in - *Elle prend un livre sur l'étagère* (She takes a book from the shelf), where it expresses a concrete action of grasping, and in - *Il prend la parole* (He speaks), where it acquires an abstract value linked to the speech act. Similarly, the verb *monter* (to go up) can signify a physical movement in - *Il monte les escaliers* (He goes up the stairs), or a more abstract operation in - *Elle monte un projet* (She is developing a project). Likewise, the verb passer (to pass on) does not have the same semantic value in the statements - *Elle passe le message à ses collègues* (She passes the message to her colleagues), where it means "to transmit," and - *Il passe à autre chose* (He moves on to something else), where it takes on the meaning of "to change the subject" or "to abandon". Similarly, the verb *jouer* (to play) refers to distinct interpretations in *L'enfant joue au ballon* (The child plays with the ball) and *Il joue un rôle important dans l'entreprise* (He plays an important role in the company).

The above examples demonstrate a type of syntactically conditioned polysemy in which the verb's meaning cannot be understood without regard to its syntactic context. The semantic scope of a word therefore seems dynamic and contextual, changing itself in response to the grammar and discourse combinations where it is used. This structure-dependent meaning confirms the view that lexical interpretation is the outcome of a tight interplay between formal features and semantic processes.

Additionally, in French, the precise position of the adjectives can play a significant role in disambiguation. The classic contrast between proposed and postposed adjectives vividly displays this phenomenon. Adjectives also exhibit position-dependent ambiguity in French: *un ancien professeur* → a former teacher; *un professeur ancien* → an elderly or long-established teacher.

The position of the adjective is therefore not purely stylistic: it has a semantic function. Pre-positioning tends to favor a more subjective, evaluative, or figurative interpretation, while post-positioning more often refers to a descriptive and objective property. The distinction between relational adjectives (political, academic, medical) and qualifying adjectives also plays a role in disambiguation. Similarly: *un débat politique* → relating to politics; *un homme politique* → a politician.

Some adjectives can shift from one category to another depending on the context, creating a temporary ambiguity that the reader or listener resolves thanks to the overall context.

Disambiguation does not rely solely on syntax or vocabulary. The discursive context and the situation of utterance are often decisive: *un silence lourd* → heavy with tension; *un sac lourd* → physically heavy.

It is knowledge of the world, pragmatic expectations, and the framework of utterance that allow us to dismiss irrelevant interpretations. The meaning of polysemous adjectives can only be determined in relation to the noun they modify. For example, the adjective *grand* (big) does not refer to the same property depending on the noun it modifies: *un grand homme* → moral grandeur, symbolic importance; *un homme grand* → physical size. Here, it is the noun context that resolves ambiguity. The adjective does not carry a stable meaning in itself: it activates a particular value depending on the noun it modifies.

Thus, lexical semantics traditionally distinguish between polysemy and homonymy. Polysemy corresponds to a set of meanings linked by semantic or conceptual relationships. For example, the word *tête* (head) can refer to a part of the body, a person at the head of a group, or the end of an object. These meanings are perceived as belonging to the same lexical unit (Kleiber, 1999). Homonymy, on the other hand, concerns identical forms associated with distinct and unrelated meanings, such as *avocat* (lawyer/fruit). In this case, some linguists consider these to be two different lexical units accidentally sharing the same form (Cruse, 1986). While theoretically important, this distinction can be ambiguous, especially for automated and computational systems that lack conceptual requirements for establishing semantic links between meanings.

3 The Role of Context in Lexical Disambiguation

Linguistic context is important in interpreting ambiguous words because it allows us to reduce the range of potential interpretations. Lexical co-occurrences, syntactic structures, and grammatical linkages are important indications for disambiguation. In other words, the meaning is carried by the word embedded in a certain linguistic arrangement rather than the word on its own. According to Rastier (2009), the meaning of a word cannot be understood without regard for its textual context. This interpretative paradigm of meaning emphasizes signification's dynamic and contextual aspect. For example: *L'avocat a plaidé au tribunal, Il s'est assis sur le banc du parc, Le banc de poissons scintillait au soleil alors qu'ils nageaient en harmonie.*

The verbs and complements in the phrase L'avocat a plaidé au tribunal (The lawyer pleaded in court) instantly point the reader to the legal interpretation. The term "bench," when used alone, can refer to a piece of furniture, a group of fish, or a quantity of material creating a more or less horizontal layer, such as a sandbank or a snowbank. However, in a sentence like "He sat on the park bench," the terms "sat" and "park" instantly drive the reading toward the physical sense of a seat. In contrast, in "The school of fish sparkled in the sun as they swam in harmony," the lexical co-occurrences and grammatical structure clearly lead to a nautical interpretation.

The language context therefore serves as a semantic filter, filtering out unrelated meanings. Besides mere language context, interpreting meaning requires encyclopedic and practical expertise. For instance, comprehending the line "He looks tired" requires acknowledging that "mine" could be referring to a facial feature rather than a mining spot. The concept of meaning, according to this viewpoint, is the outcome of a dynamic interpretative process, rather than a fixed description of the word. As an example of a highly polysemous adjective, *dur* (hard), whose uses are listed in dictionaries, we can group its meanings around several main values:

1. Resistant to touch, solid, not soft: hard stone, hard ground.
2. Painful, difficult to endure: a hard ordeal, a hard life.
3. Severe, strict, inflexible (behavior, attitude): a harsh teacher, a harsh judgment.
4. Violent, brutal (action, shock): a hard blow.
5. Difficult to understand or accept intellectually: a hard line of reasoning.
6. Lacking sensitivity or compassion: a hard man.

These different meanings are linked by a central notion of resistance or opposition. In the physical realm, a hard object resists pressure (1). This idea of resistance extends to the psychological and moral level: a hard person (3 or 6) resists emotion, pity, or indulgence, while a hard situation (2) imposes a strong constraint on the person experiencing it. Some uses can mobilize several values simultaneously. Thus, a hard blow is both violent (4) and difficult to bear emotionally (2). The two interpretations coexist without being truly dissociable.

On the other hand, *un matériau dur* (a solid material) refers exclusively to a physical property (1), while a hard man refers, depending on the context, either to a stern person or an insensitive person, without these two traits necessarily being present together. In this case, the context allows for the selection of a precise interpretation. *Clair* (Clear) is a highly polysemous adjective. Dictionaries attribute many meanings to it, which can be grouped around several main values:

1. Easy to understand, explicit: *une explication claire, un raisonnement clair* → a clear explanation, a clear line of reasoning.
2. Bright, well-lit: *une pièce claire, un ciel clair* → a bright room, a clear sky.
3. Limpid, transparent: *de l'eau claire* → clear water.
4. Sharp, clear (sound): *une voix claire* → a clear voice.
5. Obvious, obvious: *il est clair qu'il viendra* → it is clear that he will come.
6. Pale or light in color: *une robe claire* → a light-colored dress.

These different meanings are not entirely independent: they maintain a semantic continuity. In most uses, we find the general idea of visibility or the absence of darkness, whether physical or intellectual. Thus, *eau claire* (clear water) is transparent (3), which allows one to see clearly through it, while *idée claire* (a clear idea) (1) is an idea "without gray areas," easy to grasp. Similarly, *pièce claire* (a clear room) (2) and *voix claire* (a clear voice) (4) share the idea of clarity and distinctness. Some uses can activate several values at once. For example, *réponse claire* (a clear answer) is both intelligible (1) and obvious (5). In this case, it is not an ambiguity in the strict sense, because the meanings do not exclude each other: they overlap.

In contrast, *une couleur claire* (in a bright color), the chromatic meaning (6) takes dominance over all other meanings. The context then makes it possible to choose a specific meaning without uncertainty. This extralinguistic information poses a significant barrier to AI systems, which lack human-level world experience. This conception of meaning contrasts with computational approaches, in which words are represented as numerical vectors derived from their co-occurrences in large corpora (Mikolov et al., 2013). However, as Rastier (2009) points out, a purely distributional approach struggles to account for the conceptual relationships between the meanings of the same word and their interpretation in complex discursive contexts. These representations, although effective for many tasks, raise the question of their ability to account for the semantic complexity of the lexicon.

4 Representation of Meaning in Artificial Intelligence

4.1 Distributional Approaches

Contemporary language models largely rely on a distributional approach to meaning, inherited from the principle first formulated by Zellig Harris (1968), according to which "words that appear in similar contexts have similar meanings" or "the semantic proximity between words can be calculated based on their distributional proximity". Harris shows that these distributions can be formalized mathematically, opening the way to a rigorous analysis of polysemy.

Emerging in the United States in the 1930s under the influence of Leonard Bloomfield, distributional analysis was subsequently developed and formalized by Zellig S. Harris, a representative of the American school of structuralism. It rests on the idea that linguistic units are defined primarily by their positions within the discourse chain. Thus, a phrase is understood in relation to the sentence, a morpheme in relation to the phrase, and a phoneme in relation to the morpheme. Linguistic units can be described based on their distribution within the corpus, independently of any psychological or cognitive reference (Harris, 1954). In other words, linguistic analysis must be based on formal regularities observable in corpora, without resorting to unobservable mental notions. Two linguistic forms sharing similar environments can be considered as belonging to the same class. Applied to the lexicon, this approach makes it possible to identify semantic regularities based on lexical co-occurrences, thus paving the way for a formal analysis of meaning.

By observing, at each level of analysis, the contexts in which a given element appears, it becomes possible to determine its distribution. Comparing these distributions then allows us to group units sharing similar environments within the same distributional class: for example, determiners are distinguished by the fact that they generally precede nouns. Vector representation models of meaning, such as Word2Vec or GloVe at BERT, can be considered direct heirs of distributionalism. They translate the Harrisian idea into computational terms, representing words as vectors in a high-dimensional space (Mikolov et al., 2013).

Later, distributional analysis was reformulated by Firth (1957) as follows: "You shall know a word by the company it keeps." From this perspective, the meaning of a word is deduced from the contexts in which it appears. Modern contextual models (embeddings, transformers), such as BERT (Bidirectional Encoder Representations from Transformers), allow for the capture of semantic regularities on a large scale. These models and their results have demonstrated remarkable effectiveness for numerous linguistic tasks and expanded the range of tools available to linguists for exploring word meaning based on their usage in corpora. They use words as vectors where the meaning is dynamically constructed according to context.

Contemporary vector models can be seen as a mathematical formalization of this principle: they translate the linguistic distributions observed in corpora into distances and directions in a geometric space. Vector space, based on comparison and proximity, materializes this idea: a word does not have meaning, but in relation to others.

4.2 Vectorization as Semantic Translation

Language models interpret ambiguous words based on their contextual environment, relying on vector representations of meaning. Vectorization involves translating complex information, such as human language, into machine-readable numerical representations called vectors. Without vectorization, artificial intelligence would be unable to process the majority of unstructured textual data. Large language models (LLMs) rely entirely on tensors for mathematical generalizations of vectors to perform their calculations. Vectorization thus forms the foundation upon which any attempt at automated language understanding rests. It is more than just randomly associating a word with a sequence of numbers; it aims to represent both its meaning and its associations with other words in diverse situations. In other words, it converts linguistic meaning into mathematical coordinates, allowing algorithms to analyze, compare, and interpret text at a vast scale. This resembles a procedure in which each word or thought has a fixed location. Terms with comparable semantic features, such as banque (bank) (financial institution) and crédit (credit), are reasonably near, but terms with no direct semantic relationship, such as banque, carotte (bank, carrot), are far. This process is necessary since machine learning models do not directly handle words, but rather abstract mathematical constructs. This geometric approximation utilized by AI is frequently enough for practical purposes, but it does not correlate to great comprehension. In the context of lexical ambiguity, vectorization pursues many crucial goals:

- **Quantify meaning**: transform the plurality of possible meanings of a word into comparable mathematical representations.
- **Capture context**: integrate contextual clues that guide the interpretation of an ambiguous word toward one meaning rather than another.
- **Enable semantic comparison**: measure, using tools such as cosine similarity, the proximity between different occurrences of the same word in distinct contexts.

- **Reduce lexical complexity**: represent a rich and polysemous vocabulary in a dense vectorial space and a fixed dimension, compatible with the computational constraints of the models.

This portrayal takes on added significance. An imprecise vectorization that fails to effectively integrate contextual variables leads to erroneous interpretations: misunderstanding between homonyms, inadequate selection of the appropriate meaning, or semantic overgeneralization. As we have seen, this approach also has limitations: when contexts are sparse, ambiguous, or culturally marked, the resulting vectors may remain too similar, making the distinction between meanings uncertain. Thus, even the most efficient models remain dependent on how the meaning is initially encoded.

A polysemous word like *souris* (mouse) (animal or computer peripheral) does not have a single meaning. Vector models then attempt to distinguish these meanings according to the contexts in which they appear: *cliquer avec la souris* (clicking with the mouse) will not be vectorized in the same way as "*une souris grise dans la cuisine* (a gray mouse in the kitchen)".

Let's take another example, the word *banc* (bench). In *un banc public* (a public bench), the term refers to a physical object, whereas in *un banc de poissons* (a bench of fish), it designates a group of living beings. From a vector perspective, these two uses correspond to distinct areas of semantic space, constructed from different co-occurrences (sit, park, place on one side and *fish, sea, swimming* on the other). Contextualized models therefore do not store a single vector for bench but generate representations that vary according to the lexical environment.

The case of homonymy offers an even sharper contrast. The word *verre* (glass) can refer to a container (*boire un verre* - to drink a glass), a material (*verre trempé* - tempered glass), or correspond phonetically to "worm" or "green." For humans, disambiguation relies on syntax, the context of utterance, and sometimes the oral or written modality. For AI, however, only contextual distribution allows these forms to be distinguished. In the vector space, *verre* (glass) - (container) will be associated with "cup" or "drink," while "green" will be associated with *couleur* (color) or *écologique* (ecological), thus illustrating a geometric separation of meaning. Thus, it is not the isolated word that carries the meaning, but rather its place within a network of co-occurrences.

Each linguistic unit—word, phrase, sentence, or document—is represented by a vector in a high-dimensional space. The distance or proximity between these vectors reflects a calculable semantic similarity. This modeling allows algorithms to perform mathematical operations on meaning, a necessary condition for any attempt at automatic disambiguation.

For example, the word *clé* (key) can refer to an object used to open a door but also to an explanatory element, as in the expression *la clé du problème* (the key to the problem). Similarly, the word *feuille* (leaf) can refer to a part of a plant, a sheet of paper, or even a sheet of metal. In these cases, the same word is used to express several meanings. This example explains how each word develops meaning based on the existence of other parts in the speech. If one attempts to rigidly implement the concept of compositionality by gradually giving meaning to each unit, one rapidly finds a methodological dead end: the number of viable interpretations becomes significant, if not unmanageable. Ambiguity works in the opposite way that synonymy does, by using many words to describe the same concept. As Kleiber (1999) points out, lexical meaning cannot be reduced to a set of fixed definitions; it evolves via usage, at the confluence of context and common knowledge. This apparent performance does not imply a deep knowledge of lexical meaning, especially when the words employed are polysemous or homophones (Bende & Koller, 2020). This dynamic view of meaning is critical to understanding why lexical ambiguity is so difficult to comprehend mechanically.

Through a theoretical examination and investigation of specific situations, we find that present AI systems mainly inherit a distributional understanding of meaning, while also confronting the restrictions emphasized by generative linguistics in terms of linguistic comprehension. This technique rapidly displays its limits, as the analyzed corpus by definition incomplete and contingent cannot account for the whole range of utterances that a language is capable of creating. This deficiency has resulted in the development of an alternate, hypothetico-deductive paradigm.

4.3 The Use of Metaphors: A Major Challenge in Natural Language Processing

The analysis of metaphors is a particularly revealing area for understanding the capabilities and limitations of artificial intelligence systems in natural language processing. Many metaphorical idioms in French are based on semantic transfers across disparate conceptual domains, such as exploser de rire (to burst out laughing), être noyé sous le travail (to be drowned in labor), or le temps qui file. These statements show how a tangible idea (physical explosion, immersion in water, fast movement) is projected onto an abstract or subjective experience. These linkages are based on commonly recognized cognitive processes and an inherent grasp of the world in humans, but in AI models, they must be taught from statistical regularities found in corpora. Thus, in il a été noyé par la vague (he was drowned by the sea), the vector linked with drowned goes closer to physical settings, but in elle est noyée sous les emails (she is drowning in emails), it transfers to semantic domains relating to work or information overload.

Despite these gains, mistakes continue to occur, especially when the training data contains unclear, sarcastic, or rare contexts. Similarly, the sentence Cet étudiant est une éponge (this student is a sponge) is a nominal metaphor based on a comparison between a tangible thing, the sponge, and a person. A sponge is a physical entity that can absorb liquids; nevertheless, in this context, the term denotes a semantic shift toward the cognitive realm. The learner is viewed as an entity capable of rapidly and efficiently acquiring information or knowledge. The parallel is based on a shared functional trait: the ability to absorb. The term "sponge" has a semantic neighborhood related to the physical domain: water, absorption, washing, liquid, and porous.

When it appears in an educational environment (student, learning, course, etc.), the contextualized vector switches to a semantic area associated with quick learning, memory, and cognitive aptitude. "Transformer" models dynamically change this representation based on environmental inputs, allowing for a believable metaphorical interpretation. In the example *Cet étudiant est un robot* (this student is a robot), the metaphor can activate several interpretations depending on the context: **lack of emotion, mechanical efficiency, automated repetition, strict obedience**.

Unlike "*sponge*," which emphasizes cognitive absorption, "*robot*" evokes an analogy with machines and automation. As for its vector representation in a semantic space: *robot est proche de*: *machine, automatisé, programme* (robot is close to machine, automated, program). In an educational context, its contextualized vector shifts toward *intense work, rigor, and repetition*. In this context, the challenge for AI lies in the high degree of ambiguity: depending on the context, the metaphor can be positive (effective) or negative (dehumanizing).

From a linguistic point of view, metaphors introduce a form of ambiguity. The same expression has: a literal interpretation and a figurative interpretation. In the example, *Il a dévoré le livre* (he devoured the book). The literal meaning is an **improbable physical action** and the metaphorical meaning is **rapid and passionate reading**.

The correct interpretation depends on the discursive context, world knowledge, and cultural conventions. The words *avaler* (swallow) and *dévorer* (devour) frequently appear with food → dominant literal meaning. When they co-appear with *livre* (book), *roman* (novel), *dossier* (file), this indicates an increased probability of metaphorical use. This approach allows artificial intelligence to detect semantic patterns without direct access to a human understanding of the world. These models do not possess symbolic or intentional understanding, but they manage to identify metaphorical uses by relying on statistical regularities. Thus, AI's "understanding" corresponds more to probabilistic optimization than to conscious interpretation. In approaches based on vector representations of language, each word or expression is represented by a vector in a multidimensional semantic space.

The basic assumption, borrowed from distributionalism, is that contexts of usage can approximate meaning: statements that arise in comparable circumstances hold near locations in this space. As a result, the expression exploser de rire (bursting with laughter) can be associated with contexts similar to rire énormément (laughing out loud), être plié de rire (rolling with laughter), or mourir de rire (dying of laughter), allowing the model to capture its semantic intensity despite the non-literal nature of the verb exploser. Similarly, noyé sous le travail (overwhelmed by work) is vectorially related to phrases relating to work overload or stress, whereas le temps qui file (time flies) is part of a lexical network associated with speed, movement, or transience.

Unlike literal utterances, which may typically be understood using compositional approaches or by studying syntactic and distributional relationships, metaphors require a semantic shift based on conceptual similarities across distinct areas of experience. This analogical projection complicates the job of computer models since metaphorical meaning is the result of an interplay between language context, world knowledge, and cultural traditions, rather than just the sum of individual lexical meanings.

Modern machine learning techniques, notably vector representation-based language models (word embeddings, contextual embeddings, transformers), rely heavily on statistical analysis of massive text corpora to grasp metaphors. These models learn to correlate words and sentences based on their use settings, in line with the distributional principle, which states that language units appearing in comparable situations have certain semantic qualities. Thus, some traditional metaphors, which are commonly present in the data, can be accurately understood due to the vectorial closeness of the ideas involved. For example, statements like un argument solide (a solid argument) or une idée lumineuse (a brilliant idea) might be partially captured by the models' contextual associations.

Despite this improvement, there are still significant limits. First, AI models prefer to repeat the corpus' statistical regularities rather than gaining profound conceptual comprehension. They may identify repeating metaphorical patterns without necessarily understanding the underlying cognitive mechanisms that enable humans to generate and comprehend new metaphors. Second, innovative, unusual, or highly context-dependent metaphors provide a unique difficulty because they need inferential and analogical thinking abilities that frequently extend beyond basic distributional correlations.

From a theoretical standpoint, this problem shows the contrast between distributional techniques inherited from Harris' work and cognitive linguistic viewpoints, notably those of Lakoff and Johnson, who believe metaphors organize cognition itself. While AI algorithms detect statistical similarities, human cognition uses embodied conceptual schemas and extralinguistic information to quickly grasp fresh figurative language. Thus, the study of metaphors is a preferred subject for comparing human and machine modalities of meaning representation, as well as criticizing the existing boundaries of language models' attempts to replicate human semantic comprehension.

5 Empirical Discussion and Results

In order to empirically validate the theoretical propositions discussed in this study, we developed a dataset comprising 150 French ambiguous sentences, categorized into polysemy, metaphor, homonymy, and adjective position. For our evaluation, we employed three specific language models including CamemBERT-base (a model tailored specifically for the French language), mBERT-base (a multilingual equivalent), and Qwen2.5-0.5B-Instruct (a modern, compact generative LLM). Although large SOTA LLMs can effectively resolve contextual ambiguities, we deliberately chose small, lightweight models, as the widespread deployment of large LLMs is often cost-prohibitive due to significant computational and financial demands, as well as environmental concerns. Consequently, the industry increasingly relies on smaller, resource-efficient models to empower daily Natural Language Processing (NLP) tasks. The detailed performance of these models on our dataset is illustrated in Fig. 1. According to the results, homonymy emerges

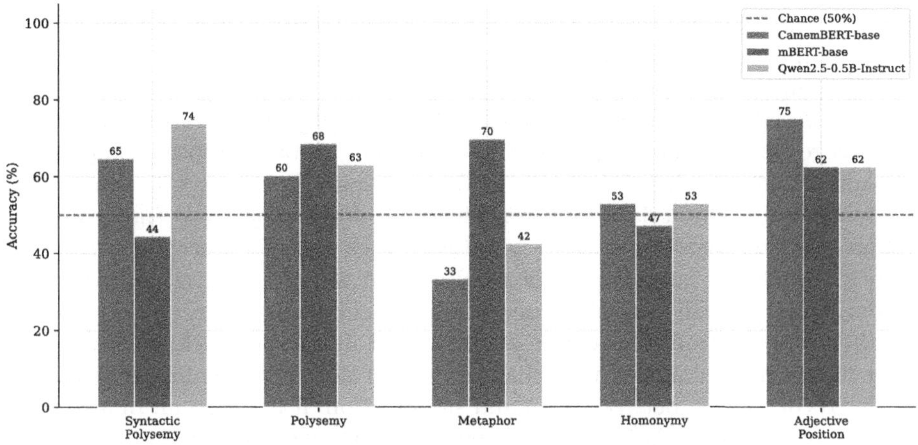

Fig. 1. Ambiguity accuracies across LLMs

as a consistently resilient challenge that all evaluated models struggle to identify. Furthermore, while the CamemBERT model achieves decent overall performance, it faces specific hardships in processing both homonyms and metaphors. Similarly, although the Qwen model demonstrates outstanding capability in resolving syntactic polysemy, it yields remarkably poor results against metaphors and homonyms. Overall, our testing demonstrates that employing compact LLMs introduces severe risks to Natural Language Understanding (NLU), necessitating targeted improvements for tiny models. One highly promising solution to mitigate this is Knowledge Distillation, as massive LLMs demonstrate superior performance in understanding semantic ambiguities and they can be utilized to fine-tune tiny, resource-efficient LLMs.

6 Conclusion

Lexical ambiguity in French is quickly developing as an important field for studying the interaction of theoretical linguistics and artificial intelligence. Rather than being a barrier to communication, it demonstrates language's cognitive flexibility and how meaning is dynamically generated from context. Contemporary computer techniques, based on distributionalism and deep learning, have significantly improved our capacity to represent semantic variability. AI models can capture some contextual regularities and provide reasonable interpretations by learning from vector representations produced from vast corpora, which represents a substantial advancement in automatic lexical disambiguation.

Nevertheless, these technological developments pose basic problems regarding the nature of meaning and the context under which it is interpreted. While neural models are excellent at identifying statistical relationships, they frequently rely on reported distributions and can struggle with unusual, implicit, or culturally particular circumstances. This border distinguishes between a human-centered view that incorporates communication intents, world knowledge, and pragmatics, and a computational interpretation that is largely data-driven.

Our empirical evaluation of three lightweight models across 150 French ambiguous sentences confirms this divide: homonymy emerged as the most persistently resistant category across all models, while metaphor posed compounded difficulties for both CamemBERT and Qwen despite their otherwise competitive performance in syntactic polysemy. These results demonstrate that compact, resource-efficient models, while pragmatically advantageous, introduce meaningful risks to natural language understanding that require targeted intervention.

Polysemy, homonymy, and contextual dependency of meaning reveal the dynamic and non-univocal character of lexical units in French, as they do in other languages. Analyzing these instances indicates that meaning is not a simple, fixed correlation between form and signification: it is the result of a continuous interplay between language structures, discursive context, world knowledge, and pragmatic judgments.

From this standpoint, the study of lexical ambiguity encourages us to go beyond the divide between linguistics and artificial intelligence and adopt a truly multidisciplinary approach. Future studies might investigate hybrid models that include distributional representations, symbolic structures, and pragmatic mechanisms to better align automatic interpretation processes with human cognitive strategies. Knowledge Distillation represents one particularly promising direction, allowing compact models to be fine-tuned on the superior semantic reasoning of large LLMs without incurring their computational and environmental costs. The integration of conceptual information, broader discursive contexts, and explanatory processes is also a viable option for increasing system resilience and transparency.

In addition to technological uses, lexical ambiguity raises deeper concerns regarding our understanding of language itself. The endeavor to teach robots how to manage numerous meanings forces us to reexamine the underlying mechanics of human interpretation. As a result, artificial intelligence does more than just duplicate language events; it also serves as a heuristic tool for testing theoretical ideas and renewing semantic viewpoints. Lexical ambiguity is consequently viewed not as a problem to be solved, but as a crucial key to comprehending the interaction of language, cognition, and technology.

References

Asgarov, V.: Gérondif, Participe Présent. PAF (Presses Académiques Francophones), Saarbrücken, Adjectif Verbal (2016)

Aurnague, M., Borillo, A.: La polysémie lexicale. Cah. Lexicol., (1995)

Auroux, S.: La Philosophie du langage. PUF, Paris (1998)

Bender, E.M., Koller, A.: Climbing Towards NLU: On Meaning, Form, and Understanding in the Age of Data, p. ACL. (2020)

Bird, S., Klein, E., Loper, E.: Natural Language Processing with Python. O'Reilly, Sebastopol (2009)

Brown, T.: Language Models are Few-Shot Learners. NeurIPS, (2020)

Collobert, R., et al.: Natural Language Processing (Almost) from Scratch. J. Mach. Learn. Res. **12**, 2493–2537 (2011)
Cruse, D.A.: Lexical Semantics. Cambridge University Press, Cambridge (1986)
Devlin, J., Chang, M.W., Lee, K., Toutanova, K.: BERT: Pre-training of Deep Bidirectional Transformers for Language Understanding. In: NAACL-HLT (2019)
Firth, J.R.: A Synopsis of Linguistic Theory 1930–1955. In: Studies in Linguistic Analysis. Blackwell, Oxford (1957)
Fillmore, C.J.: Frame Semantics and the Nature of Language, Ohio State University Working Papers in Linguistics (1975)
Geeraerts, D.: Theories of Lexical Semantics. Oxford University Press, Oxford (2010)
Harris, Z.: Distributional Structure. Word **10**(2–3), 146–162 (1954)
Hug, J.: La polysémie lexicale: approches théoriques et applications. Artois Presses Université, Arras (2015)
Jurafsky, D., Martin, J.H.: Speech and Language Processing, (2023). 3rd ed. draft
Kleiber, G.: Prototype, Stéréotype: un air de famille? DRLAV **38**, 1–6 (1988)
Kleiber, G.: Problèmes de sémantique lexicale. Presses Universitaires du Septentrion, Villeneuve d'Ascq (1999)
Kleiber, G.: Prototypes, stéréotypes et catégorisation. PUF, Paris (1999)
Lakoff, G., Johnson, M.: Metaphors We Live By. University of Chicago Press, Chicago (1980)
Marcus, G.: Deep Learning: A Critical Appraisal, (2018). arXiv:1801.00631
Manning, C.D., Schütze, H.: Foundations of Statistical Natural Language Processing. MIT Press, Cambridge (1999)
Mellet, S.: Ambiguïté lexicale et polysémie dans les langues romanes. Presses Universitaires de Rennes, Rennes (2002)
Mikolov, T., Chen, K., Corrado, G., Dean, J.: Efficient Estimation of Word Representations in Vector Space, (2013). arXiv:1301.3781
Mikolov, T.: Distributed Representations of Words and Phrases. NeurIPS, (2013)
Mitchell, T.M.: Machine Learning. McGraw Hill, New York (1997)
Pennington, J., Socher, R., Manning, C.: GloVe: Global Vectors for Word Representation. EMNLP, (2014)
Prieto, L.J., Corréard, G., Michelena, L., Waldo, G.S., Austerlitz, R.: Reviews. Word **14**(1), 84–107 (1958)
Pustejovsky, J.: The Generative Lexicon. MIT Press, Cambridge (1995)
Rastier, F.: Sémantique interprétative. Presses Universitaires de France, Paris (2009)
Riegel, M., Pellat, J.C., Rioul, R.: Grammaire méthodique du français. PUF, Paris (2014)
de Saussure, F.: Cours de linguistique générale. Payot, Paris (1972)
Sperber, D., Wilson, D.: Relevance: Communication and Cognition. Blackwell, Oxford (1986)

Improving Named Entity Recognition for the Azerbaijani Language Using Neural and Conditional Random Fields-Based Models

Tofig Aliyev[1(✉)] and Samir Rustamov[2,3]

[1] Department of General and Applied Mathematics, Azerbaijan State Oil and Industry University, Baku, Azerbaijan
aliyev.tofiq@asoiu.edu.az

[2] School of IT and Engineering, ADA University, Baku, Azerbaijan
srustamov@ada.edu.az

[3] MegaSec AI Company, Baku, Azerbaijan

Abstract. In this paper, we are examining existing Named Entity Recognizer (NER) algorithms for the Azerbaijani language and comparing traditional methodologies to state-of-the-art neural network based methodologies. In terms of morphology and availability of labeled linguistic resources, the Azerbaijani language has both corresponding unique benefits and challenges. To provide a means for experimentation within the NER domain, we created a silver standard NER corpus for the Azerbaijani language through a cross-lingual annotation methodology that has resulted in over 90,000 sentences and 1.17 million tokens annotated for 37 different entity types. We created five models: a baseline using Conditional Random Fields (CRFs) with manually created features, a BiLSTM-CRF model using FastText embeddings, a BERT-BiLSTM-CRF model, an ALBERTo-BiLSTM-CRF model, and an XLM-RoBERTa-BiLSTM-CRF model. Of all the models tested the XLM-RoBERTa based BiLSTM-CRF model produced the best performance with an F1 score of 0.7918, precision of 0.8505, recall of 0.7574, and token based accuracy of 0.7823. These results show that multilingual pretrained transformer architectures present a promising opportunity for future NER research in low resource languages. In addition, we addressed issues related to morphological forms of rare entities, morphological variability, and issues related to noise from projecting a cross-lingual projection epistemologically. The direction for future work includes incorporation of morphological information, consideration of data augmentation techniques, and creation of gold standard NER corpora.

Keywords: Named entity recognition · Azerbaijani · Conditional random fields · BiLSTM · BERT · XLM-RoBERTa · low resource languages · cross-lingual annotation · transformer architectures

C. Zanni-Merk et al. (Eds.): AIDT 2026, CCIS 3023, pp. 205–219, 2026.
https://doi.org/10.1007/978-3-032-31319-5_14

1 Introduction

Named Entity Recognition (NER) refers to the capability of automatically identifying and classifying words that represent people, organizations, places, dates, etc. in unstructured text. NER has traditionally relied on statistical modeling (e.g., Conditional Random Fields (CRFs)), which can capture sequential dependencies of data using a set of hand-crafted linguistic features [1]. The emergence of neural NER models, such as BiLSTM-CRF, have improved upon the use of raw data by allowing for the generation of progressively richer contextual representations [2]. Most recently, the most advanced forms of NER include transformer-based models (e.g., BERT, multilingual BERT), which have been pre-trained on various types of information and computational complexity [3].

Despite these advances, much of the research being done within the NER community has focused on high-resource languages (e.g., English, Chinese, German) compared to low-resource languages (e.g., Azerbaijani). Specifically, due to the multi-syntactic nature of the highly agglutinative Turkic language, Azerbajani NER presents unique challenges due to the presence of multiple possible suffixes for a single root word. Furthermore, the highly agglutinative nature of the language, combined with a lack of publicly available high-quality annotated data sets, is expected to increase the level of difficulty when trying to create a suitable NER system in Azerbaijani [4,7].

To address the gaps in the literature surrounding NER of Azerbaijani, this study will review both traditional and modern methodologies of NER. In addition, we created an annotated silver-standard data set consisting of over 90,000 Azerbaijani sentences, 1.17 million tokens, and 37 different entities using a semi-automated annotation process. Through the creation of this annotated data set, we were able to compare the five models that were created (CRF baseline, BiLSTM-CRF with FastText embeddings, BERT-BiLSTM-CRF, ALBERTo-BiLSTM-CRF, and XLM-RoBERTa-BiLSTM-CRF) to one another.

Furthermore, the following contributions to the field will be presented in this article, in addition to how NER can be utilized for Azerbaijani, such as:

- The comparative performance of five different models (neural and non-neural), all of which use the same Azerbaijani NER data set;
- The highest-performance NER model, XLM-RoBERTa-BiLSTM-CRF model, received a F1-score of 0.7918, precision of 0.8505, recall of 0.7574, and token accuracy of 0.7823;
- The impact that rare entity types and morphological variation as well as translation noise had on model performance will be analyzed in-depth;
- Suggestions for future research will include using better morphological analysis, external language resources, and cross-linguistic transfer learning.

Overall, the findings of this article will assist in establishing NER systems in Azerbaijani, as well as providing an overarching insight for other morphologically rich, low-resource languages. The structure and comprehensive comparison of five architecturally distinct systems will serve as a supportive reference for future studies. Before starting our work on Azerbaijani NER, we reviewed recent Azerbaijani NLP studies produced within ADA University research groups, including prior work on Azerbaijani part-of-speech tagging and lemmatization, which highlight the practical impact of agglutinative morphology on token-level modeling [12–14]. These findings informed our preprocessing decisions and motivated the use of robust sequence labeling architectures for downstream NER. The remainder of this article will cover previous literature relevant to this study including background, methodologies, experimental results, and future recommendations.

2 Related Work

NER has evolved from early statistical sequence models (e.g., HMMs and MEMMs) to Conditional Random Fields (CRFs), which address label bias and support rich feature sets [1,6,10,11]. Feature-based systems such as Stanford NER demonstrated strong performance in well-resourced languages using handcrafted lexical and orthographic features [1].

Neural NER reduced reliance on manual feature engineering through architectures such as BiLSTM–CRF [2]. Transformer encoders (e.g., BERT) further improved sequence labeling by providing contextual representations, and multilingual variants (mBERT, XLM-RoBERTa) enable transfer to low-resource languages [3,9]. **Azerbaijani NER context.** Azerbaijani remains under-resourced for NER, with limited publicly available labeled corpora and only a small number of prior studies [4]. This motivates silver-standard data creation and benchmarking across classical and transformer-based architectures.

Regarding low-resource language processing, researchers typically use a variety of "external" resources and semi-supervised methods in order to address their training/test data problems in low-resource environments. Although Das et al. (2020) found that external knowledge sources (e.g. Wikipedia categories) provide opportunities for improving the performance of NER for low-resource languages, they have also made it possible to work with these sources more easily. An excellent example of this is the WikiAnn corpus (Huang et al. 2020), which is a new multilingual corpus that consists of automatically annotated data (including Azerbaijani) and provides a great source of low-resource training/test data. However, automatic annotations may also introduce much noise into a dataset, which could add incentive to researchers as they are continuing to work towards finding a reasonable volume of human-collaboratively collected data that reflects what each central linguist chose as a placeholder, type, etc., for the entity tags used in the data.

In some respects, this research applied learnings from other research studies to achieve the overall approach; in other words, this research combines the utilization of CRFs, which are one of the more traditional methods in experimental systems with the use of BiLSTM-CRF and BERT, and the newest neural architecture, XLM-RoBERTa, to develop completely new reference benchmarks to allow for the direct emulation of current benchmarks contained within previous research publications in the fields of NER. In this study, a primary objective is to expedite the process of collecting appropriate ETL standards for the generation of performance benchmarks and the provision of established benchmarks to enable comprehensively measuring NER in order to have a basis for continued research in NER.

3 Research Methodology

In the current research, we evaluate five Named Entity Recognition (NER) models (a CRF model and four based on neural networks) for the Azerbaijani language. Each of the models uses the IOB2 tag format, meaning that every token developed an indicator to show whether it was at the Beginning (B), Inside (I), or Outside (O) of a Named Entity boundary. All the models were applied to the same annotated dataset and were assessed using the same evaluation methodologies to give fair evaluation criteria.

3.1 Conditional Random Field (CRF) Baseline

As the CRF model does not use neural networks to generate features, it can be thought of as a non-neural model and uses many different features which have been manually generated. The CRF model itself is a linear-chain probabilistic model, which allows it to consider the sequential dependencies present among the output label predictions when generating the predicted most likely label sequence. The features we created and used in training the model are:

Orthographic features: Pattern of capitalization of words, as well as the number and presence of numbers in the token could assist the CRF model to identify proper nouns in the Azerbaijani Language.

Character n-grams: The use of 2–4 character prefixes and suffixes because the agglutinative morphological forms of the Azerbaijani language do not produce segments of the same character writing.

Contextual features: The context of the current token, including the words preceding and following the token, their casing, and frequency of use may aid in establishing local syntactic rules.

Word identity: Word frequencies (name and keyword frequencies) were obtained from the dataset. For example, "Sinan" would be a name associated with a frequency greater than 1 and when used in association with the keyword "Universitet."

POS Tags: The lack of a reliable POS Tagging system for the Azerbaijani Language has prevented the use of POS Tagging for feature building; however, it is conceivable that if we had yet another reliable POS tagging system for Turkish, we could have considered adopting and implementing it as a part of our model architecture.

Feature Engineering. All CRF features are manually designed and computed at token level (casing patterns, digit flags, character n-grams, and local context windows); no neural embeddings are used in this baseline.

Training was conducted using stochastic gradient descent with L2 regularization; once locked, the model features could no longer be modified. The CRF model does not use pre-trained embeddings or language-contextually-derived knowledge; however, it remains a far less complex feature-based model compared to the other feature-based models, and thus provides a strong baseline by providing an easily interpreted framework for model development.

CRF Objective. Given an input token sequence x and label sequence y, the linear-chain CRF defines

$$p(y \mid x) = \frac{\exp\left(\sum_t s_t(y_t, x) + \sum_t A_{y_{t-1}, y_t}\right)}{Z(x)}, \tag{1}$$

where $s_t(\cdot)$ are emission scores, A are transition scores, and $Z(x)$ is the partition function. Decoding is performed via Viterbi:

$$\hat{y} = \arg\max_y \left(\sum_t s_t(y_t, x) + \sum_t A_{y_{t-1}, y_t}\right). \tag{2}$$

3.2 Neural Network Models

The last four neural architectures use a deep contextual pre-training architecture combined with a CRF decoding layer (CRF + decoding). They include (1) BiLSTM-CRF with fasttext word embeddings, (2) BERT-BiLSTM-CRF, (3) ALBERTo-BiLSTM-CRF, and (4) XLM-RoBERTa-BiLSTM-CRF (the best performing model).

1. BiLSTM-CRF with FastText: This model uses static FastText word embeddings (300-dimensional Azerbaijan FastText vectors were trained with data from both the Wikipedia and Common Crawl Corpus [8]) to populate the input of a bidirectional LSTM. The bidirectional LSTM captures the context around an entity via prior and subsequent tokens. The FastText embeddings can be frozen during training to ensure that only noise from additional data used for model training, does not cause overfitting on the few samples that were available to train with this model. The BiLSTM uses 128 hidden units to encode a sequence of entities in each direction (a total of 256 hidden units). The output of the BiLSTM is then passed to a CRF layer, which decodes the valid entity tag sequences as the output of the CRF. To account for misclassifying an entity (causing a dropout), a dropout rate of 0.3 will occur, thereby increasing generalization capabilities in addition to addressing overfitting.

2. BERT-BiLSTM-CRF: This model combines the use of multilingual BERT (mBERT) in the embedding layer as the embedding for as fine-tuned Azerbaijani NER. Contextual tokenized representations (768-D) produced by mBERT are used as the input to a BiLSTM, and the outputs of BERT subword tokens are either averaged together to aggregate the weightings of the scores for the respective tokens or the first subword token from each output is mapped to a corresponding token level embedding. The output of the BERT will be passed to

a BiLSTM and then to a CRF for decoding. BERT was fine-tuned using 2e−5 as the learning rate, while the BiLSTM and CRF were both fine-tuned using 1e−3 as the learning rate. A dropout rate of 0.1 applied during training and fine-tuning, had the same intention as described for the first model (Fig. 1).

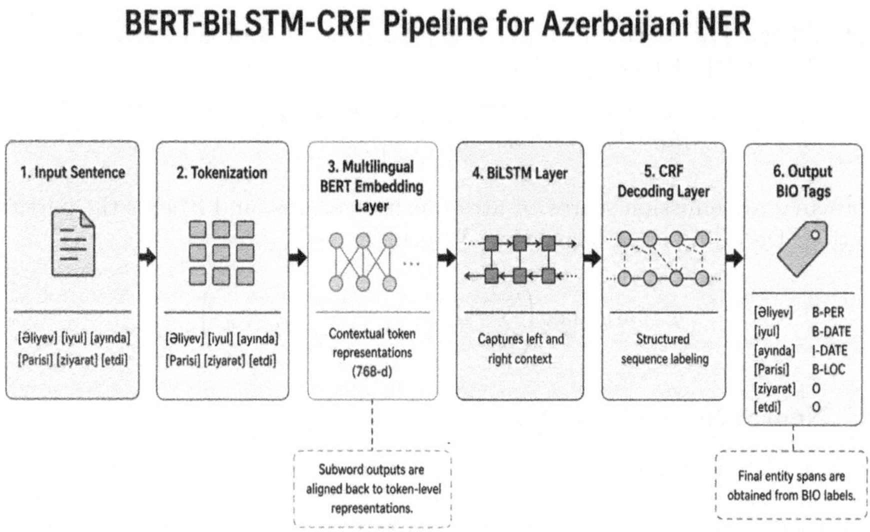

Fig. 1. BERT–BiLSTM–CRF sequence labeling pipeline used for Azerbaijani NER.

3. ALBERTo-BiLSTM-CRF: The ALBERTo-BiLSTM-CRF model uses the ALBERTo embedding parameter that is a representative sampling of Italian as a cross-lingual baseline. Although it is architecturally similar to BERT, no linguistically aligned variants have been made available for Azerbaijani. Consequently, being unable to select the appropriate variant resulted in the ALBERTo-BiLSTM-CRF model performing the worst and confirming the benefits of variant selection.

4. XLM-RoBERTa-BiLSTM-CRF (Best Model): This model achieved the highest overall performance in this study and is also the model noted as the best. The rationale for XLM-RoBERTa's success is the supply of multilingual contextual embeddings (trained in 100 languages), which are provided to a BiLSTM and then to a CRF for decoding. The XLM-RoBERTa-BiLSTM-CRF model's overall performance was as follows:

- F1 Score: 0.7918
- Precision: 0.8505
- Recall: 0.7574
- Token Accuracy: 0.7823

3.3 Dataset Preparation and Preprocessing

An Azerbaijani NER corpus featuring a large publicly accessible annotated NER dataset does not currently exist. The custom dataset was created with the 90,685 sentences (which consist of 37 distinct entity types). The new dataset will have entity tags associated with cross-linguistic resources. In this project, the Azerbaijani sentences were machine translated into English to produce quality translations and then automatically annotated for named entities (NER) using the spaCy English NER model to tag the entities and re-align the identified entity spans back to the original Azerbaijani sentences using token level alignments. Through this semi-automated procedure, a silver standard is developed for the Azerbaijani NER corpus without the need for extensive manual artistic annotation. The IOB2 and annotation tagging schemes were used for every sentence in the dataset and included the basic named entity types (person, location, organization, and date) as well as additional granularity (cardinal, ordinal, money, time, quantity, and percentage), and followed a Universal Dependencies-style tokenization discipline while adopting the WikiAnn/Wikipedia-derived NER label conventions [7]. Furthermore, to ensure text was normalized, a standard normalization (Unicode NFC) was utilized to maintain the integrity of the unique Azerbaijani specific Latin characters (ə,ı,ö,ü,ç,ş). Here, tokenization was executed through the use of whitespace and punctuation as boundaries. Capitalization was maintained to provide important cues in order to accurately identify named entities. To maintain lexical integrity, no stemming or case folding was performed on the dataset by entity types.

The dataset was divided into two distinct subsets (80% for training [72,500 sentences] and 20% for validation [18,100 sentences]). The sentences in each subset were shuffled chronologically and ordered to prevent repeated topics and to create approximately equal numbers of each of the five main topics in both the training and the validation sets. The performance of the model was assessed only with the validation set. We did not use gazetteers, rule-based systems or any external linguistic resources, other than pre-trained embeddings. Only the CRF used the hand-crafted features; the neural models obtained FastText, multilingual BERT, ALBERTo or XLM-RoBERTa token embeddings. A consistent model setup enabled fair comparison of these models using the same data and processing methods.

We collected Azerbaijani sentences primarily from Wikipedia and online news, then removed duplicates and applied basic cleaning to obtain over 90k sentences with naturally occurring entities.

We implemented a cross-lingual projection pipeline to achieve token level named entity labels while avoiding the cost of manual annotation. This process involved translating each of the Azerbaijani sentences into English, using the English NER tagger to label the entity spans and types of the translated sentences, and then using word alignment to return the entity spans to the original Azerbaijani tokens. As translation errors and word alignment errors can introduce a fair amount of noise into the final output of these projections, the process is both successful and can result in a large-scale silver-standard corpus for low-resource NER.

Manual Validation of Projected Annotations: We manually validated a random subset of 50 sentences to estimate the noise level introduced by cross-lingual projection. Using span-level exact-match metrics, the projected annotations reached Precision = 0.9048, Recall = 0.8313, and F1 = 0.8664, while token-level accuracy was 0.9666. These results indicate a reliable silver-standard signal; remaining errors mostly come from boundary mismatches and rare classes.

We employed IOB2 tagging; therefore, every token will be tagged as either: B-XXX (beginning), I-XXX (inside), O (outside). The dataset provides support for eighteen different entity types with a combination of B and I forms plus O tags would provide a total of thirty-seven distinct tags. The many standard entity types include PERSON, ORG, LOC, DATE and GPE. There are also many finer-grained entity types such as EVENT, WORK_OF_ART, LAW, LANGUAGE that provide for more varied benchmarking within the Azerbaijani language (Table 1).

Table 1. Entity labels and definitions used in the Azerbaijani NER dataset

Entity Label	Definition
CARDINAL	Numerals that do not fall under another type
DATE	Absolute or relative dates or periods
EVENT	Named hurricanes, battles, wars, sports events, etc.
FAC	Buildings, airports, highways, bridges, etc.
GPE	Countries, cities, states
LANGUAGE	Any named language
LAW	Named documents made into laws
LOC	Non-GPE locations, mountain ranges, bodies of water
MONEY	Monetary values, including unit
NORP	Nationalities or religious or political groups
ORDINAL	"First", "Second", etc.
ORG	Companies, agencies, institutions, etc.
PERCENT	Percentage, including the "%" symbol
PERSON	People, including fictional characters
PRODUCT	Objects, vehicles, foods, etc. (not services)
QUANTITY	Measurements such as weight or distance
TIME	Times smaller than a day (e.g., 5 p.m., noon)
WORK_OF_ART	Titles of books, songs, artworks, etc.

Once we finished projecting our data, we performed standardization and cleansing of all projected data using normalization and removal of extraneous whitespace/punctuation (e.g., NFC Unicode Normalization for Azerbaijani characters) as well as an automated tokenization process using basic rules for separating out multi-token expressions (e.g., dates). Very little manual filtering occurred, and we kept every single automatically annotated sentence that belonged to our projected data. In the end, we produced a corpus that is composed of 90,865 sentences and roughly 1.17 million tokens, allowing for extensive training of modern NER systems.

Dataset Characteristics. Entities range from single-token mentions to multi-token expressions. The corpus contains residual noise from translation/alignment, while punctuation and capitalization are preserved. We keep surface forms (no stemming or lowercasing) to retain informative morphological suffixes for NER.

3.4 Model Training and Evaluation

To be able to compare our models' performance in an equitable manner, we built all of them in Python using the same training/evaluation framework.

All models were trained using the same train/validation split and evaluated with Precision, Recall, span-level F1, and Token Accuracy. Early stopping was applied for neural models to reduce overfitting and training time.

The output layer of the CRF model produced structured tags corresponding to the predicted outputs. Evaluation Metrics included Precision, Recall, F1-score, and Token-Level Accuracy. The F1-score was chosen as the main evaluation metric because it alone requires that the predicted entity span and label match exactly to the true values.

3.5 Implementation and Hyperparameter Settings

In order to conduct a valid comparison between the various approaches, we utilized the same data splits, preprocessing methods, and evaluation scripts across all models. Model training utilized early stopping, based on the validation F1-score where appropriate, to limit overfitting and to maximize efficiency.

- Conditional Random Field (CRF) baseline: Trained via L-BFGS optimization using L2 regularization and maximum iterations. The same features were used as previously described in Sect. 3.1.
- BiLSTM-CRF (FastText): Trained for 50 epochs with Adam (learning rate 0.001) and utilized early stopping (5 epochs). Gradient clipping was implemented in order to stabilize training of the recurrent model.
- Transformer-based models (BERT, ALBERTo, and XLM-RoBERTa): Fine-tuned using AdamW with a small learning rate on the transformer parameters (2e−5) and a larger learning rate on the newly added BiLSTM/CRF layers (1e−3). The batch size of 64 was constrained by the available GPU memory. Both a warm-up schedule and dropout-based regularization were implemented to allow for stable fine-tuning.

4 Experiments and Results

Five different models were evaluated on the Custom Azerbaijani NER dataset. These models included: 1.) CRF (non-Neural baseline), 2.) BiLSTM-CRF with FastText, 3.) BERT-BiLSTM-CRF, 4.) ALBERTo-BiLSTM-CRF and 5.) XLM-RoBERTa-BiLSTM-CRF.

4.1 Overall Model Comparison

The evaluation of these five models can be found in Table 2. The results shown (Precision, Recall, F1 Score, Token Accuracy) there is a significant performance difference between traditional sequential models and those using transformer-based architectures. Overall, the models using contextual representations produced greater F-1 scores on the Azerbaijani validation dataset than did traditional sequential models.

As an example, the F1 Score of the CRF baseline model is very low due to missed entities and partially detected entities, despite it having a high Token accuracy (because "O" dominates in terms of the number of tokens). The BiLSTM-CRF with FastText provided a good improvement in Precision, however still conservative with Recall. Alternatively, the BERT based and XLM-RoBERTa based models achieve a better balance in Precision and Recall with respect to the NER task on the Azerbaijani language due to the pre-training on multilingual sources and the ability to obtain contextual NER information.

Table 2. Performance of NER Models on Azerbaijani Dataset (Results on Validation Set)

Model	Precision	Recall	F1-score	Token Accuracy
CRF (baseline)	0.73	0.67	0.70	0.92
BiLSTM-CRF (FastText)	0.86	0.60	0.70	0.61
BERT-BiLSTM-CRF	0.80	0.66	0.72	0.69
ALBERTo-BiLSTM-CRF	0.72	0.37	0.49	0.41
XLM-RoBERTa-BiLSTM-CRF	0.85	0.76	0.79	0.78

Comparison with Turkish NER. Turkish has several public NER resources and benchmarks, and prior work shows that performance can drop substantially on noisy Turkish text compared to cleaner domains [5]. Compared to these results, our best model (XLM-RoBERTa-BiLSTM-CRF, F1 = 0.7918) is competitive given that our corpus is silver-standard and produced via cross-lingual projection, which introduces boundary and type noise.

Token Accuracy vs. Span F1. Token accuracy is the fraction of tokens whose predicted BIO label matches the gold label. In NER, the label `O` typically dominates, so a model can obtain high token accuracy by predicting `O` frequently, while still missing entity spans. Span-level F1 is stricter: an entity is counted as correct only if both its boundaries and its type match exactly. *Example.* For the sequence "Aliyev visited Paris", a model that predicts `O O O` achieves 1/3 token accuracy if the gold is `B-PER O B-LOC`, but it gets zero span-level true positives because it detects no entities.

Overall, XLM-RoBERTa–BiLSTM–CRF achieved the best balance of precision and recall (F1 = 0.7918), while the CRF baseline shows inflated token accuracy due to the dominance of the `O` label. ALBERTo underperforms because it is pretrained on Italian and is not linguistically aligned with Azerbaijani.

4.2 Entity-Wise Performance of the Best Model

While overall performance metrics are interesting, it is important to know which entity types the best-performing models benefit the most from being used with multilingual contextual embeddings. The table below, Table 3, shows the best-performing XLM-RoBERTa-BiLSTM-CRF architecture/model performance metrics for each of the named entity "taxonomy" types. In addition to per-class performance, Table 3 reports the relative frequency of entity types in the train and validation splits to contextualize the metrics, especially for rare classes.

The model has high F1 effectiveness (i.e. the highest F1-scores) for many frequent and structured classes such Example Classes DATE, ORDINAL, CARDINAL. However, performance reduces for classes with low quantities, such as LAW and PRODUCT, which are likely to have fewer training examples and increased ambiguity due to fewer examples of class training. The entity-level analysis complements the aggregate results, and directly informs methods of developing targeted strategies for improving the prediction of rare-class examples as dataset development progresses.

4.3 Error Analysis

While the XLM-RoBERTa-BiLSTM-CRF model delivers the strongest overall performance, examining individual predictions and performing qualitative assessment reveals systematic errors, which are useful in improving Azerbaijani NER in future efforts.

Confusions between semantically related entity-types: Confusion among the Entity Types of GPE (Geopolitical entity) and LOC (Location) can occur when the mention refers to a Region, Administrative Division or Colloquial denotes a place-name. The need for more contextual information or more clear annotator guidelines is apparent for borderline classification of GPE, LOC, ORG, and PERSON entities.

- The confusion of GPE and LOC can occur when referencing to a Region, Administrative Division or Colloquial denotes a place-name.
- The confusion of ORG and PERSON occur when organization names contain an individual name as part of the organization's name (i.e., named after an individual).

Table 3. Entity-wise precision, recall, F1-score, and entity distribution across the train/validation splits for XLM-RoBERTa-BiLSTM-CRF

Entity Type	Precision	Recall	F1-Score	Train (%)	Val (%)
CARDINAL	0.92	0.84	0.88	11.28	11.24
DATE	0.94	0.92	0.93	16.65	16.45
EVENT	0.70	0.41	0.52	0.52	0.52
FAC	0.57	0.21	0.31	0.53	0.52
GPE	0.87	0.73	0.80	15.36	15.36
LANGUAGE	0.63	0.72	0.67	0.40	0.37
LAW	0.25	0.19	0.22	0.19	0.18
LOC	0.56	0.53	0.55	1.85	1.80
MONEY	0.69	0.51	0.59	0.25	0.28
NORP	0.89	0.80	0.84	13.35	13.23
ORDINAL	0.96	0.91	0.93	4.86	4.83
ORG	0.82	0.47	0.60	11.51	11.41
PERCENT	0.92	0.90	0.91	0.57	0.60
PERSON	0.79	0.84	0.81	19.33	19.90
PRODUCT	0.00	0.00	0.00	0.33	0.36
QUANTITY	0.77	0.84	0.80	1.25	1.24
TIME	0.81	0.70	0.75	1.09	1.03
WORK_OF_ART	0.61	0.32	0.42	0.70	0.71

- The confusion of EVENT type entities and WORK_OF_ART type entities occurs when materials designed as cultural artifacts (i.e., works of art) and events are coded similarly or display similar surface patterns.

Impact of rare and low frequency classes: Targeted data augmentation efforts, use of class-aware loss functions, or acquiring more quality annotations of infrequent categories is likely necessary to address these issues.

- Several low frequency categories (e.g., PRODUCT, LAW, WORK_OF_ART) produced significantly lower F1-scores compared to other frequent categories. The low F1-scores suggest a lack of sufficient training examples for models tasked with general prediction across categories.
- Rare or low frequency mentions may be mistakenly coded as a false negative (i.e., missed) and/or partial or incomplete detections of rare or low frequency mentions will occur, which is adversely affected by approximate span boundary-based evaluation models.

Errors identifying span boundaries: In general, difficulty in finding span boundaries of TEST entities may result from the nature of tokenization of Azerbaijani entities and implementations for suffixing. This additional Complexity suggests potential benefits from developing span-aware models to better utilize spacing between suffix tokens and subword-type representations in models for NER.

- Multi-token entities may be truncated at the first token if the first token is classified correctly and the subsequent tokens are classified incorrectly as O (i.e., not an entity).
- The misclassification of long organization and facility names will also be adversely affected by span matching models.

Inconsistent tagging between B-tag and I-tag: Differences in tagging of Entities by utilizing B-XXX and I-XXX vs. using an alternative BIOES tagging format, will reduce fragmentation of Multi-entities:

- Generally, B-XXX tags show a higher degree of correct prediction than I-XXX tags leading to multi-token entity fragmentation.
- Predicting the second token of a two-token entity as a new, beginning entity-token of a two-token entity produces fragmented entities B-tag/I-tag.

Precision/Recall behavior: Future studies can investigate the use of calibration methods, semi-supervised learning methods, and/or thresholding techniques (which will improve recall while producing a high level of precision).

- For the models included, precision values are generally greater than recall values indicating a conservative prediction method yielding a low number of false positives, however more entities that are missed.
- Under-prediction of rare or contextually dependent entities will generate more pronounced evidence.

5 Conclusion and Future Work

In this research, an extensive evaluation has been performed on different methods/models that can perform NER (Named Entity Recognition) in the Azerbaijani language, as well as how well they perform when compared to each other using both traditional and modern methods (i.e. artificial neural networks). A variety of models/methods were tested including: a baseline CRF model, a BiLSTM-Combined CRF model trained on FastText word embeddings, a BERT-BiLSTM-CFR model and an XLM-RoBERTa-BiLSTM-CFR model. Out of all the above models/methods tested, the XLM-RoBERTa-BiLSTM-CFR model had the best overall performance with an F1 score of .7918, precision of .8505, recall of .7574 and was better than all of the other models tested across all performance metrics measured.

The implications of this work show that contextual embeddings from multilingual transformers such as BERT and XLM-RoBERTa provide a substantial benefit to NER tasks in Azerbaijani which is a morphologically rich low resource language. Previous models such as the CRF and BiLSTM allowed for some level accuracy within Azerbaijani NER tasks, but were limited due a lack of recall and/or a lack of contextual information. The proposed transformer based architecture was shown to provide a level of accuracy that was more balanced in terms of precision and recall thereby demonstrating its value and efficacy within relevant NER tasks in Azerbaijani NER tasks.

Future efforts will focus on addressing the limitations identified in this work by increasing the amount of available training data via different semi-supervised annotation methods, crowdsourcing and/or data augmentation. As well, further developing models that will be able to generalize across different languages will be accomplished through cross-lingual transfer learning by using Turkish (which is related) as a reference point. In addition, an external knowledge base (such as a gazetteer or industry specific lexicons) can be used to enhance recall for entities that are uncommon or otherwise complex. Further exploration of the use of pure transformer based sequence labeling methods will be carried out (e.g., XLM-RoBERTa-CRF or BERT in variants that are specific to Azerbaijani). Finally, there will be an evaluation of the distributed accuracy of trained models in noisy/informal environments such as social media (which will include an assessment of this NER work for real-world applications).

This work has provided the foundation to establish strong baseline NER performance metrics for Azerbaijani, and also has shown support for the continued use of multilingual transformers for low-resource languages. **Data availability:** The Azerbaijani NER silver-standard corpus and related resources are available at https://github.com/Tofig999/azerbaijani-ner-silver-corpus.

LLM Baseline (Future Work). Recent instruction-tuned LLMs were not evaluated in this study. As future work, we plan to benchmark prompt-based and lightweight fine-tuning approaches (e.g., multilingual LLMs) on the proposed corpus to compare task-specific models with large generalist models.

Acknowledgments. This work was developed within the Center of Data Analytics Research (CeDAR) and the MegaSec AI Laboratory at ADA University.

References

1. Lafferty, J., McCallum, A., Pereira, F.C.: Conditional random fields: probabilistic models for segmenting and labeling sequence data. In: Proceedings 18th International Conference on Machine Learning (ICML), pp. 282–289 (2001)
2. Huang, Z., Xu, W., Yu, K.: Bidirectional LSTM-CRF models for sequence tagging. arXiv preprint arXiv:1508.01991 (2015)
3. Devlin, J., Chang, M.-W., Lee, K., Toutanova, K.: BERT: pre-training of deep bidirectional transformers for language understanding. In: Proceedings NAACL-HLT, Minneapolis, MN, USA, pp. 4171–4186 (2019)
4. Akhundova, N.: Named entity recognition for the Azerbaijani language. In: Proceedings 15th International Conference on Application of Information and Communication Technologies (AICT), pp. 1–7 (2021)

5. Akkaya, E.K., Can, B.: Transfer learning for Turkish named entity recognition on noisy text. Nat. Lang. Eng. **27**(1), 35–64 (2021)
6. Das, A., Ganguly, D., Garain, U.: Named entity recognition with word embeddings and Wikipedia categories for low-resource language. ACM Trans. Asian Low-Resource Lang. Inf. Process. **16**(3), 1–19 (2017)
7. Pan, X., Zhang, B., May, J., Nothman, J., Knight, K., Ji, H.: Cross-lingual name tagging and linking for 282 languages. In: Proceedings 55th Annual Meeting of the ACL (Short Papers), pp. 194–200 (2017)
8. Bojanowski, P., Grave, E., Joulin, A., Mikolov, T.: Enriching word vectors with subword information. Trans. Assoc. Comput. Linguistics **5**, 135–146 (2017)
9. Conneau, A., et al.: Unsupervised cross-lingual representation learning at scale. In: Proceedings 58th Annual Meeting of the ACL, pp. 8440–8451 (2020)
10. Cotterell, R., Duh, K.: Low-resource named entity recognition with cross-lingual, character-level neural conditional random fields. In: Proceedings of the Eighth International Joint Conference on Natural Language Processing (IJCNLP 2017), Volume 2: Short Papers, Taipei, Taiwan, pp. 91–96 (2017)
11. Dong, L., Qiu, Y., Zhang, W.: Enhancing named entity recognition in low-resource scenarios using diffusion variational autoencoder and knowledge graphs. J. King Saud Univ. Comput. Inf. Sci. **37**, 285 (2025). https://doi.org/10.1007/s44443-025-00314-w
12. Mammadov, S., Mustafali, A., Mammadov, Z., Rustamov, S.: Part-of-speech tagging for Azerbaijani language. In: 12th International Conference on Application of Information and Communication Technologies (AICT). IEEE (2018). https://doi.org/10.1109/ICAICT.2018.8747154
13. Gulu-Zada, A., Ahmadova, H., Kazimov, E., Rustamov, S.: A hybrid approach to Azerbaijani lemmatization integrating rule-based methods with machine learning. In: 18th International Conference on Application of Information and Communication Technologies (AICT). IEEE (2024). https://doi.org/10.1109/AICT61888.2024.10740428
14. Mammadli, S., Huseynov, S., Alkaramov, H., Jafarli, U., Suleymanov, U., Rustamov, S.: Sentiment polarity detection in Azerbaijani social news articles. In: Proceedings RANLP 2019, pp. 703–710 (2019). https://doi.org/10.26615/978-954-452-056-4082

An Approach for Evaluating Recommender Systems: Assessing Usefulness and Coherence Through Item Relationships

Marie Griffon[1,2](✉), Nicolas Delestre[1], and Maxime Guériau[1]

[1] INSA Rouen Normandie, Univ Rouen Normandie, Université Le Havre Normandie, Normandie Univ, LITIS UR 4108, F-76000 Rouen, France
{marie.griffon,nicolas.delestre,maxime.gueriau}@insa-rouen.fr
[2] TraceParts, Parc Éco Normandie, F-76430 Saint-Romain-de-Colbosc, France
mgriffon@traceparts.com

Abstract. Recommender systems (RSs) have become essential components of online platforms. As their economic impact continues to grow, evaluating their performance can no longer be limited to accuracy-based metrics alone. In particular, existing offline evaluation protocols often overlook the usefulness of recommendations from the user's perspective. In this paper, we highlight the importance of usefulness-oriented evaluation for RSs and propose two new offline evaluation metrics: Usefulness and Intra-List Relationships (ILR). Usefulness evaluates how well recommendations align with the user's session context, whereas ILR measures the relational coherence among recommendations. These metrics are based on the assumption that user interactions within a session reflect a decision-making process in which items may act as substitutes for one another or, conversely, complement each other in order to satisfy a given need. To model this behavior, we formalize the notion of session context and propose two data-driven scores that quantify substitutability and complementarity relationships between items. Experiments conducted on a public dataset show that our scores capture meaningful item relationships with performance comparable to existing approaches, and that both metrics exhibit behavior consistent with their definition, decreasing monotonically as the relational coherence of recommendations degrades. Overall, the proposed approach provides a reproducible way to assess the usefulness and relational coherence of recommendations.

Keywords: Recommender systems · Offline evaluation metric · Item relationships

1 Introduction

Recommender systems (RSs) have become central to the commercial success of modern online platforms. Widely deployed in e-commerce, content streaming,

C. Zanni-Merk et al. (Eds.): AIDT 2026, CCIS 3023, pp. 220–235, 2026.
https://doi.org/10.1007/978-3-032-31319-5_15

online advertising, and digital services, they help users navigate large item catalogs while stimulating engagement and sales. Major companies such as Amazon and Netflix attribute a substantial share of interactions and revenues to personalized recommendations [3,10]. Consequently, RSs have evolved from auxiliary tools to core components of online business strategies [4].

As RSs increasingly shape sales, consumption, and user experience, their evaluation has become a central issue for both research and industry. Three main evaluation paradigms are commonly distinguished: offline evaluations, user studies, and online evaluations. Offline evaluations rely on historical datasets containing explicit (*e.g.*, ratings) or implicit (*e.g.*, purchases, views) feedback. User studies involve controlled experiments with a limited number of participants, requiring substantial resources and potentially introducing behavioral biases such as the "Hawthorne effect" [9]. Online evaluations deploy the RS in real-world conditions [22] and provide realistic behavioral estimates, but they entail significant risks. In particular, exposing users to irrelevant or incoherent recommendations may rapidly erode trust, with long-lasting negative effects [3]. Consequently, robust offline evaluation protocols are essential to assess both recommendation accuracy and overall quality before real-world deployment.

Such an approach requires evaluating an RS using multiple complementary metrics, as relying on a single measure provides a biased and incomplete view of performance. Nevertheless, most RS studies still prioritize accuracy-oriented metrics in offline settings. These metrics assess the system's ability to predict past user interactions and include prediction errors such as MAE and RMSE, usage-based measures such as Precision@k and Recall@k, and ranking metrics such as NDCG and MRR [22].

Although accuracy has traditionally been the primary objective of RSs, it is widely recognized as insufficient to fully assess recommendation quality [5,7,11]. Prior work emphasizes the importance of considering additional factors and evaluating recommendation lists as coherent sets rather than isolated items. Consequently, complementary objectives such as diversity, serendipity, novelty, and coverage have been introduced [5,7,22]. Unfortunately, the metrics commonly used in offline evaluations do not explicitly capture two essential aspects: (1) whether recommendations address the user's current needs; and (2) whether recommendations form a collection of related items that may be perceived as a bundle by the user.

In this paper, we propose two new offline evaluation metrics for RSs: the Usefulness metric to assess the first aspect, and the Intra-List Relationships metric to assess the second. To construct these metrics, we propose two scores designed to identify and quantify substitutability and complementarity relationships between items based on observed user session interactions. The use of substitutability and complementarity scores is motivated by the observation that interactions within a session reflect a goal-oriented decision-making process, leading users to explore substitutable and complementary items.

The remainder of this paper is organized as follows. Section 2 reviews related work on RS evaluation and on extracting substitutability and complementarity relationships. Section 3 presents our approach, Sect. 4 reports the experimental results, and Sect. 5 concludes with a discussion and future research directions.

2 Related Work

A substantial body of literature has addressed the evaluation of RSs, defining and categorizing the associated evaluation metrics. Survey papers provide overviews of commonly used metrics, grouping them into broad categories such as prediction accuracy, ranking, and business-oriented metrics [3,22], while others focus on specific subsets of metrics [5,7,8].

However, none of the reviewed offline metrics explicitly assesses whether recommendations are useful to the user or whether they can be perceived as a coherent bundle. Accuracy-oriented metrics evaluate items with available feedback but provide no insight into the relevance of the remaining recommended items. To address these limitations, we propose new offline evaluation metrics grounded in item relationships.

To ground evaluation in item relationships, it is necessary to understand how such relationships can be modeled. Several studies have investigated substitutability and complementarity between items in RSs, motivated by their importance in economics, marketing, and nutrition.

Tian *et al.* [19] propose a graph-based method to extract substitutability and complementarity from transaction data by modeling products and transactions as a bipartite graph. Relationships are inferred from co-occurrence patterns: complementary items co-occur more frequently than expected, whereas substitutable items rarely co-occur but exhibit similar purchasing patterns. While we adopt similar assumptions to characterize item relationships (*cf.* Sect. 3.3), their bipartite graph modeling increases implementation complexity and may limit scalability in large-scale settings.

Item co-occurrence (IC) has been proposed by Sun *et al.* [18] to quantify substitutability and complementarity through pairwise co-occurrence probabilities. It is integrated into a hierarchical feature-based recommendation framework to improve accuracy and interpretability. However, this approach does not account for session structure or interaction context.

Akkoyunlu *et al.* [2] investigate item substitutability in food recommendation by introducing the notions of dietary context (co-consumed items) and food intake context (*e.g.*, meal type or participants). Two items are considered substitutable if they appear in similar dietary contexts, with substitutability modulated by the food intake context. To control combinatorial complexity, meals are modeled as graphs and comparisons are restricted to meals differing by a single item. While effective, this strategy increases implementation complexity, omits certain possible contexts, and focuses solely on substitutability without addressing complementarity.

3 Approach

In this section, we introduce the notations, the formal framework, and the recommendation problem under study. These elements constitute the foundation of the proposed approach and enable the definition of the scores and metrics.

3.1 Notations and Problem Statement

Let I denote the set of items and S the set of sessions. We define a session as a set of items with which a user has interacted, delimited by boundaries defining the beginning and the end of a given event (*e.g.* a transaction or a consumption episode) [20]. The recommendation task considered in this work is the following: given a session $Q \in S$, generate a set of items $R \subseteq I$, referred to as recommendations, that are likely to interest the user. Our approach is guided by two objectives. First, we aim to evaluate the usefulness of the recommendations R with respect to the session Q, that is, to determine to what extent the recommendations address the user's needs. Second, we aim to evaluate the internal coherence of R by analyzing whether the recommendations form a coherent set of items connected through meaningful relationships.

Within a session, users tend to explore strongly related items that share a common intention [21]. These interactions reflect a decision-making process driven by one or multiple short-term goals. In this context, a recommendation is considered useful if it maintains a significant relationship with at least one item already present in the session, in accordance with the intention underlying the session.

In economics, relationships between two items are traditionally characterized through substitutability and complementarity [13,18]. These relationships are not binary: different degrees of substitutability and complementarity may arise depending on usage and context. Consider, for example, *peanut butter* and *jam* consumption habits at breakfast. In the United States, these two products are frequently consumed together on *bread*, indicating a strong complementarity relationship. In contrast, in France, *peanut butter* and *jam* are more often consumed alternatively on *bread*, corresponding to a substitutability relationship. This example highlights that relationships between two items may be complementary in certain contexts and substitutable in others.

Our approach therefore consists in quantifying the degree of substitutability and complementarity between two items. Given a session database, we aim to identify these relationships from observed interactions. To this end, we rely exclusively on contextual information derived from sessions, drawing inspiration from Akkoyunlu *et al.* [2].

3.2 Defining Context

We adapt the two notions of context introduced by Akkoyunlu *et al.* [2] to our application setting. We thus define two distinct types of context: the *item context within a session* and the *session context*.

The item context within a session corresponds to the set of other items with which the user interacted during that session. For instance, if the breakfast {*coffee, bread, butter, jam*} is modeled as a session, the item context of *coffee* within that session is {*bread, butter, jam*}. Formally, for a session $s \in S$ and an item $i \in s$, the item context of i in session s is defined as $c_i(s) = s \backslash \{i\}$, where $c_i(s) \subset I$.

The session context characterizes the setting in which the session occurred. It may include, for instance, user-related characteristics (age, location) and the user's goals during the session (having breakfast, getting dressed, decorating, *etc.*) [1,21].

Both the item context within a session and the session context are essential for modeling item substitutability. An item can replace another item only if (i) the two items exhibit similar interactions with other items, i.e., they share common complementary items [19], and (ii) they are associated with the same contextual factors [2].

3.3 Substitutability and Complementarity Scores

Two items are said to be substitutable when they serve the same purpose and can be used *in place of one another*, whereas two items are complementary when they are used *jointly* to satisfy the same need.

Our hypotheses, inspired by Tian *et al.* [19] and Akkoyunlu *et al.* [2], assume that substitutability and complementarity relationships between items can be inferred from observed session interactions. Intuitively, substitutability is associated with interactions occurring in similar contexts, whereas complementarity manifests itself through frequent co-interactions within the same session. However, co-interaction between two items reduces the likelihood that they are substitutable, since the items may instead be complementary. Conversely, when two items consistently share identical item contexts within sessions, their degree of complementarity tends to decrease, suggesting instead a substitutability relationship. Consider, for example, five breakfasts: {*bread, butter, coffee*}, {*bread, butter, tea*}, {*bread, chocolate spread, coffee*}, {*bread, butter, juice*}, and {*bread, chocolate spread, tea*}. The co-interaction between *butter* and *bread* is observed in three breakfasts. This repeated co-occurrence suggests a complementarity relationship between these two items, with *butter* enriching the consumption of *bread* during breakfast. *Butter* and *chocolate spread* appear in two shared item contexts, {*bread, coffee*} and {*bread, tea*}, while rarely co-occurring within the same breakfast. This contextual similarity, combined with low co-interaction, indicates a substitutability relationship: the two items fulfill a comparable role and are used as substitutes for one another.

We therefore propose scores that quantify the degree of substitutability and complementarity between two items while accounting for the opposing influence of the other relationship. In this framework, we define substitutability and complementarity scores that satisfy the following hypotheses:

(1) Two items are highly substitutable when their item contexts within sessions are identical;

(2) Two items are less substitutable when users frequently interact with both items within the same session;
(3) Substitutability is a symmetric relationship;
(4) Two items are highly complementary when users frequently interact with both items within the same session;
(5) Two items are less complementary when their item contexts within sessions are identical;
(6) Complementarity is a symmetric relationship.
(7) The degree of substitutability and complementarity between two items depends on the session context in which they occur.

First, we define the set C_i as the set of item contexts within sessions associated with item i, that is:

$$C_i = \{c_i(s) \mid s \in S, i \in s\} \tag{1}$$

For two items $q, r \in I$, the hypotheses (1) and (5) are captured by the intersection of C_q and C_r. If $|C_q \cap C_r|$ is large, then q and r frequently occur in identical item contexts across sessions.

Next, we denote by $A_{q:r}$ the set of contexts of q where r appears:

$$A_{q:r} = \{c \in C_q \mid r \in c\} \tag{2}$$

The cardinality of $A_{q:r}$ quantifies how often r appears in the item contexts of q.

Taking these considerations into account and building upon Akkoyunlu *et al.* [2], we define the substitutability score between two items q and r as follows:

$$subst(q,r) = \frac{|C_q \cap C_r|}{|C_q \cup C_r| + |A_{q:r}| + |A_{r:q}|} \tag{3}$$

The substitutability score equals 1 when the item contexts of q and r are identical across all sessions in which they appear and when $A_{q:r} = A_{r:q} = \emptyset$. If users never interact with q and r within the same contexts, the score reduces to 0. The larger $|A_{q:r}| + |A_{r:q}|$, the lower the score, reflecting hypothesis (2).

For the complementarity score, we first introduce an association score that estimates the probability that item r appears in an item context of q, with a penalty based on the intersection of C_q and C_r:

$$assoc(q,r) = \frac{|A_{q:r}|}{|C_q| + |C_q \cap C_r|} \tag{4}$$

The association defined above is asymmetric. However, we aim to compute a complementarity score that quantifies the degree of mutual association between two items. We therefore combine the two asymmetric association scores into a symmetric complementarity score that attains a high value only when both asymmetric components are high. In the literature, the harmonic mean is a commonly used measure of central tendency for combining rates and is well suited for score aggregation [14]. It is also sensitive to large imbalances between aggregated values, meaning that a strong asymmetry in association between two items lowers the final score.

We thus define the complementarity score as the harmonic mean of the two association scores:

$$compl(q,r) = \frac{2 \times assoc(q,r) \times assoc(r,q)}{assoc(q,r) + assoc(r,q)} \tag{5}$$

The complementarity score of q and r equals 1 when r appears in all item contexts of q, q appears in all item contexts of r, and $C_q \cap C_r = \emptyset$. If users never interact with q and r within the same session, the score is 0. The larger $|C_q \cap C_r|$, the higher the degree of substitutability between q and r, and consequently the lower the complementarity score, reflecting hypothesis (5).

3.4 Usefulness and Intra-List Relationships Metrics

We consider the offline evaluation setting of an RS. Our first objective is to evaluate the usefulness of R with respect to Q to determine the extent to which the recommendations address users' needs. We consider R to be useful with respect to Q if (i) every recommendation maintains at least one strong relationship with an item in the session, and (ii) every item in the session maintains at least one strong relationship with a recommendation.

Equation 6 captures condition (i), and Eq. 7 captures condition (ii):

$$u(R,Q) = \frac{1}{|R|} \sum_{r \in R} \max_{q \in Q}\big(\max\big(subst(q,r), compl(q,r)\big)\big) \tag{6}$$

$$u(Q,R) = \frac{1}{|Q|} \sum_{q \in Q} \max_{r \in R}\big(\max\big(subst(q,r), compl(q,r)\big)\big) \tag{7}$$

where $\max\big(subst(q,r), compl(q,r)\big)$ extracts the strongest relationship between q and r, whether substitutability or complementarity.

To obtain the overall usefulness of R with respect to Q, we compute the harmonic mean of $u(Q,R)$ and $u(R,Q)$:

$$Usefulness(Q,R) = \frac{2 \times u(Q,R) \times u(R,Q)}{u(Q,R) + u(R,Q)} \tag{8}$$

The Usefulness metric ranges from 0 (no relationships between recommendations and session items) to 1 (maximal relational strength between recommendations and session items), ensuring that the recommendations address the user's current needs.

The second objective is to evaluate the internal coherence of R by analyzing whether the recommendations form a consistent set from the perspective of their relationships. Inspired by Intra-List Diversity [8], we propose the Intra-List Relationships (ILR) metric, which measures the average relational strength among all recommendations.

Formally, for a recommendation set $R = \{r_1, r_2, \ldots, r_k\}$, where k denotes the number of recommended items, ILR is defined as:

$$ILR(R) = \frac{2}{k(k-1)} \sum_{i=1}^{k-1} \sum_{j=i+1}^{k} \max\big(subst(r_i, r_j), compl(r_i, r_j)\big) \quad (9)$$

ILR ranges from 0 (no relationships between recommended items) to 1 (maximal relational strength), ensuring that users are exposed to recommendation sets that can naturally be experienced as a bundle.

4 Experiments

This section reports the experimental evaluation conducted on the BundleRec dataset. We designed two experiments to assess the proposed scores and metrics. The code used to reproduce these experiments is publicly available[1]. The first experiment evaluates the ability of our substitutability and complementarity scores to recover item relationships annotated by humans. The second experiment investigates the effectiveness of our metrics in capturing relational consistency, by analyzing their behavior under progressive degradation of item relationships.

The BundleRec dataset[2] was introduced by Sun *et al.* [17] in the context of bundle recommendation. A bundle refers to a set of items offered or sold together to fulfill a specific objective (*e.g.*, dressing for a particular event or completing a gaming setup). Bundle recommendation aims to automatically suggest to a user a bundle that aligns with their purchasing intent. In [17], the authors designed a crowdsourcing task to annotate candidate bundles and the associated objectives implicitly expressed in user sessions. The user sessions are derived from the Amazon datasets [6] and span three domains: electronics, clothing, and food.

Each domain-specific dataset is first randomly partitioned into training and test subsets using an 80/20 session-level split, followed by a constrained session-swapping procedure to improve the balance of unique item occurrences across subsets while preserving the original split sizes. The training split is used to compute the substitutability and complementarity scores between items, while the remaining 20% is reserved for evaluation. Descriptive statistics of these datasets are reported in Table 1.

4.1 Evaluation of Substitutability and Complementarity Scores

The objective of this experiment is to determine whether the proposed scores identify meaningful relationships of substitutability and complementarity based on user sessions. The identified relationships are compared against a ground truth constructed from the human annotations provided in Sun *et al.* [17].

[1] https://github.com/MadPuppyForce/evaluating-recsys-item-relationships.

[2] https://github.com/BundleRec/bundle_recommendation.

Table 1. Descriptive statistics of the BundleRec datasets across domains and train/test splits.

	Electronic		Clothing		Food	
	Train	Test	Train	Test	Train	Test
#Items	4 142	1 435	5 261	1 452	4 283	1 424
#Sessions	916	229	944	237	928	233
#Bundles	1 330	420	1 515	395	1 354	430
Average bundle size	3.47	3.69	3.31	3.32	3.52	3.80
#Session-item interactions	4 947	1 592	5 328	1 465	5 018	1 529
#Bundle-item interactions	4 616	1 549	5 016	1 310	4 760	1 635
Density of session-item interactions	0.13%	0.48%	0.11%	0.43%	0.13%	0.46%
Density of bundle-item interactions	0.08%	0.26%	0.06%	0.23%	0.08%	0.27%

In this experiment, we consider individual items rather than item categories, and we exploit the training sessions of each domain to compute substitutability and complementarity scores between items within that domain. The ground truth consists of the set of unique item pairs for each domain, extracted from the annotated bundles associated with the training sessions.

Methodology. We compare two approaches: (1) the Item Co-occurrence (IC) method proposed by Sun *et al.* [18], and (2) our substitutability and complementarity scores.

Precision and recall are used as evaluation metrics to measure how many item pairs are correctly predicted in each domain:

$$Precision = \frac{\#Correct.\ pred.\ pairs}{\#Pred.\ pairs} \qquad Recall = \frac{\#Correct.\ pred.\ pairs}{\#GT\ pairs}$$

where $\#GT\ pairs$ denotes the number of pairs in the ground truth, $\#Pred.\ pairs$ the number of predicted pairs, and $\#Correct.\ pred.\ pairs$ the number of pairs in the intersection between the predicted pairs and the ground-truth pairs.

A high precision indicates that few non-annotated or spurious relationships are identified, whereas a high recall indicates that most of the annotated relationships are successfully recovered.

Results. The results obtained for the three domains are reported in Table 2. Both approaches identify the same complementarity relationships, which can be explained by the fact that our complementarity score and the IC are both based on item co-occurrence within sessions. However, our approach additionally identifies substitutability relationships, namely 5 in the electronics domain, 5 in the food domain, and 1 in the clothing domain.

The maximum recall achieved by both approaches indicates that all ground-truth pairs are successfully identified. Nevertheless, the precision values reveal

Table 2. Precision and recall obtained by IC [18] and by our scores in each domain.

Domain	Method	#GT pairs	#Pred. pairs	#Correct. pred. pairs	Precision	Recall
Electronic	IC [18]	6 841	13 534	6 841	50.55%	100%
	Our scores		13 539	6 841	50.53%	100%
Clothing	IC [18]	6 744	15 201	6 744	44.37%	100%
	Our scores		15 202	6 744	44.36%	100%
Food	IC [18]	6 950	13 396	6 950	51.88%	100%
	Our scores		13 400	6 950	51.87%	100%

that approximately half of the predicted pairs are not present in the annotations or are not considered validated relationships. These results show that our approach achieves performance comparable to IC in terms of precision and recall, while offering the additional advantage of identifying substitutability relationships between items.

In addition, a complementary experiment conducted in a separate study[3] shows that the proposed substitutability score identifies meaningful relationships comparable to those obtained with the IC [18] and the substitutability score proposed by Akkoyunlu *et al.* [2]. This experiment highlights the importance of incorporating the notion of item context within sessions to improve the identification of substitutable items, as well as the influence of session context on the relationships between items.

4.2 Evaluation of Usefulness and ILR

This experiment evaluates the Usefulness and ILR metrics in order to analyze their behavior on real-world data. A distinct experimental protocol is designed for each metric to assess its specific properties. Both evaluations follow a stress-test protocol, in which structured bundles associated with the test sessions are progressively corrupted to examine the robustness and sensitivity of the proposed metrics under controlled perturbations [16]. The computation of Usefulness and ILR relies exclusively on the substitutability and complementarity scores previously estimated from the training sessions of each domain.

Experimental Protocol for ILR. The objective of this protocol is to assess whether ILR is sensitive to the internal relational coherence of recommendation sets. If ILR correctly captures internal structure, its value should exhibit an overall decreasing trend as coherent bundles are progressively corrupted through random item replacement.

[3] Accepted for publication at INFORSID 2026, URL: https://inforsid2026.sciencesconf.org/resource/page/id/41.

Let $B = \{b_1, \ldots, b_k\}$ denote an original bundle of size k, assumed to exhibit strong internal relational coherence. To simulate progressive degradation, we introduce controlled corruption by replacing $m \in \{0, \ldots, k\}$ items.

For a given corruption level m, a subset of indices $W \subset \{1, \ldots, k\}$ with $|W| = m$ is sampled uniformly without replacement. Each selected item b_i, $i \in W$, is replaced by an item drawn uniformly at random from the catalog $C \setminus B$, also without replacement to avoid duplicates. This produces a corrupted bundle $B^{(m)}$ of identical size k containing exactly m randomly injected items.

To reduce sampling variance, the corruption process is repeated R times independently for each pair (B, m). We set the number of repetitions to $R = 100$ to obtain stable Monte Carlo estimates of the expected ILR under random corruption, ensuring a reliable trade-off between estimation precision and computational cost [15].

Since bundle sizes vary ($k \in [2, 10]$), corruption intensity is normalized using the relative ratio $\rho = m/k \in [0, 1]$, enabling aggregation across bundles of different sizes. For each corruption level, the expected ILR is estimated by averaging across repetitions and bundles, yielding the degradation curve $\mathbb{E}[ILR \mid \rho]$.

Experimental Protocol for Usefulness. The objective of this protocol is to evaluate whether Usefulness captures relational alignment between an input set and an output set. If the metric is well-calibrated, its value should exhibit an overall decreasing trend as the relational consistency between input and output items is progressively disrupted.

Each bundle is partitioned into an input set IP and an output set O using a fixed 60/40 ratio. The input size is defined as $|IP| = \max(1, \lfloor 0.6k \rfloor)$ and the output size as $|O| = k - |IP|$. The partition is performed deterministically according to the bundle ordering.

To simulate progressive misalignment, the output set is corrupted while keeping the input set fixed. For a given corruption level $m \in \{0, \ldots, |O|\}$, m items are randomly selected from O without replacement and replaced by m items drawn uniformly from the catalog $C \setminus (IP \cup O)$, also without replacement. This yields a corrupted output set $O^{(m)}$ of unchanged cardinality.

Corruption intensity is normalized using $\rho_O = m/|O| \in [0, 1]$. The procedure is repeated $R = 100$ times for each pair (IP, O, m). For each corruption level, the expected Usefulness score is estimated by averaging across repetitions and bundles, producing the degradation curve $\mathbb{E}[Usefulness \mid \rho_O]$.

Degradation Indicators. To assess the robustness and discriminative behavior of the evaluated metrics under progressive corruption, two complementary indicators are computed.

First, the Spearman rank correlation coefficient between the corruption ratio and the metric value evaluates whether the metric exhibits an overall monotonic decreasing trend. A strong negative correlation indicates consistent degradation. Following commonly used interpretation guidelines [12], absolute correlation values below 0.20 are considered very weak, values in [0.20, 0.39] weak, [0.40, 0.59] moderate, [0.60, 0.79] strong, and values above 0.80 very strong. These thresholds provide a standardized basis for assessing the strength of the monotonic association.

Second, the total drop Δ is defined as the difference between the expected metric value at zero corruption and at maximum corruption. This quantity measures the discriminative capacity of the metric between fully coherent and fully corrupted configurations. Since both Usefulness and ILR are bounded in the interval $[0, 1]$, the delta drop is necessarily constrained to the same range. As it captures the absolute amplitude of degradation, its interpretation is inherently scale-dependent: larger values indicate higher sensitivity to corruption intensity, whereas smaller values suggest limited discriminative power.

Results. Figure 1 reports the degradation curve of Usefulness under progressive output corruption. As the normalized corruption ratio ρ_O increases, the mean Usefulness score exhibits an overall decreasing trend. The Spearman correlation between ρ_O and Usefulness equals -0.7857. The total drop between the fully coherent configuration ($\rho_O = 0$) and the fully corrupted one ($\rho_O = 1$) is $\Delta = 0.0011$. Figure 2 illustrates the behavior of ILR under progressive bundle corruption. As ρ increases, the mean ILR score also exhibits an overall decreasing trend. The Spearman correlation between ρ and ILR equals -0.8155. The total drop between $\rho = 0$ and $\rho = 1$ is $\Delta = 0.0030$.

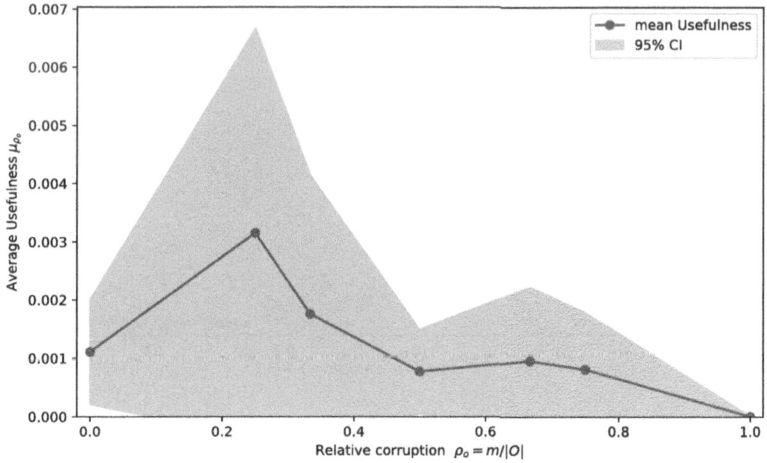

Fig. 1. Mean Usefulness as a function of the normalized output corruption ratio $\rho_O = m/|O|$. Results are averaged across bundles, repetitions, and domains.

The negative Spearman coefficients obtained for both Usefulness and ILR indicate a consistent decreasing tendency as corruption intensity increases. Moreover, the magnitude of the correlations reflects a strong to very strong negative association, according to standard interpretation guidelines. This result demonstrates that the degradation of both metrics is highly monotonic. The absolute magnitude of Δ remains small for both metrics, indicating limited discriminative power under high sparsity conditions.

In fact, a substantial proportion of test bundle items does not appear in the training sessions (53.85% in electronics, 81.43% in clothing, and 55.59% in food), meaning that no learning about potential relationships between those unseen items and others in the dataset is done. As a result, both ILR and Usefulness often evaluate to zero, since complementarity and substitutability are themselves null, thereby weakening the rank correlation and diminishing the overall drop magnitude.

Despite the sparsity-induced attenuation, both metrics demonstrate monotonic decreasing trends under controlled random perturbation. This behavior supports their theoretical validity as relational consistency measures. However, their discriminative capacity fundamentally depends on the coverage and density of item relationships in the dataset. When such relationships are sparse, either due to insufficient item coverage or because items are intrinsically weakly related, the substitutability and complementarity structure becomes underdetermined, which in turn limits the practical relevance of the proposed metrics.

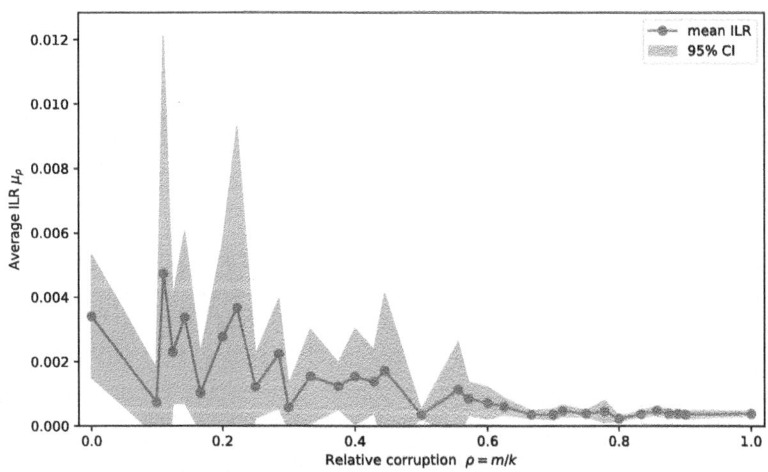

Fig. 2. Mean ILR as a function of the normalized corruption ratio $\rho = m/k$. Results are averaged across bundles, repetitions, and domains.

5 Conclusion

This paper introduced an approach to evaluating RSs, addressing limitations of traditional offline evaluation protocols that primarily rely on accuracy-based metrics and overlook both the usefulness of recommendations from the user's perspective and the coherence among recommended items. We proposed two complementary metrics, Usefulness and Intra-List Relationships (ILR), designed to capture respectively the alignment of recommendations with the user's session context and their internal relational coherence. Both metrics rely on item relationships, modeled through substitutability and complementarity scores derived from user session interactions.

The proposed approach shifts evaluation from isolated item relevance to relational consistency. Usefulness assesses recommendation relevance without requiring feedback by leveraging contextual relationships, while ILR evaluates whether recommendations form a coherent set that can be naturally perceived as a bundle, particularly in multi-item consumption scenarios. These metrics complement conventional evaluation measures by capturing distinct aspects of recommendation quality not reflected by traditional metrics, and can be used independently or jointly to support a multi-dimensional view of RS performance.

Experimental results on the BundleRec dataset show that the proposed scores capture meaningful item relationships with performance comparable to existing approaches. Under controlled corruption protocols, both Usefulness and ILR exhibit consistent monotonic degradation, supported by strong negative Spearman correlations, confirming their validity as indicators of relational consistency. However, the magnitude of score variation remains limited, highlighting a key limitation: the effectiveness of the proposed approach depends on the density and coverage of item relationships. In sparse settings, where many items lack sufficient interaction history, the substitutability and complementarity structure become underdetermined, reducing the discriminative power of the metrics.

Despite this limitation, the proposed approach is particularly relevant in domains where items are intrinsically related and user interactions reflect structured, goal-oriented behavior, such as bundle or session-based recommendation scenarios. In such contexts, relational evaluation provides insights that cannot be captured by accuracy, diversity, or novelty metrics alone. More broadly, this work suggests that incorporating relational structure into evaluation protocols is a necessary step toward more realistic and user-centered assessment of RSs.

Future work will focus on improving the robustness of the proposed scores through richer contextual modeling, such as filtering or weighting mechanisms. Another important direction is to conduct experimental evaluations of RSs using the proposed metrics on additional session-based datasets (e.g., YOOCHOOSE, Diginetica, Tmall). By combining Usefulness and ILR with established evaluation measures from the literature (e.g., accuracy, diversity, novelty, coverage), such experiments would enable a comparative analysis of models, providing insights into the practical impact of the proposed metrics and their ability to discriminate between RSs.

Acknowledgments. This work was funded by the French National Association for Research and Technology (ANRT) under a CIFRE doctoral fellowship agreement (No. 2024/1103).

Disclosure of Interests. The authors have no competing interests to declare that are relevant to the content of this article.

References

1. Adomavicius, G., Tuzhilin, A.: Context-aware recommender systems. In: Recommender Systems Handbook, pp. 217–253. Springer, Boston, MA (2010)
2. Akkoyunlu, S., Manfredotti, C., Cornuéjols, A., Darcel, N., Delaere, F.: Investigating substitutability of food items in consumption data. In: HealthRecSys. vol. 5. Como, Italy (2017)
3. Chen, M., Liu, P.: Performance evaluation of recommender systems. IJPE **13**(8), 1246 (2017)
4. Choi, J., Park, H.Y.: Usage complementarity vs. basket co-occurrence: discount depth reliance in digitally personalized product recommendations. J. Retail. **101**(2), 177–196 (2025)
5. Ge, M., Delgado-Battenfeld, C., Jannach, D.: Beyond accuracy: evaluating recommender systems by coverage and serendipity. In: RecSys 2010, pp. 257–260. ACM, New York, NY, USA (2010), barcelona, Spain
6. He, R., McAuley, J.: Ups and downs: Modeling the visual evolution of fashion trends with one-class collaborative filtering. In: WWW 2016, pp. 507–517. WWW 2016, International World Wide Web Conferences Steering Committee, Republic and Canton of Geneva, CHE (2016)
7. Kaminskas, M., Bridge, D.: Diversity, serendipity, novelty, and coverage: a survey and empirical analysis of beyond-accuracy objectives in recommender systems. ACM TiiS **7**(1), 42 (2016)
8. Kunaver, M., Pozrl, T.: Diversity in recommender systems, a survey. Knowl.-Based Syst. **123**, 154–162 (2017)
9. Landsberger, H.A.: Hawthorne Revisited: Management and the Worker, Its Critics, and Developments in Human Relations in Industry. Cornell University Press, Ithaca, N.Y. (1958)
10. MacKenzie, I., Meyer, C., Noble, S.: How retailers can keep up with consumers. McKinsey & Company **18**(1), 1–10 (2013)
11. McNee, S.M., Riedl, J., Konstan, J.A.: Being accurate is not enough: how accuracy metrics have hurt recommender systems. In: CHI 2006 Extended Abstracts on Human Factors in Computing Systems. pp. 1097–1101. ACM, New York, NY, USA (2006), montréal, Québec, Canada
12. Mukaka, M.M.: A guide to appropriate use of correlation coefficient in medical research. Malawi Med. J. **24**(3), 69–71 (2012)
13. Nicholson, W., Snyder, C.: Demand relationships among goods, pp. 187–205. Joe Sabatino, 11th edn. (2012)
14. Ravana, S.D., Moffat, A.: Score aggregation techniques in retrieval experimentation. In: CRPIT Series, vol. 92, pp. 59–67 (2009)
15. Robert, C.P., Casella, G.: Monte Carlo Statistical Methods. Springer, New York, NY (2004)
16. Saltelli, A., et al.: Experimental Designs, chap. 2, pp. 53–107. Wiley (2007)
17. Sun, Z., et al.: Revisiting bundle recommendation for intent-aware product bundling. ACM TORS **2**(3), 34 (2024)
18. Sun, Z., Yang, J., Zhang, J., Bozzon, A.: Exploiting both vertical and horizontal dimensions of feature hierarchy for effective recommendation. In: Proceedings of the AAAI Conference on Artificial Intelligence, vol. 31, no. 1 (2017)

19. Tian, Y., Lautz, S., Wallis, A.O., Lambiotte, R.: Extracting complements and substitutes from sales data: a network perspective. EPJ Data Sci. **10**(45), (2021)
20. Wang, S., Cao, L., Wang, Y., Sheng, Q.Z., Orgun, M.A., Lian, D.: A survey on session-based recommender systems. ACM CSUR **54**(7), 38 (2021)
21. Wang, S., Hu, L., Wang, Y., Sheng, Q.Z., Orgun, M., Cao, L.: Modeling multi-purpose sessions for next-item recommendations via mixture-channel purpose routing networks. In: IJCAI-19. pp. 3771–3777. International Joint Conferences on Artificial Intelligence Organization (7 2019)
22. Zangerle, E., Bauer, C.: Evaluating recommender systems: survey and framework. ACM CSUR **55**(8), 38 (2022)

Deep Features, Classical Models: Binary vs. Multiclass Deepfake Detection

Abbas Aliyev[1(✉)] and Samir Rustamov[1,2]

[1] ADA University, Baku, Azerbaijan
{ashaliyev,srustamov}@ada.edu.az
[2] MegaSec AI Company, Baku, Azerbaijan

Abstract. The proliferation of highly realistic deepfakes poses a significant threat to digital media integrity. While deep learning models have achieved remarkable success in deepfake detection, they typically rely on computationally expensive end-to-end fine-tuning and formulate the task strictly as a binary classification problem (real vs. fake). In this paper, we present a systematic comparative study of frame-level deepfake detection and method attribution using frozen, pre-trained Convolutional Neural Network (CNN) embeddings paired with lightweight classical classifiers. We exhaustively benchmark 12 embedding configurations—derived from MobileNetV2, EfficientNetB0, and ResNet50 using various pooling and fusion strategies—coupled with Support Vector Machines (SVM), XGBoost, Multilayer Perceptrons (MLP), and an AUC-weighted ensemble. Evaluated on the FaceForensics++ dataset, our approach achieves highly competitive results without requiring backbone fine-tuning. For binary detection, the optimal configuration achieves an accuracy of 92.53% and an AUC of 0.9547. When extending the pipeline to a challenging 6-class method attribution task, the model maintains a strong accuracy of 89.36% and an AUC-OVR of 0.9795. Our contributions are threefold: (1) we present the first matched binary vs. 6-class deepfake comparison on identical frozen-feature pipelines on FF++-c23 including FaceShifter; (2) we conduct a systematic 12-configuration backbone × pooling × fusion ablation—the first such ablation for frozen features in this task; and (3) we quantify the accuracy–training-time Pareto front for forensic deployment with statistical significance testing and cross-dataset validation on Celeb-DF-v2.

Keywords: Deepfake Detection · Transfer Learning · Feature Extraction · Multiclass Classification · Ensemble Learning

1 Introduction

There is a notable gap in the literature regarding the systematic evaluation of lightweight, frozen-feature approaches at the frame level, and an even greater scarcity of studies directly comparing binary detection against multiclass attribution using the exact same feature pipeline. Addressing this gap, we make three contributions:

C. Zanni-Merk et al. (Eds.): AIDT 2026, CCIS 3023, pp. 236–250, 2026.
https://doi.org/10.1007/978-3-032-31319-5_16

1. **Matched binary vs. multiclass comparison.** We present the first direct comparison of binary deepfake detection and 6-class method attribution on identical frozen-feature pipelines, evaluated on FaceForensics++ (c23) [19] with the addition of FaceShifter [12].
2. **Systematic frozen-feature ablation.** We exhaustively benchmark 12 embedding configurations spanning three CNN backbones (MobileNetV2, EfficientNetB0, ResNet50), two pooling strategies, and three fusion levels—the most comprehensive such ablation for frozen features in deepfake detection to date.
3. **Deployment-oriented efficiency analysis.** We quantify the accuracy–training-time Pareto front across four classifier paradigms, demonstrating that frozen embeddings paired with SVMs achieve 92.53% binary and 89.36% multiclass accuracy without any backbone fine-tuning, validated with bootstrap confidence intervals, McNemar tests, and cross-dataset evaluation on Celeb-DF-v2 [11].

2 Related Work

CNNs form the backbone of visual deepfake detection [7,17]. Multi-branch and attention-guided architectures isolate manipulation artifacts by suppressing high-level face content [14,23,26], while hybrid models combine spatial CNN extractors with temporal modules such as LSTMs or Vision Transformers to capture long-range dependencies [9,18,20,24,25]. Although these end-to-end approaches achieve high accuracy, fine-tuning millions of parameters demands immense computational resources, limiting real-time deployment.

Transfer learning with frozen CNN features offers a lighter alternative. Pretrained MobileNetV2 [5] and EfficientNet [4] strike an effective speed–accuracy balance, and feature stacking further improves performance: Naskar et al. [15] combined Xception and EfficientNet-B7 embeddings, Ismail et al. [6] fed InceptionResNetV2 features into XGBoost, and Kong et al. [13] applied mixture-of-experts adapters on frozen ViTs. Our work extends this paradigm by systematically ablating pooling and fusion strategies across three backbones.

While the vast majority of deepfake studies treat detection as a binary task, determining the specific generative algorithm is crucial for forensic analysis. Multiclass attribution has been explored with patch-wise ViTs [2], CNN ensembles [16], and cross-dataset generalization frameworks [21,22]. However, systematic comparisons of binary detection against multiclass attribution on identical frozen-feature pipelines remain scarce—a gap our work directly addresses.

3 Methodology

To systematically evaluate the efficacy of frozen CNN embeddings for both binary detection and multiclass attribution, we designed a modular, frame-level feature extraction and classification pipeline. The architecture is illustrated in Fig. 1. The pipeline consists of three primary stages: (1) pre-trained feature extraction using distinct CNN backbones and pooling strategies, (2) embedding fusion, and (3) lightweight classification and ensembling.

3.1 Feature Extraction Pipeline

We selected three diverse CNN architectures, pre-trained on ImageNet, to serve as our frozen feature extractors. These models were chosen to represent a spectrum of depth, parameter efficiency, and architectural paradigms:

- **MobileNetV2:** A lightweight architecture utilizing depthwise separable convolutions, highly optimized for mobile and edge deployment.
- **EfficientNetB0:** A network designed using compound scaling that balances depth, width, and resolution, offering robust feature representation with relatively low parameter counts.
- **ResNet50:** A deeper, widely adopted architecture relying on residual connections to capture highly complex, hierarchical spatial patterns.

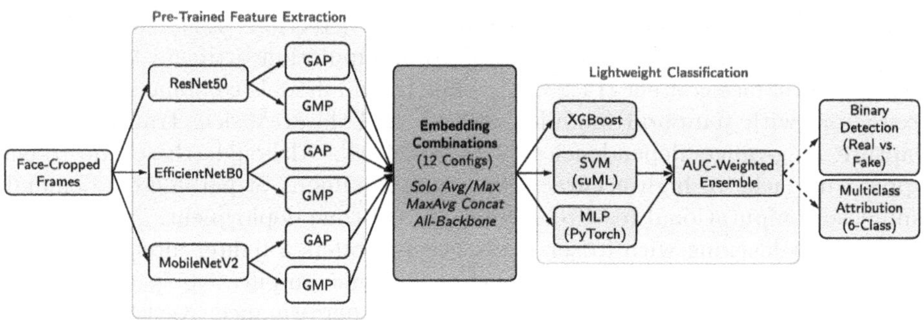

Fig. 1. Overview of the frame-level feature extraction and classification pipeline. Pre-trained CNN backbones extract spatial features, which are pooled, fused into 12 distinct configurations, and evaluated using an ensemble of lightweight classifiers for binary detection and multiclass attribution.

For each input frame, the face is cropped and resized to 224×224 pixels. The frame is passed through the frozen convolutional base of each model, bypassing the original classification heads. To collapse the resulting 2D spatial feature maps into 1D embedding vectors, we evaluate two distinct pooling strategies:

1. **Global Average Pooling (GAP):** Averages the spatial dimensions, capturing the overall, diffuse statistical distribution of the feature map.
2. **Global Max Pooling (GMP):** Extracts the maximum value across the spatial dimensions, highlighting the most salient, localized artifact activations.

The baseline embedding dimensions yielded by these pooling operations are 1280 for MobileNetV2, 1280 for EfficientNetB0, and 2048 for ResNet50.

3.2 Embedding Combinations

To thoroughly ablate the impact of backbone choice and pooling strategy, we construct 12 distinct embedding configurations for evaluation:

- **Solo Max/Solo Avg (6 configurations):** Independent embeddings from each of the three backbones using either GMP or GAP (e.g., *ResNet50-Avg*).
- **Intra-Model MaxAvg Concat (3 configurations):** For a single backbone, the GMP and GAP vectors are concatenated along the feature dimension (e.g., *EfficientNet-MaxAvg*, yielding a 2560-dimensional vector). This aims to capture both diffuse and salient artifacts simultaneously.
- **All-Backbone Fusion (3 configurations):** Embeddings from all three backbones are concatenated. *All-Max* and *All-Avg* yield 4608-dimensional vectors, while *All-MaxAvg* concatenates both pooling types from all three models, resulting in a dense 9216-dimensional representation.

3.3 Classification Models

Rather than fine-tuning the deep backbones, we feed the extracted embeddings into highly optimized, lightweight classifiers. We evaluate three distinct paradigms: XGBoost, Support Vector Machine (SVM, accelerated via RAPIDS cuML), and a PyTorch-based Multi-Layer Perceptron (MLP).[1] Prior to classification, all embeddings undergo standardization (zero mean, unit variance) using parameters fit strictly on the training folds to prevent data leakage.

3.4 Task Formulations

We structure our experiments to evaluate the exact same feature combinations on two distinct forensic tasks:

Binary Classification Setup. The objective is to determine whether a given frame is pristine or manipulated. To understand model sensitivity to specific forgery methods, we divide the data into isolated method folds: *Real vs. Deepfakes*, *Real vs. Face2Face*, *Real vs. FaceSwap*, and *Real vs. NeuralTextures*. Finally, we evaluate the comprehensive *Real vs. ALL* fold, combining all manipulation methods into a single "Fake" class.

[1] All hyperparameters were selected via grid search on the validation set. The resulting configurations are: XGBoost with 200 estimators, max depth 6, and a learning rate of 0.04; SVM with an RBF kernel ($C = 1.0$, gamma='scale'); MLP with two hidden layers (256 and 128 neurons), Batch Normalization, ReLU activations, a 0.3 dropout rate, trained using Adam ($lr = 10^{-3}$) with early stopping. Class weights are applied across all models to handle dataset imbalance.

Multiclass Classification Setup. In this formulation, the model must explicitly attribute the frame to its source. The task is formulated as a 6-class problem: Original (0), Deepfakes (1), Face2Face (2), FaceSwap (3), NeuralTextures (4), and FaceShifter (5). The MLP utilizes a softmax output layer, the XGBoost model utilizes the `multi:softprob` objective, and the SVM employs a One-Vs-Rest (OVR) strategy.

3.5 Ensemble Strategy

To maximize predictive robustness, we propose a soft-voting ensemble mechanism. Rather than a simple average, the ensemble aggregates the predicted probabilities from the SVM, XGBoost, and MLP models, weighted by their respective validation AUC scores. Let $p_{i,c}$ be the predicted probability of class c from model i, and w_i be the validation AUC of model i. The ensembled probability P_c for class c is calculated as:

$$P_c = \frac{\sum_i w_i \cdot p_{i,c}}{\sum_i w_i} \tag{1}$$

For binary classification, the final prediction is positive if $P_1 \geq 0.5$. For the multiclass formulation, the prediction is assigned to $\arg\max_c P_c$.

4 Experimental Setup

To ensure a rigorous and reproducible evaluation of the proposed feature extraction and classification pipeline, we established a standardized experimental protocol covering dataset preparation, model training, and performance metrics.

4.1 Dataset

All experiments use the widely adopted **FaceForensics++** (FF++) dataset [19], a large-scale forensics benchmark of 1,000 pristine YouTube sequences and 4,000 manipulated videos produced by four methods: two identity-swapping techniques (*Deepfakes*, *FaceSwap*) and two expression-reenactment techniques (*Face2Face*, *NeuralTextures*). We additionally incorporate *FaceShifter* [12], a high-fidelity identity-swapping method from a Kaggle-hosted derivative [3], extending multiclass attribution to six categories.

To simulate realistic conditions commonly encountered on social media platforms, we utilized the *c23* (light compression) variant of the dataset, encoded using the H.264 codec. Operating directly on full video frames often introduces background noise and irrelevant spatial artifacts that can severely degrade the learning process [8]. The Kaggle-hosted derivative [3] addresses this by providing pre-extracted, face-cropped frames resized to a standard resolution of 224×224 pixels, matching the expected input dimensions of the pre-trained CNN backbones.

To prevent data leakage and ensure that our models generalize to unseen identities, the dataset was strictly split at the video level prior to frame extraction. The extracted frame-level dataset was partitioned into **70% training**, **15% validation**, and **15% testing** subsets. All splits were stratified based on the target labels to maintain consistent class distributions across the subsets.

4.2 Implementation Details

The entire pipeline was implemented in Python. Feature extraction and the Multi-Layer Perceptron (MLP) were built using the `PyTorch` deep learning framework. To dramatically accelerate the training of traditional machine learning algorithms on high-dimensional embeddings, we utilized GPU-accelerated libraries. Specifically, the Support Vector Machine (SVM) was implemented using the `RAPIDS cuML` library, and the XGBoost classifier was configured to run on the GPU via the `device="cuda"` parameter. All experiments were executed on a high-performance compute node equipped with a dedicated NVIDIA GPU.

To prepare the embeddings for classification, the extracted feature vectors were standardized to have a zero mean and unit variance. The `StandardScaler` was fitted exclusively on the training fold to prevent information leakage, and subsequently used to transform the validation and test folds. To address class imbalance—particularly in the *Real vs. ALL* binary task, where the number of manipulated frames outnumbers pristine frames by a ratio of 4:1—we computed balanced class weights. These weights were applied directly to the loss functions of the MLP and XGBoost models, and used to conduct stratified resampling for the SVM.

Deep learning and tree-based models can be sensitive to weight initialization. To ensure robustness and reproducibility, all classifiers were trained across three distinct random seeds (42, 123, and 7). For each classifier, the model instance that achieved the highest Area Under the Curve (AUC) on the validation set was selected and ultimately evaluated on the hold-out test set.

4.3 Evaluation Metrics

For binary tasks (Real vs. Specific Method, and Real vs. ALL), we report Accuracy, Area Under the ROC Curve (AUC), Macro-F1, and Matthews Correlation Coefficient (MCC). For the 6-class attribution task, we report Accuracy, Macro-F1, MCC, and AUC computed using the One-Vs-Rest (AUC-OVR) strategy, which averages per-class AUC scores to measure discriminative power across all manipulation methods.

Table 1. Top-5 binary classification results on the `real_vs_ALL` fold with 95% bootstrap confidence intervals. Best value per metric in **bold**.

Embedding	Model	Accuracy (%)	AUC (%)
All-MaxAvg	SVM	**92.53** [92.22, 92.84]	95.32 [95.02, 95.62]
All-Avg	SVM	92.41 [92.11, 92.73]	**95.47** [95.18, 95.76]
All-MaxAvg	Ensemble	92.19 [91.88, 92.50]	92.47 [92.01, 92.93]
All-Avg	Ensemble	91.93 [91.62, 92.24]	93.24 [92.81, 93.66]
All-Max	SVM	91.13 [90.80, 91.46]	93.68 [93.31, 94.04]

5 Results and Discussion

We present results for binary detection, multiclass attribution, a direct comparison of the two paradigms, computational efficiency, state-of-the-art positioning, ensemble ablation, and cross-dataset robustness.

5.1 Binary Classification Results

Table 1 reports the top-5 binary classification configurations on the `real_vs_ALL` fold with 95% bootstrap confidence intervals. Fusing features from all three backbones yields strictly superior performance (Fig. 2): the highest accuracy (92.53%) is achieved by *All-MaxAvg* + SVM, while the highest AUC (0.9547) belongs to *All-Avg* + SVM. The AUC-weighted Ensemble performs closely to SVM (AUC 0.9324 on *All-Avg*), but the SVM's RBF kernel proves most effective in this frozen feature space. Global Average Pooling generally outperforms Global Max Pooling, as deepfake artifacts are often diffuse rather than localized; concatenating both (*MaxAvg*) provides marginal accuracy gains (+0.12%) at the cost of doubled dimensionality.

All reported differences are statistically significant. Bootstrap 95% confidence intervals on the top-5 configurations are narrow (±0.3 pp for both accuracy and AUC), reflecting the large test set (28,733 samples). A McNemar paired test confirms that SVM significantly outperforms both MLP ($p < 10^{-300}$; 8,033 vs. 868 discordant pairs) and the Ensemble ($p = 1.68 \times 10^{-9}$) on binary hard accuracy. Results are also robust to random seed choice: SVM and Ensemble metrics vary by fewer than 0.2 pp across three seeds, while MLP shows the largest sensitivity (up to 3 pp standard deviation), confirming that performance differences reflect genuine capability gaps.

Per-method Detection Difficulty. To understand the nuances of the binary detection task, we evaluated the models on specific forgery methods. Table 2 highlights a clear hierarchy of difficulty. Methods that execute full identity swaps, such as *Deepfakes* (AUC: 0.9956) and *FaceShifter* (AUC: 0.9924), introduce significant structural and blending artifacts that the frozen CNNs readily capture. Conversely, expression-reenactment techniques like *NeuralTextures* (AUC:

0.9450, F1: 0.8702) and *Face2Face* (AUC: 0.9763) prove much harder to detect. These techniques modify localized regions (e.g., the mouth) while leaving the overall identity untouched, resulting in spatial feature distributions that remain much closer to pristine frames in the pre-trained embedding space.

Table 2. Per-method binary detection difficulty (best config per method, All-Avg/SVM unless noted). Ranked by AUC.

Forgery Method	AUC	Acc	F1	MCC
Deepfakes	0.9956	0.9706	0.9706	0.9415
FaceShifter	0.9924	0.9648	0.9648	0.9298
FaceSwap	0.9888	0.9476	0.9476	0.8958
Face2Face†	0.9763	0.9190	0.9190	0.8399
NeuralTextures	0.9450	0.8703	0.8702	0.7418
ALL (combined)	0.9547	0.9241	0.8599	0.7205

†All-MaxAvg + SVM.

Table 3. Top-5 multiclass (6-class) results on the `real_vs_ALL` fold with 95% bootstrap confidence intervals. Best value per metric in **bold**.

Embedding	Model	Accuracy (%)	AUC-OVR (%)
All-Avg	SVM	**89.36** [89.01, 89.72]	97.09 [96.96, 97.22]
All-MaxAvg	SVM	89.17 [88.81, 89.52]	97.03 [96.89, 97.16]
All-Avg	Ensemble	89.01 [88.64, 89.37]	**97.95** [97.85, 98.05]
All-MaxAvg	Ensemble	88.83 [88.45, 89.19]	97.85 [97.74, 97.95]
All-Max	SVM	86.82 [86.43, 87.22]	96.19 [96.04, 96.34]

5.2 Multiclass Attribution Results

Transitioning from binary detection to a 6-class method attribution task fundamentally changes the learning dynamics. As shown in Table 3, the *All-Avg* embedding remains optimal, achieving an accuracy of 89.36% with the SVM and the highest AUC-OVR (0.9795) via the Ensemble.

The normalized confusion matrix for the best multiclass model (Fig. 4) reveals that *Deepfakes* (95.7%) and *FaceShifter* (95.9%) are accurately attributed, while the *Original* class exhibits the lowest accuracy (76.3%).

In a multiclass formulation, pristine frames lack the distinct generative "fingerprints" present in the four fake classes. Consequently, "Original" becomes a

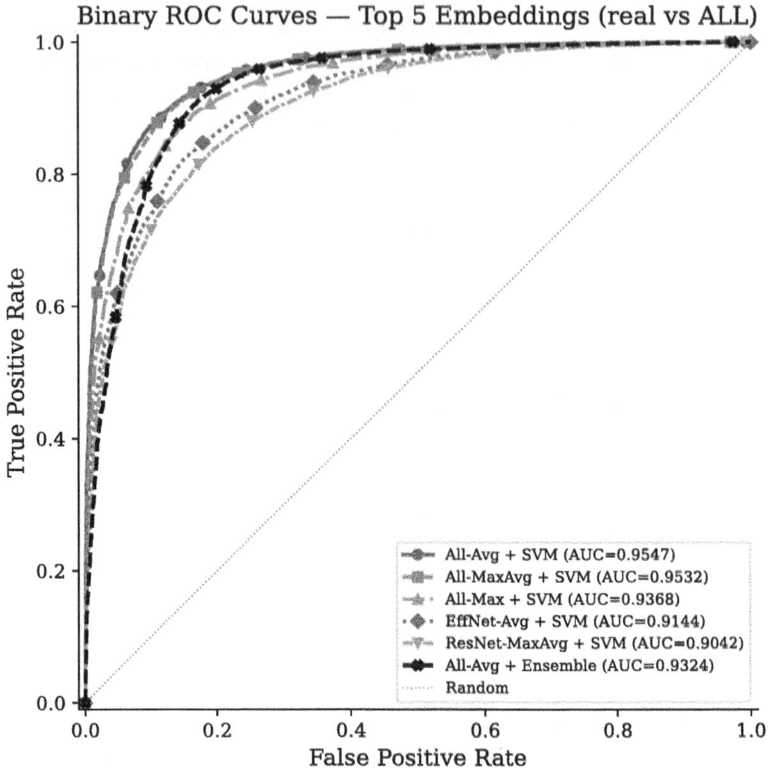

Fig. 2. ROC curves for the top-5 binary classification configurations on the `real_vs_ALL` fold. The separation between fused embeddings and single-backbone embeddings demonstrates the clear benefit of combining diverse spatial features.

default category for subtle, hard-to-detect forgeries. Specifically, 10.3% of *Original* frames are misclassified as *NeuralTextures*, and 8.3% of *NeuralTextures* are misclassified as *Original*. This bidirectional confusion reinforces our finding that texture-based reenactment leaves the faintest spatial footprints. Beyond this primary axis, Original frames are misclassified as Face2Face at a 1.73× higher rate than the reverse (337 vs. 195 errors), and the strongest asymmetry occurs for FaceShifter, where real→FaceShifter errors outnumber the reverse by 3.12× (156 vs. 50). This suggests that FaceShifter produces distinctive artifacts, yet certain pristine facial features (e.g., specular highlights, skin smoothness) trigger the same CNN activations.

Figure 3 presents the highest-confidence misclassified frames along the dominant Original ↔ NeuralTextures confusion axis. The top row shows pristine frames whose soft-focus textures or compression artifacts mimic neural rendering; the bottom row shows NeuralTextures manipulations that are nearly imperceptible, with artifacts confined to sub-pixel intensity shifts around the mouth region.

Fig. 3. Highest-confidence misclassifications along the Original ↔ NeuralTextures axis. Top: real frames predicted as NeuralTextures. Bottom: NeuralTextures predicted as Original.

5.3 Binary vs. Multiclass Comparison

Table 4 quantifies the cost of method attribution. On average, transitioning from binary to 6-class attribution causes a 5.62 pp accuracy drop, while AUC and F1 *increase*—a metric-aggregation effect: the five highly separable manipulation classes elevate the macro-averaged AUC-OVR, compensating for the weaker Original class. For the optimal *All-Avg* configuration the accuracy drop is only 3.05 pp (92.41% → 89.36%), an acceptable cost for the added attribution capability.

5.4 Computational Efficiency and Embedding Analysis

Table 6 maps the computational efficiency of our classifiers. The *All-MaxAvg* embedding (9,216 dimensions) pushes SVM multiclass training to nearly 14 min (829.1 s) due to the $\mathcal{O}(d{\cdot}n^2)$ complexity of kernel methods. XGBoost, by contrast, maintains low training times (14.0 s for binary *All-Avg*; 58.7 s for multiclass) regardless of embedding width. While SVM provides a +3–10% accuracy edge, XGBoost offers a scalable near-real-time alternative, and the marginal accuracy gains of *MaxAvg* concatenation do not justify the 2× increase in dimensionality.

5.5 Comparison with State-of-the-Art

As shown in Table 5, fine-tuned methods achieve marginal accuracy leads, but our frozen-feature approach remains competitive (92.41% binary, 89.36% multiclass) and surpasses similar extraction pipelines such as Ismail et al. [6] by +1.68% accuracy and +4.85% AUC. The true advantage lies in operational efficiency: frozen ImageNet representations bypass the gradient computations and

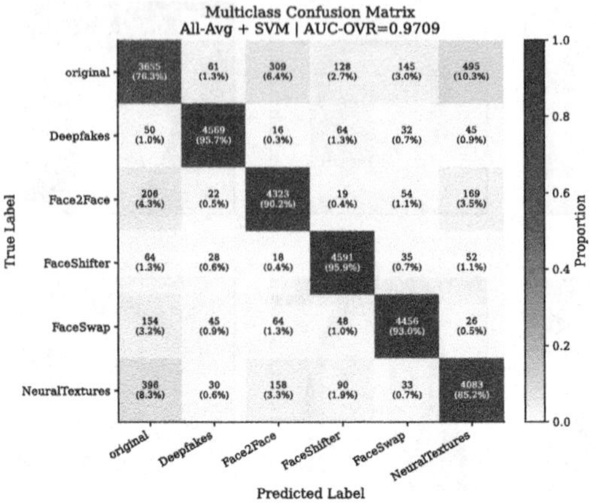

Fig. 4. Normalized confusion matrix for the best multiclass model (All-Avg + SVM). Pristine frames act as an attractor for misclassifications, particularly from NeuralTextures and Face2Face.

Table 4. Binary vs. multiclass comparison (best model per embedding). Δ: pp change.

Embedding	ΔAcc	ΔAUC	ΔF1
MNet-Avg	−7.8	+4.0	+1.7
ENet-Avg	−7.9	+2.5	+0.3
RN50-MxAv	−7.3	+3.9	+2.5
All-Max	−4.3	+2.5	+4.2
All-Avg	−3.1	+1.6	+3.3
All-MxAv	−3.4	+1.7	+3.4
Average	−5.6	+2.7	+2.6

Table 5. Comparison with state of the art on FF++ (c23). Our method uses frozen CNNs.

Method	FT?	Acc	AUC	Task
Rössler [19]	Yes	.957	–	Bin
Ismail [6]	No	.907	.906	Bin
Naskar [15]	No	.980	.999	Bin
Ours (SVM)	No	.924	.955	Bin
Arshed [2]	Yes	.999	–	MC
Omar [16]	Yes	.917	–	MC
Ours (SVM)	No	.894	.971	MC
Ours (Ens)	No	.890	**.980**	MC

prolonged training required to fine-tune architectures like Xception or Vision Transformers.

5.6 Ensemble Ablation

To justify our AUC-weighted soft-voting strategy, Table 7 compares five aggregation rules applied to the same base classifiers (SVM, XGBoost, MLP) on the *All-Avg* embedding. The best single classifier (SVM) achieves the highest accuracy on both tasks, while the ensemble improves multiclass AUC from 0.9709 to 0.9795 through better probability calibration. Hard majority voting is the

Table 6. Training time (seconds) for the top embedding configurations.

Embedding	Binary			Multiclass		
	SVM	XGB	MLP	SVM	XGB	MLP
ResNet50-MaxAvg	316.7	12.2	302.9	408.4	51.9	521.1
All-Max	359.1	13.3	482.0	461.6	60.1	443.9
All-Avg	338.1	14.0	591.3	434.8	58.7	449.3
All-MaxAvg	710.1	30.2	405.1	829.1	116.7	399.1

weakest strategy (85.01% binary) because XGBoost and MLP frequently outvote SVM with incorrect predictions. Following Kuncheva's framework for classifier combination [10], AUC-weighted averaging acts as maximum-likelihood pooling under the assumption that per-classifier AUC proxies discriminative calibration—it preserves ranking quality even when individual classifiers have heterogeneous accuracy.

5.7 Cross-Dataset Robustness

To evaluate generalization beyond FF++, we applied our best binary models, trained on FF++ c23 only, to Celeb-DF-v2 [11] without retraining (Table 8). Celeb-DF-v2 contains 104,440 face crops (14,216 real, 90,224 fake; 6.3:1 class imbalance) with identical preprocessing (backbone-specific ImageNet normalization on 224×224 crops), so the evaluation isolates content and style shift with no geometry confound. The best AUC (0.6518, All-Avg + Ensemble) represents a 33 pp drop from the in-domain result (0.9547), confirming that frozen-feature pipelines trained on a single dataset do not generalize out of the box—a known limitation of approaches relying on dataset-specific cues. The ensemble's AUC advantage over SVM widens under shift (0.6518 vs. 0.6204), suggesting that soft averaging is most beneficial when the decision boundary is uncertain. Accuracy (0.77–0.79) is inflated by class imbalance; low MCC values (0.12–0.17) confirm near-chance balanced performance.

6 Ethical Considerations and Limitations

Dataset Bias. FF++ is sourced from YouTube actors whose demographic distribution is undocumented; frozen ImageNet features may inherit pre-training biases. Fairness audits across protected groups are essential before deployment.

Adversarial Robustness. Frozen backbones cannot adapt to adversarial perturbations targeting the ImageNet feature manifold. We recommend pairing frozen-feature detectors with adversarial-augmented retraining or input-preprocessing defenses.

Table 7. Ensemble ablation on the *All-Avg* embedding (`real_vs_ALL` fold). Accuracy (%) for binary and 6-class tasks. The best single classifier (SVM) leads on accuracy; soft-voting ensembles improve multiclass AUC (0.9795 vs. 0.9709) through better probability calibration.

Strategy	Binary Acc (%)	Multiclass Acc (%)
Best single (SVM)	92.41	89.36
Hard majority vote	85.01	86.10
Uniform soft avg	91.66	88.98
Val-acc-weighted soft	91.93	89.01
AUC-weighted soft (ours)	**91.93**	89.01

Table 8. Cross-dataset evaluation on Celeb-DF-v2. Models trained on FF++-c23 (binary), evaluated without retraining. Class imbalance: 6.3:1 fake : real.

Configuration	Accuracy	AUC	Macro-F1	MCC
All-Avg + SVM	0.7727	0.6204	0.5576	0.1176
All-Avg + Ensemble	0.7754	0.6518	0.5640	0.1307
All-MaxAvg + SVM	0.7699	0.6257	0.5621	0.1281

Deployment Guardrails. The system is a forensic screening aid, not a sole decision-maker. Human-in-the-loop review, calibrated confidence thresholds, and audit logging should accompany deployment; the tool should not serve as criminal evidence without corroborating signals.

Scope and Temporal Extension. Frame-level analysis ignores temporal artifacts that can aid detection. Our companion work [1] extends this pipeline to video-level aggregation with temporal-difference features, confirming that inter-frame dynamics provide complementary discriminative signal. Both FF++ and Celeb-DF-v2 are publicly available under academic licenses; all processing was performed locally.

7 Conclusion

This work addressed three objectives. First, we provided the first matched binary vs. 6-class deepfake comparison on identical frozen-feature pipelines: multi-backbone fusion (*All-Avg* + SVM) achieved 92.53% binary accuracy [92.11, 92.73] and 89.36% multiclass accuracy [89.01, 89.72] on FF++-c23, with all differences confirmed significant via bootstrap CIs and McNemar tests. Second, our 12-configuration ablation revealed that fusing all three backbones with average pooling consistently dominates, while MaxAvg concatenation doubles dimensionality for marginal (<0.2 pp) accuracy gain. Third, the accuracy–training-time Pareto front shows SVMs maximize forensic accuracy, while XGBoost offers 20× faster training at competitive performance—and cross-dataset evaluation on Celeb-DF-v2 exposes a 33 pp AUC drop, underscoring the need for domain adaptation.

Future work will pursue domain-adaptation techniques to close the cross-dataset gap, incorporate temporal features following our companion video-level study [1], and extend toward a holistic cross-modal forensic system capable of tracing both visual and acoustic generative fingerprints.

Acknowledgements. This work was developed within the Center of Data Analytics Research (CeDAR) and the MegaSec AI Laboratory at ADA University.

References

1. Aliyev, A., Rustamov, S.: An ensemble approach to deepfake video detection using pre-trained CNNs and statistical feature aggregation. In: ITTA 2026, Communications in Computer and Information Science. Springer, to appear (2026)
2. Arshed, M.A., et al.: Multiclass AI-generated deepfake face detection using patch-wise deep learning model. Computers **13**(1), 31 (2024)
3. Adham 7elmy: FaceForensics++ extracted frames [Data set]. Kaggle (2025). https://doi.org/10.34740/KAGGLE/DSV/12366330
4. Deng, L., Suo, H., Li, D.: Deepfake video detection based on EfficientNet-V2 network. Comput. Intell. Neurosci. **2022**, 3441549 (2022)
5. Elhassan, A., et al.: DFT-MF: enhanced deepfake detection using mouth movement and transfer learning. SoftwareX **19**, 101115 (2022)
6. Ismail, A., Elpeltagy, M., Zaki, M.S., Eldahshan, K.: A new deep learning-based methodology for video deepfake detection using XGBoost. Sensors **21**(16), 5413 (2021)
7. Jheelan, J., Pudaruth, S.: Using deep learning to identify deepfakes created using generative adversarial networks. Computers **14**(2), 60 (2025)
8. Kaddar, B., et al.: Deepfake detection using spatiotemporal transformer. ACM Trans. Multimedia Comput. Commun. Appl. **20**(11), 345 (2024)
9. Khan, S.A., Dang-Nguyen, D.T.: Hybrid transformer network for deepfake detection. In: Proceedings of the CBMI, pp. 1–7 (2022)
10. Kuncheva, L.I.: Combining Pattern Classifiers: Methods and Algorithms, 2nd edn. Wiley, Hoboken (2014)
11. Li, Y., Yang, X., Sun, P., Qi, H., Lyu, S.: Celeb-DF: a large-scale challenging dataset for deepfake forensics, pp. 3207–3216. CVPR (2020)
12. Li, L.: FaceShifter: towards high fidelity and occlusion aware face swapping, pp. 12–21. CVPR (2020)
13. Kong, C., et al.: MoE-FFD: mixture of experts for generalized and parameter-efficient face forgery detection. IEEE Trans. Dependable Secure Comput. **23**(1), 82–96 (2026)
14. Masi, I., et al.: Two-branch recurrent network for isolating deepfakes in videos. In: ECCV, pp. 667–684. Springer (2020)
15. Naskar, G., et al.: Deepfake detection using deep feature stacking and meta-learning. Heliyon **10**(4), e25933 (2024)
16. Omar, K., Sakr, R.H., Alrahmawy, M.F.: An ensemble of CNNs with self-attention for DeepFake video detection. Neural Comput. Appl. **36**, 2749–2765 (2024)

17. Pan, D.: Deepfake detection through deep learning. In: IEEE/ACM BDCAT, pp. 134–143 (2020)
18. Petmezas, G., et al.: Video deepfake detection using a hybrid CNN-LSTM-transformer model. Multimed. Tools Appl. **84**, 40617–40636 (2025)
19. Rössler, A., et al.: FaceForensics++: learning to detect manipulated facial images. In: ICCV, pp. 1–11 (2019)
20. Saikia, P.: A hybrid CNN-LSTM model for video deepfake detection by leveraging optical flow features. In: IJCNN, pp. 1–8. IEEE (2022)
21. Tolosana, R., et al.: DeepFakes detection across generations: analysis of facial regions, fusion, and performance evaluation. Eng. Appl. Artif. Intell. **110**, 104673 (2022)
22. Tran, V.N., et al.: MEViT: generalization of deepfake detection with meta-learning EfficientNet vision transformer. IEEE Open J. Comput. Soc. **6**, 789–800 (2025)
23. Yadav, A., Vishwakarma, D.K.: AW-MSA: adaptively weighted multi-scale attentional features for DeepFake detection. Eng. Appl. Artif. Intell. **127**, 107443 (2024)
24. Yadav, S., Mangalampalli, S.S.: Deepfake defense: Combining spatial and temporal cues with CNN-BiLSTM-transformer architecture. PLoS ONE **20**(11), e0334980 (2025)
25. Zafar, F., et al.: A hybrid deep learning framework for deepfake detection using temporal and spatial features. IEEE Access **13**, 79560–79570 (2025)
26. Zhao, H.: Multi-attentional deepfake detection. In: CVPR, pp. 2185–2194 (2021)

Can We Trust LLMs for Mental Health-Based Decisions? A Causality Aware Reliability Analysis

Yusif Ibrahimov[1(✉)], Turan Mutallimov[2], Seymour Mirzabayov[1], and Elgun Hasanov[3]

[1] French-Azerbaijani University Under Azerbaijan State Oil and Industry University, Baku, Azerbaijan
{yusif.ibrahimov,seymour.mirzabayov}@ufaz.az
[2] University of Aberdeen, Aberdeen, UK
t.mutallimov.22@abdn.ac.uk
[3] Institut Polytechnique De Paris, Paris, France
elgun.hasanov@polytechnique.edu

Abstract. Mental health concerns constitute a global burden as they affect individuals' functional, behavioral, and cognitive abilities. Unfortunately, people who suffer from mental health issues do not get enough professional help, which necessitates the use of automatic tools for early risk identification. Large Language Models (LLMs), as emerging models, have a broad usage including mental health screening. However, there is not enough research about the trustworthiness and responsibility of these systems. In this research, we investigate whether, although LLMs have high predictive performance, these models follow clinical reasoning or if there are mainly spurious correlations. We analyse Qwen2.5-7B on depression, stress, and suicidal ideation detection to investigate if its internal attention mechanism reflects clinical reasoning. We further identify and correct a theoretical flaw in attention based causal discovery for mental health by introducing contrastive TF-IDF and directed Pointwise Mutual Information (PMI) and validated by three independent LLM judges across all three disorders.

Keywords: Large language models · Mental health · Reliability · Causal discovery · Attention analysis

1 Introduction

Mental health challenges such as depression, stress, and suicidal ideation affect millions of people and cause a global burden. However, only a few portion of the affected people seek for the professional help, especially on the early stages, which worsens the situation [5]. While LLMs have shown promising prediction performance [11,12], there is strong need of the ethical analysis of the decision mechanisms of these models, which is usually performed via explainability and causality [1,3].

C. Zanni-Merk et al. (Eds.): AIDT 2026, CCIS 3023, pp. 251–263, 2026.
https://doi.org/10.1007/978-3-032-31319-5_17

Attention values are widely argued as a potential explainability method [6, 10], and recent work has shown that self-attention can be employed to reveal the structural causal path, enabling zero-shot causal discovery from pre-trained transformers [8]. However, this approach assumes that token representations are mutually independent by avoiding the role of the external factors behind the token generation, such as latent mental disorder (which could be totally different than the annotated label).

We address this gap through an attention value analysis of Qwen2.5-7B models across depression, stress, and suicidal ideation. We empirically demonstrate that attention scores carry correlational information regarding the annotated labels, however, there is not enough evidence for the causality, due to the underestimation of the causal factors behind the tokens of the user posts. To correct the theoretical limitation above, we propose Domain-Infused Attention Correction, injecting disorder-relevance information by adopting matrix Φ^{sym} built from contrastive TF-IDF and directed PMI. Evaluation strongly confirms statistically consistent improvement over the baselines. In this study, we aim to validate causal-graph plausibility and explainability, rather than precise clinical causality.

The remainder of the paper is organised as follows. Section 2 presents the causal model and the proposed correction. Section 3 describes the experimental settings, datasets and results. Finally, Sect. 4 concludes the paper.

2 Methodology

2.1 Causal Model for Mental Health Screening with Proxy Signals

The main objective of this study is try to imitate the therapy processes conducting screening by inferring the presence of the latent conditions from linguistic proxy signals. However, in practice, the true disorder is generally unobservable. Therefore, clinicians and automated systems use observable symptoms and proxy signals do perform the screening. Therefore, we formulate this task using Structural Causal Model (SCM) as illustrated in Fig. 1.

Exogenous Variables

$$\mathbf{U} = \underbrace{\{E\}}_{\text{environment}} \cup \underbrace{\{S_1, \ldots, S_m\}}_{\text{style / tone / word frequency / others}}$$

Endogenous Variables

$$\mathbf{X} = \underbrace{\{T_1, \ldots, T_k\}}_{\text{traumatic events}} \cup \underbrace{\{D\}}_{\text{latent disorder}} \cup \underbrace{\{W_1, \ldots, W_m\}}_{\text{linguistic proxy variables}} \cup \underbrace{\{Y\}}_{\text{model prediction}}$$

Traumatic events arise from exogenous systems, such as environmental conditions and previous life events.

$$\mathrm{Pa}(T_j) \subseteq \{E\} \cup \{T_1, \ldots, T_{j-1}\}$$

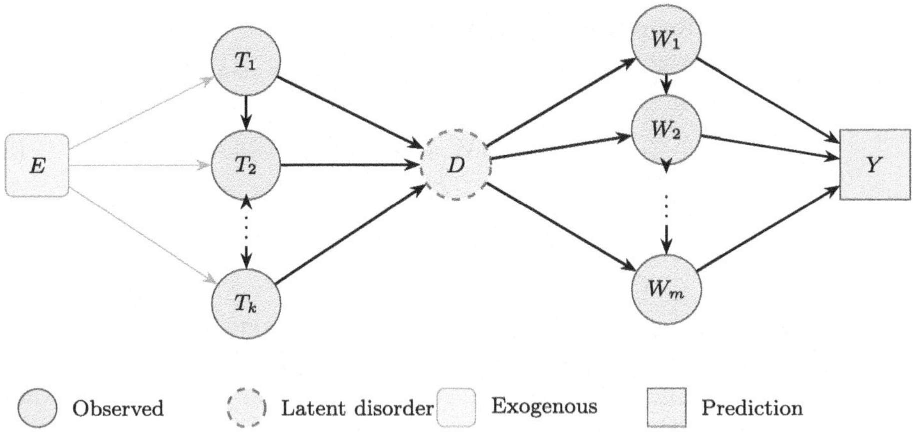

Fig. 1. Causal DAG for the mental health screening process.

$$T_j \leftarrow f_{T_j}(\mathrm{Pa}(T_j)), \qquad j = 1, \ldots, k \tag{1}$$

The true mental health condition is directly influenced by environmental factors and traumatic events as well, which has a latent nature:

$$\mathrm{Pa}(D) \subseteq \{E\} \cup \{T_1, \ldots, T_k\}$$

$$D \leftarrow f_D(\mathrm{Pa}(D)) \tag{2}$$

Although the language of the patient carry the valuable information regarding the mental state, they are still proxy variables, and they are influenced by other factors such as tone, writing style, word frequency and even the previously used words, therefore we adopt autoregression for modelling the language.

$$\mathrm{Pa}(W_i) \subseteq \{W_1, \ldots, W_{i-1}\} \cup \{D\} \cup \{S_i\}$$

$$W_i \leftarrow f_{W_i}(\mathrm{Pa}(W_i)), \qquad i = 1, \ldots, m \tag{3}$$

Finally, the screening process produce prediction by employing the language based proxy variables:

$$\mathrm{Pa}(Y) = \{W_1, \ldots, W_m\}$$

$$Y \leftarrow f_Y(W_1, \ldots, W_m) \tag{4}$$

$$Y \approx P(D \mid W_1, \ldots, W_m)$$

2.2 Disorder-Informed Causal Discovery

Inspired by Rohekar [8], we model the screening process using Gaussian Markov DAGs, in the form of:

$$\mathbf{W} = \mathbf{G}\mathbf{W} + \Lambda D \tag{5}$$

$$\mathbf{W} = (\mathbf{I} - \mathbf{G})^{-1}\Lambda D \tag{6}$$

The theoretical causal covariance $\mathbf{C}_W$ of the observed words is:

$$\mathbf{C}_W = (\mathbf{I} - \mathbf{G})^{-1}\mathbf{C}_D(\mathbf{I} - \mathbf{G})^{-1})^\top \tag{7}$$

The Attention based representations generated by Transformer model is $Z = \mathbf{AV}$, which represents that projections among the input features as $\mathbf{G}$, however, original CLEANN method [8] relies on the assumption that the value projections $\mathbf{V}$ are independent and identically distributed (i.i.d.). This is fundamentally flawed in mental healthcare, as the words tokens are not independently distributed ans effected by the latent disorder, which furthermore affects their embedding representations. Therefore, we can not assure $\mathbf{C}_V \approx I$, and therefore, $\mathbf{C}_Z \neq \mathbf{AA}^\top$, therefore, attention mechanism is insufficient alone for causality. Based on that, we propose the Domain-Infused Attention Injection which restores the missing clinical covariance, which can be defined as $V \propto \gamma D + \epsilon$,where γ defines the psycholinguistic load. As a final result,

$$\mathbf{C}_Z \propto \mathbf{AA}^\top + \rho(\mathbf{A}\gamma)(\mathbf{A}\gamma)^\top = \mathbf{AA}^\top + \rho\mathbf{A}\gamma\gamma^\top\mathbf{A}^\top = AA^\top + \rho A\Phi^{\text{sym}}A^\top. \tag{8}$$

where the ρ was empirically set to 2.0 and Φ^{sym} represent symmetrized disorder-relevance matrix, which is further calculated by:

$$\boldsymbol{\Phi}^{\text{sym}} = \frac{1}{2}(\Phi + \Phi^\top) \tag{9}$$

where $\Phi_{ij} = \gamma_i\gamma_j\text{PMI}(i \to j)$. γ_i defines the characteristic of token W_i in a document d can be calculated via contrastive TF-IDF method:

$$\gamma_i = \underbrace{\frac{\text{count}(W_i, d)}{|d|}}_{\text{TF}} \times \underbrace{\left[\log\frac{|\mathcal{D}^+| + 1}{\text{df}^+(W_i) + 1} - \log\frac{|\mathcal{D}^-| + 1}{\text{df}^-(W_i) + 1}\right]}_{\text{Contrastive IDF}} \tag{10}$$

Moreover, to ensure the clinical flow of the tokens, we calculate the Directed Pointwise Mutual Information (PMI) within $\mathcal{D}^+$ which calculates the likelihood that the term i precedes term j.

$$\text{PMI}(i \to j) = \log\frac{P(W_i \text{ precedes } W_j \mid \mathcal{D}^+)}{P(W_i \in \mathcal{D}^+)\cdot P(W_j \in \mathcal{D}^+)} \tag{11}$$

where we computed the token co-occurrences at the document level, pairing every token with all subsequent tokens within the same post. Tokenization was performed via standard whitespace and lowercase splitting. To mitigate instability due to token sparsity, a small epsilon smoothing factor (10^{-9}) was added.

Overall, this injection conducts the de-confounding the discovery process, which is helpful to isolate true causal dependencies $\mathbf{G}$.

2.3 Causal Graph Generation

Algorithm 1 represents the causal graph extraction process which originally extends CLEANN [8] with the Φ-correction. Based on the last attention layer of the pre-trained language model, it builds two covariance matrices including original ($\tilde{A}\tilde{A}^{\top}$) and corrected ($AA^{\top} + \rho\, A\Phi^{\text{sym}}A^{\top}$), are further fed into CausalDiscovery, which runs the PC algorithm to recover a causal graph (PAG) over the input tokens via partial correlation CI-tests. FindExplanation then searches this graph for the minimal potentially influencing (PI) set, indicating the smallest token subset whose removal changes the model prediction.

Algorithm 1. Φ-Corrected CLEANN: CausaL Explanations from Attention in Neural Networks

---

Require: Post $s = [s_1, \ldots, s_n, \langle\text{mask}\rangle]$, model $\mathcal{M}$ (Qwen2.5-7B), corpora $(\mathcal{D}^+, \mathcal{D}^-)$, ρ, α
Ensure: Explanation set E, alternative prediction $\tilde{s}'_{n+1}$, causal graphs $\mathcal{G}_{\text{orig}}$, $\mathcal{G}_{\text{corr}}$
function Main(s, $\mathcal{M}$, $\tilde{s}_{n+1}$)
    $\tilde{s} \leftarrow [s_1, \ldots, s_n, \tilde{s}_{n+1}]$ ▷ replace mask with prediction
    Forward pass: $\mathcal{M}(\tilde{s})$; get A, the attention matrix in the last layer
    Compute γ_i for all $s_i \in s$ via contrastive TF-IDF
    Compute $\Phi_{ij} = \gamma_i \gamma_j \,\text{PMI}(i \to j)$; $\Phi^{\text{sym}} = \frac{1}{2}(\Phi + \Phi^{\top})$
    $C_{\text{orig}} \leftarrow AA^{\top}$
    $C_{\text{corr}} \leftarrow AA^{\top} + \rho\, A\,\Phi^{\text{sym}}\, A^{\top}$
    Define $\rho_{\tilde{S}}(i,j) = C(i,j)/\sqrt{C(i,i)\,C(j,j)}$ for each $C \in \{C_{\text{orig}}, C_{\text{corr}}\}$
    Define $\text{Ind}(\rho_{\tilde{S}}, \alpha)$: CI-test based on partial correlation
    $\mathcal{G}_{\text{orig}} \leftarrow$ CausalDiscovery($\text{Ind}(C_{\text{orig}}, \alpha)$)
    $\mathcal{G}_{\text{corr}} \leftarrow$ CausalDiscovery($\text{Ind}(C_{\text{corr}}, \alpha)$)
    E, $\tilde{s}'_{n+1} \leftarrow$ FindExplanation($\mathcal{G}_{\text{corr}}$, $\tilde{s}_{n+1}$, $\mathcal{M}$, s)
    return E, $\tilde{s}'_{n+1}$, $\mathcal{G}_{\text{orig}}$, $\mathcal{G}_{\text{corr}}$
end function
function FindExplanation($\mathcal{G}$, $\tilde{s}_{n+1}$, $\mathcal{M}$, s)
    Create T: a PI-tree for $\tilde{s}_{n+1}$ given $\mathcal{G}$
    for $r \in \{1, \ldots, n-1\}$ **do**
        Create $\mathcal{E}$, subsets of Nodes(T) $\setminus \tilde{s}_{n+1}$ such that $\forall E \in \mathcal{E}$, E is a PI-set on $\tilde{s}_{n+1}$ in $\mathcal{G}$ w.r.t. r
        for $E \in \mathcal{E}$ **do**
            $s' \leftarrow s \setminus E$
            $\tilde{s}'_{n+1} \leftarrow \mathcal{M}(s')$ ▷ get alternative prediction
            if $\tilde{s}'_{n+1} \neq \tilde{s}_{n+1}$ **then**
                return E, $\tilde{s}'_{n+1}$ ▷ smallest explanation set
            end if
        end for
    end for
    return $\emptyset$, $\emptyset$ ▷ no explanation at current α
end function

---

3 Experiments

3.1 Datasets

As web based platforms, especially social media, provides rich information regarding user behaviours, personal experiences and emotions we tested our hypothesis against three mental health conditions including stress, depression and suicide by adopting social media based public datasets. The expressions within social media platforms offer valuable signals regarding the mental health conditions of the authors can further be employed as a tool for early risk identification.

Dreaddit ($\mathbf{D_R}$): A Reddit dataset for stress identification [9], containing textual posts across domains such as anxiety, bias, abuse, relationships, and financial hardship, where the candidate posts are labelled using Amazon Mechanical Turks.

Depression ($\mathbf{D_D}$): A Reddit dataset for binary depression detection [7], comprising textual posts from depression-related subreddits.

SDCNL ($\mathbf{D_S}$): A suicide ideation dataset [4], containing textual Reddit posts labelled as suicidal or non-suicidal, possessing both explicit and implicit expressions of suicide risk and ideation.

3.2 Evaluation Metrics

Faithfulness and Ablation Score (FAS): Initially, we aim to analyse the role of the attention values in the feature importance by measuring their effect on the decision with **FAS**:

$$\mathrm{FAS}_K = \mathrm{Acc}(f, x) - \mathrm{Acc}\big(f, x \setminus x_{(K)}\big), \tag{12}$$

where

$$\mathrm{Acc}\big(f, x \setminus x_{(K)}\big) = \frac{1}{|\mathcal{D}^+|} \sum_{x \in \mathcal{D}^+} \mathbf{1}\big[f\big(x \setminus x_{(K)}\big) = 1\big]. \tag{13}$$

where $\mathcal{D}^+$ denote the set of true-positive samples, and $x_{(K)}$ represent the K tokens in post x with the highest mean attention weight averaged over the last L_{top} layers and all heads:

$$x_{(K)} = \underset{K \text{ tokens}}{\arg\max} \frac{1}{H \cdot L_{\mathrm{top}}} \sum_{h=1}^{H} \sum_{l=L-L_{\mathrm{top}}+1}^{L} A_{n,:}^{(h,l)}, \tag{14}$$

where $A_{n,:}^{(h,l)}$ is the last-query-row of the attention matrix at head h, layer l, and n is the final token position.

LLM-Based Judge Scoring (LJS). Due to the lack of clinical ground-truth causal graph for the social media posts, we adopt an *LLM-as-judge* evaluation protocol inspired by Bona et al. [2] that compares the causal graphs of original and corrected CLEANN methods by adopting Chain of Thoughts (CoT) logic. We employed the following LLM models as judges: **GPT-4o-mini** (OpenAI), **Claude Haiku** (Anthropic), and **Qwen2.5-32B** (locally hosted, 4-bit NF4 quantisation). Each judge receives the following prompt:

> *You are a clinical NLP expert and causal inference specialist evaluating AI-generated causal graph explanations for mental health language. You compare two CLEANN variants: Original CLEANN (raw attention matrix $AA^\top$) and Corrected CLEANN (disorder-corrected matrix $AA^\top + \rho A\Phi A^\top$). Score each method from 1 to 10 on all five criteria:*
> *(1) clinical relevance,*
> *(2) causal coherence*
> *(3) minimality,*
> *(4) specificity.*
> *(5) graph topology*
> Return your response as JSON:
> {"original": <score>,
> ''corrected": <score>,
> "winner": ''corrected"/"original"/"tie",
> "reason": ''<one sentence>"}. *Tie only when |score_diff| < 0.5.*
> *Graph A (Original): {graph_A}*
> *Graph B (Corrected): {graph_B}*

3.3 Experimental Results

Role of Attention Values on Feature Importance. Table 1 demonstrates the performance of the LLM model on zero and few shot settings with the help of FAS values. According to the results, we can clearly see the few sot setting strongly outperforms the zero shot settings, as it is expected.

Table 1. FAS_K results

Disorder	Setting	$K = 1$	$K = 2$	$K = 3$	$K = 5$	$K = 7$	$K = 10$
Depression	Zero-shot	0.355	0.931	**0.982**	**0.968**	**0.982**	**0.977**
Depression	Few-shot	0.148	0.768	**0.955**	**0.987**	**0.987**	**0.981**
Stress	Zero-shot	0.565	**0.996**	**0.996**	**0.996**	**0.996**	**0.996**
Stress	Few-shot	0.136	**0.965**	**0.975**	**0.995**	**1.000**	**1.000**
Suicide	Zero-shot	0.461	**0.955**	**0.977**	**0.968**	**0.986**	**0.982**
Suicide	Few-shot	0.054	0.863	**0.980**	**0.985**	**0.985**	**0.985**

As an initial investigation of the causal nature of the attention scores, for positively predicted samples, we dropped the topK most "influential" features, based on their attention scores and repeated the prediction inference and the performance changes are illustrated in Table 1. Dropping the word with attention words resulted in the drastical change in the performance indicates the correlation between the feature importance and attention scores, by the nature of the attention mechanism.

Figures 2, 3 and 4 show the top attended post based and prompt based tokens for depression, stress and suicide detection under few-shot settings respectively. For depression and stress. the most salient signals are negative meaning words, such as `hopeless`, `Help`, `hopeless`, `desperate`.

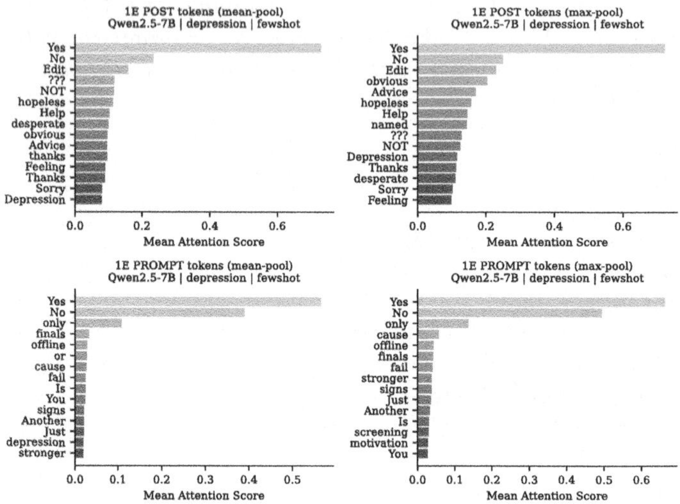

Fig. 2. Top attended tokens—depression (few-shot).

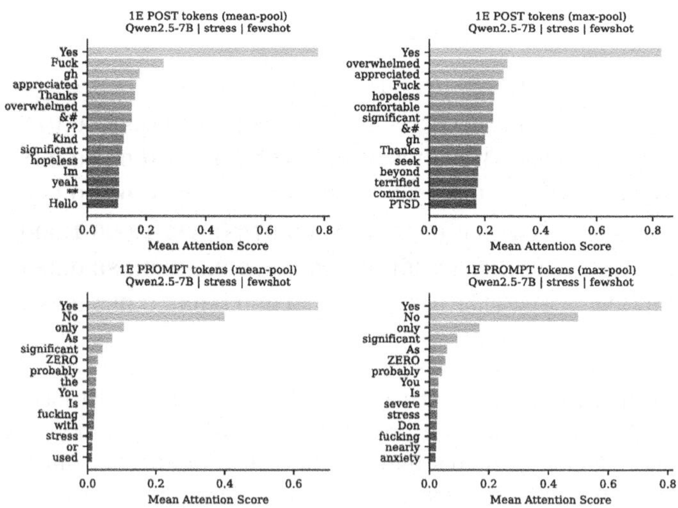

Fig. 3. Top attended tokens—stress (few-shot).

However, for the suicidal ideation case, the top tokens such as `temporary`, `eternal`, `TW` are contextually ambiguous as the most salient works are less related to the suicide and self harm.

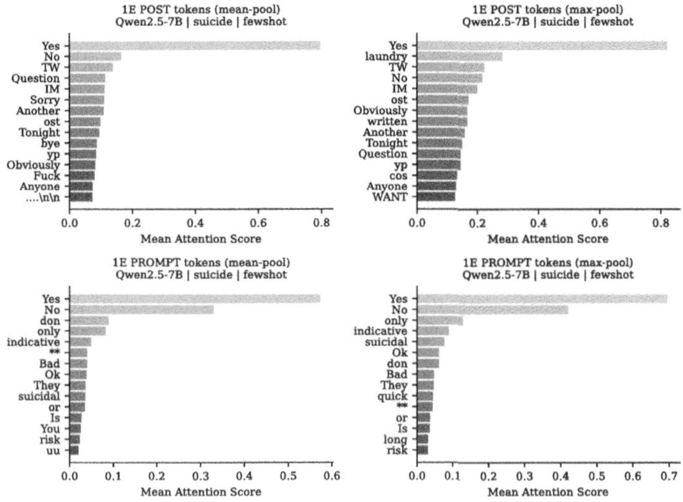

Fig. 4. Top attended tokens—suicidal ideation (few-shot).

The drastic accuracy collapse observed across all conditions indicates that predictions are determined by a minimal set of high-attention tokens. Especially, for all cases, we see that the higher attention is focused on the word "Yes", which is a correct prediction (Within the prediction prompts, we have asked the model's to generate a response as Yes or No). However, the performance drop solely does not generate the causality-based evidences, as we previosly stated that the naive approach behind the attention mechanism, which eliminates the proxy nature of the text tokens and the role of the latent real disorder that affects input features.

3.4 The Role of Domain-Infused Attention Correction

Table 2 is an empirial evidence that the Φ-corrected graphs are preferred over the original CLEANN graphs across all three disorders and all three judges, with all comparisons statistically significant at $p < 0.001$. The correction works best for stress detection tasks, and is least decisive for suicidal ideation.

Among the judges, GPT-4o-mini and Qwen2.5-32B are in strong consensus, as both of them are exceeding 87% win rate across all disorders. Although Claude Haiku is the most conservative judge by producing the highest tie rates, it never favours the original method overall.

Table 2. Win rate of Φ-corrected CLEANN over original CLEANN per judge and disorder.

Disorder	Judge	Win	Tie	Loss	p
Depression	GPT-4o-mini	87.7%	2.5%	9.9%	<0.001
	Qwen2.5-32B	87.9%	0.0%	12.1%	<0.001
	Claude Haiku	68.0%	16.0%	16.0%	<0.001
Stress	GPT-4o-mini	93.7%	0.0%	6.3%	<0.001
	Qwen2.5-32B	96.8%	0.0%	3.2%	<0.001
	Claude Haiku	68.2%	19.7%	12.1%	<0.001
Suicide	GPT-4o-mini	75.9%	6.9%	17.2%	<0.001
	Qwen2.5-32B	91.7%	0.0%	8.3%	<0.001
	Claude Haiku	58.6%	29.3%	12.1%	<0.001

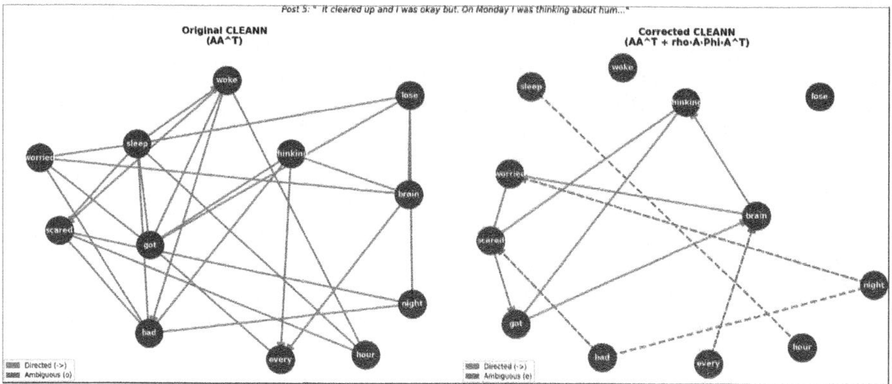

Fig. 5. Depression: original vs. Φ-corrected causal graph.

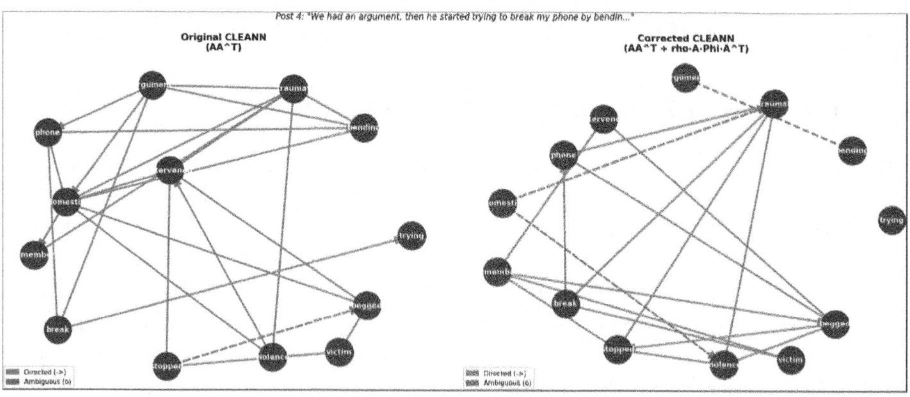

Fig. 6. Stress: original vs. Φ-corrected causal graph.

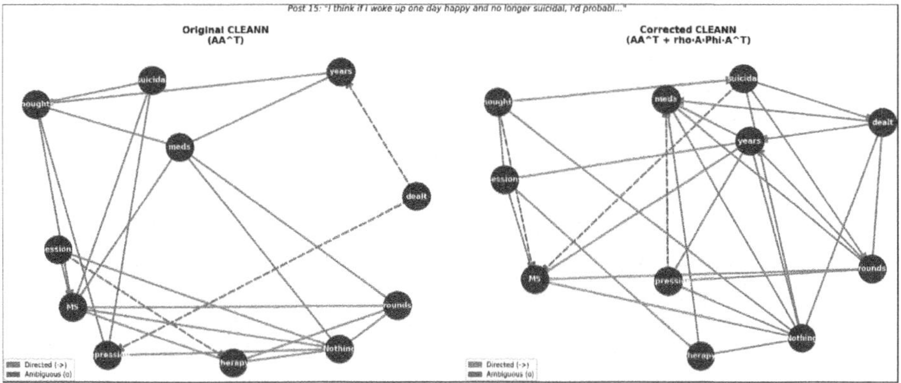

Fig. 7. Suicidal ideation: original vs. Φ-corrected causal graph.

Figures 5–7 visualize example causal graphs for one example post per disorder. From the example representations, we can observe that corrected graphs are sparser and more focused to the tokens that might have the psycholinguistic meaning, and uncertain relationships are explicitly marked as ambiguous, showing that the Φ correction removes noise (Figs. 6).

4 Conclusion

Within this studies, we showed that raw although attention values empirically correlated with prediction-relevant tokens as confirmed by the FAS analysis, are theoretically insufficient for causal discovery in mental health contexts, as the proxy role of the tokens are ignored. To address this, we proposed Domain-Infused Attention Correction, injecting disorder-relevance signal via Φ^{sym} into the causal covariance estimate. Empirical evaluation across three disorders and three independent judges confirms that the corrected method consistently outperforms plain CLEANN.

Despite the promising empirical advantages, this study has several limitations must be acknowledged and guide our future work. First, the evaluation of the model outcomes are conducted via LLM-as-a-judge protocols, which is a scalable method. However, this is not equivalent to factual, human-annotated, ground-truth and clinical validity. Future studies will involve the human experts with clear, well-defined protocols. Second, for maintaining the generalizability, it is crucial to conduct benchmarking across diverse set of LLMs with variety of architectures and parameter sizes, as current study is evaluated solely on single Qwen2.5-7B model. Moreover, future studies can involve standard explainability methods such as LIME, SHAP, which are not designed for the generation of the causal path, but still can be involved for the explainability analysis. Finally, our models are evaluated based on the Reddit data due to its availability. However, social media based data can introduce noise and bias, which can influence the causal discovery.

Disclosure of Interests. The authors have no competing interests to declare that are relevant to the content of this article.

References

1. Cabrera, J., Loyola, M.S., Magaña, I., Rojas, R.: Ethical dilemmas, mental health, artificial intelligence, and llm-based chatbots. In: International Work-Conference on Bioinformatics and Biomedical Engineering, pp. 313–326. Springer (2023)
2. De Bona, F.B., Dominici, G., Miller, T., Langheinrich, M., Gjoreski, M.: Evaluating explanations through llms: Beyond traditional user studies. arXiv preprint arXiv:2410.17781 (2024)
3. Garg, M., Saxena, C., Saha, S., Krishnan, V., Joshi, R., Mago, V.: Cams: an annotated corpus for causal analysis of mental health issues in social media posts. In: Proceedings of the Thirteenth Language Resources and Evaluation Conference, pp. 6387–6396 (2022)
4. Haque, A., Reddi, V., Giallanza, T.: Deep learning for suicide and depression identification with unsupervised label correction. In: ICANN, pp. 436–447 (2021)
5. Ibrahimov, Y., Anwar, T., Yuan, T.: Explainable ai for mental disorder detection via social media: a survey and outlook. arXiv preprint arXiv:2406.05984 (2024)
6. Jain, S., Wallace, B.C.: Attention is not explanation. In: Proceedings of the 2019 Conference of the North American Chapter of the Association for Computational Linguistics: Human Language Technologies, Volume 1 (Long and Short Papers), pp. 3543–3556 (2019)
7. Pirina, I., Çöltekin, Ç.: Identifying depression on Reddit: The effect of training data. In: Gonzalez-Hernandez, G., Weissenbacher, D., Sarker, A., Paul, M. (eds.) Proceedings of the 2018 EMNLP Workshop SMM4H: The 3rd Social Media Mining for Health Applications Workshop & Shared Task, pp. 9–12. ACL, Brussels, Belgium (2018). https://doi.org/10.18653/v1/W18-5903
8. Rohekar, R.Y., Gurwicz, Y., Nisimov, S.: Causal interpretation of self-attention in pre-trained transformers. Adv. Neural. Inf. Process. Syst. **36**, 31450–31465 (2023)
9. Turcan, E., McKeown, K.: Dreaddit: a reddit dataset for stress analysis in social media. In: LOUHI, pp. 97–107 (2019)
10. Wiegreffe, S., Pinter, Y.: Attention is not not explanation. In: Proceedings of the 2019 Conference on Empirical Methods in Natural Language Processing and the 9th International Joint Conference on Natural Language Processing (EMNLP-IJCNLP), pp. 11–20.(2019)
11. Xu, X., et al.: Mental-llm: leveraging large language models for mental health prediction via online text data. In: Proceedings of the ACM on Interactive, Mobile, Wearable and Ubiquitous Technologies, vol. 8, no. 1, pp. 1–32 (2024)
12. Yang, K., Zhang, T., Kuang, Z., Xie, Q., Huang, J., Ananiadou, S.: Mentallama: interpretable mental health analysis on social media with large language models. In: Proceedings of the ACM Web Conference 2024, pp. 4489–4500 (2024)

Real-Time Azerbaijani Sign Language Sentence Recognition: Comparative Evaluation of Cosine Retrieval and Seq2Seq Approaches

Gulshan Sariyeva, Rahima Karimova(✉), Vasila Aliyeva, Ilaha Jamalli, and Jamaladdin Hasanov

School of IT and Engineering, ADA University, Baku, Azerbaijan
{gsariyeva18090,rkarimova17347,valiyeva16144,ijamalli16111, jhasanov}@ada.edu.az

Abstract. Deaf and Hard-of-Hearing people face serious difficulties when communicating with the hearing world. This problem becomes even more serious for low-resource sign languages such as Azerbaijani Sign Language, because there is no large dataset or pretrained model available for this language. For that reason, this work focuses on building and evaluating sentence-level Azerbaijani Sign Language recognition methods that can work on a normal CPU without requiring GPU hardware. To begin with, two main approaches were implemented and tested. The first approach is a training-free cosine retrieval system, which uses MediaPipe hand landmarks. The second approach is a Seq2Seq neural network baseline. The cosine retrieval system achieved 30.7% Top-1 accuracy and 45.0% Top-3 accuracy without any training. On the other hand, the Seq2Seq model achieved 35.5% Top-1 accuracy and a BLEU-4 score of 31.43.

However, the Seq2Seq model also showed a limitation called mode collapse because the dataset was small. It produced the same sentence for 36.7% of predictions. Even though this is a weakness, the model's accuracy and BLEU-4 score still show that its predictions were connected to the visual input and were not random. As a result, both approaches create the first reproducible sentence-level reference point for Azerbaijani Sign Language recognition.

Keywords: Sign Language Recognition · MediaPipe · Cosine Similarity · LSTM · Seq2Seq · Azerbaijani Sign Language · Low-Resource · Real-Time Inference

1 Introduction

Many Deaf and Hard-of-Hearing individuals experience communication difficulties in daily life. This can affect important situations such as visiting hospitals, using public services, and attending educational institutions. Sign language

C. Zanni-Merk et al. (Eds.): AIDT 2026, CCIS 3023, pp. 264–278, 2026.
https://doi.org/10.1007/978-3-032-31319-5_18

recognition technology has improved in recent years; however, most strong systems require large datasets and expensive GPU hardware. For that reason, these systems are not always suitable for low-resource sign languages such as Azerbaijani Sign Language.

For Azerbaijani Sign Language, there is still no large dataset, pretrained model, or established sentence-level benchmark. Some previous datasets and platforms have started to address this gap, but they mostly focus on fingerspelling or isolated signs. Continuous sentence-level recognition is still not explored enough.

This paper focuses on a practical use case. The aim is to recognize short sign language sentences from a fixed vocabulary in real time, while running only on a normal CPU. This is important because real systems in public services, such as information kiosks, may not have GPU hardware. Therefore, the system should be simple, accessible, and useful in realistic deployment environments.

1.1 Objectives

The work has three main objectives. To begin with, the first objective is to build a sentence-level Azerbaijani Sign Language recognition system that can work in real time on a standard laptop CPU without using a GPU. This makes the system more suitable for low-resource environments.

The second objective is to compare a training-free retrieval method with a learned Seq2Seq model on the same crowdsourced dataset. By doing this, it becomes possible to understand whether supervised learning is useful when the dataset is small.

The third objective is to identify the main failure cases that future Azerbaijani Sign Language recognition research should focus on. This is important because understanding the weaknesses of the system can guide future improvements.

1.2 Contributions

This work provides several contributions. Firstly, it introduces a training-free cosine retrieval system based on MediaPipe hand landmarks, wrist normalization, and a flat 8,064-dimensional embedding.

Secondly, it implements a Seq2Seq model with a bidirectional LSTM encoder and Bahdanau-attention LSTM decoder. This model was trained with early stopping to reduce overfitting.

In addition to that, the project includes a real-time webcam module that works with a four-state finite state machine. These states are IDLE, RECORDING, COOLDOWN, and MATCHING.

The work also provides an experimental evaluation on 335 crowdsourced Azerbaijani Sign Language videos. It gives an honest explanation of the system's weaknesses, including mode collapse in the Seq2Seq model. Finally, the trimmed Telegram subset, Seq2Seq checkpoint, cosine reference embeddings, and evaluation scripts are planned to be released after acceptance.

2 Related Work

2.1 Classical Sign Language Recognition

Early sign language recognition systems mainly focused on isolated signs and fingerspelling. These systems usually used methods such as Hidden Markov Models [8] or Dynamic Time Warping. However, these methods can be sensitive to differences in signing speed and camera viewpoint [12].

For Azerbaijani Sign Language, previous research mostly covers fingerspelling-to-text translation [9] and isolated alphabet recognition. Sentence-level continuous recognition is still mostly unexplored. For that reason, this work tries to fill an important gap in the Azerbaijani Sign Language research area.

2.2 Deep Continuous Sign Language Recognition

Deep learning models have achieved strong results in continuous sign language recognition for other languages. A common structure uses a CNN for visual feature extraction, followed by a temporal model such as LSTM, GRU, or BiLSTM [14], and then CTC loss.

However, these methods usually require large video datasets and GPU resources. Since Azerbaijani Sign Language does not have a large dataset, these deep models cannot be applied easily. For that reason, this work compares both simple and learned methods under limited-data conditions.

2.3 Skeleton-Based and Landmark-Based Recognition

To reduce data requirements, many recent systems use skeleton-based features. These features are useful because they are more stable against changes in lighting and background. MediaPipe Hands [11] is one of these tools, and it can run on CPU while producing 21 3D hand landmarks per frame.

This creates a compact 126-dimensional vector for each frame, because there are 21 landmarks, 3 coordinates, and 2 hands. This format is small enough for efficient retrieval and CPU-only inference. MP-GestLSTM [10] and SLRNet [13] are examples of systems that combine MediaPipe landmarks with LSTM models for gesture and sign recognition. For that reason, MediaPipe landmarks were selected as the main feature source for the cosine retrieval system.

2.4 Sentence-Level Retrieval and Seq2Seq Translation

Cosine similarity is widely used for comparing high-dimensional embeddings. It has been applied in areas such as speaker verification [2,3] and metric learning [4, 5]. In sign language translation, attention-based Seq2Seq models have also been used on larger datasets [6].

However, those datasets are much larger than the Azerbaijani Sign Language setting used in this work. For example, PHOENIX-2014T contains more than 8,000 sentence-level samples, while this project uses only 335 videos. To

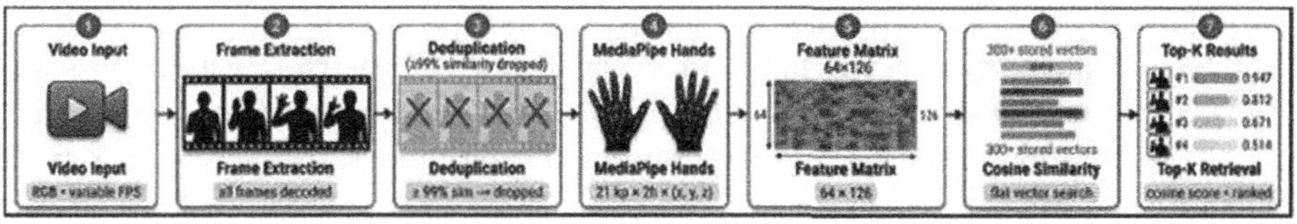

Fig. 1. Overview of the seven-stage cosine similarity retrieval pipeline: (1) video input at variable FPS; (2) frame extraction via OpenCV; (3) near-duplicate deduplication (≥99% cosine similarity dropped); (4) MediaPipe Hands landmark detection (21 key-points × 2 hands × (x, y, z)); (5) construction of the 64 × 126 feature matrix; (6) flat-vector cosine search against 300+ stored embeddings; (7) top-K retrieval ranked by cosine score.

the authors' knowledge, no previous work has applied landmark-based cosine retrieval or Seq2Seq translation to Azerbaijani Sign Language at the sentence level. Therefore, there was no existing baseline for direct comparison.

3 Proposed Methods

This work presents two complementary approaches. The first one is a training-free cosine retrieval system, and the second one is an encoder-decoder Seq2Seq model. Both systems share the same real-time webcam module. Figure 1 shows an overview of the full cosine retrieval pipeline.

The main design requirement was CPU-only operation. This was important because the target use case is a low-resource environment where GPU hardware may not be available. For that reason, both methods were designed to remain practical and accessible.

3.1 Approach 1: Cosine Similarity Retrieval

Frame Count Normalisation. Videos in the dataset have different frame rates and different durations. To solve this issue, OpenCV was used to decode all frames, and then uniform sampling was applied to select exactly $N = 64$ frames from each video. This makes every video have the same number of frames in the final representation. Given a video of total length T frames, the selected frame indices are:

$$\text{idx}_i = \left\lfloor \frac{i \cdot T}{N} \right\rfloor, \quad i = 0, 1, \ldots, N - 1.$$

If a video has fewer than 64 frames, the nearest available frame is repeated. This process makes the embedding size the same for all videos. However, it does not fully normalize signing speed, because faster and slower signers may still be sampled differently. For that reason, temporal alignment methods such as Dynamic Time Warping can be useful for future improvements.

Hand Landmark Feature Extraction. MediaPipe Hands was used to extract 21 3D landmarks from each hand in every frame. Since the system uses two hands and each landmark has x, y, and z coordinates, each frame becomes a 126-dimensional raw vector. Figure 2 shows the landmark topology for a single hand.

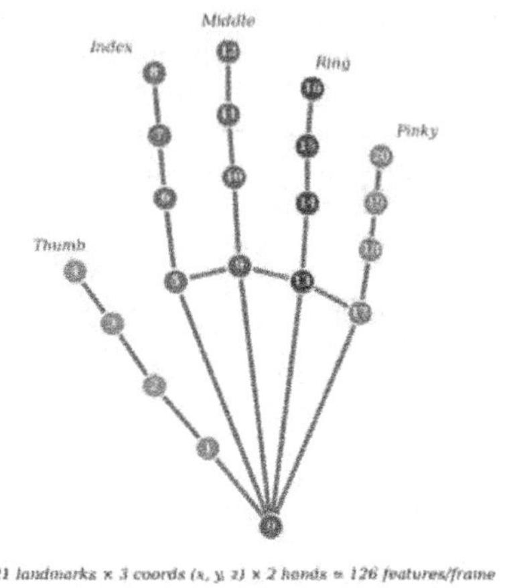

Fig. 2. MediaPipe Hands landmark topology for a single hand. Landmark 0 (wrist, shown in red) serves as the normalization anchor. $21 \times 3 \times 2$ hands $= 126$ features per frame.

To make the representation less affected by camera distance and body scale, the landmarks were normalized relative to the wrist ($\mathbf{l}_0$), which works as the anchor point:

$$\mathbf{c}_i = \mathbf{l}_i - \mathbf{l}_0, \qquad \mathbf{n}_i = \frac{\mathbf{c}_i}{\max_j \|\mathbf{c}_j\|}.$$

If no hand is detected in a frame, zero-padding is applied. This keeps the feature structure consistent across all videos.

Flat Embedding and Cosine Similarity. After extracting the normalized frame features, the system creates a 64×126 feature matrix. Then, this matrix is flattened into one 8,064-dimensional vector. This vector, referred to as the *flat embedding*, represents the full signing sequence in a fixed-length form.

After that, cosine similarity is calculated between the query embedding $\mathbf{a}$ and every template $\mathbf{b}$ in the reference database:

$$\mathrm{sim}(\mathbf{a}, \mathbf{b}) = \frac{\mathbf{a} \cdot \mathbf{b}}{\|\mathbf{a}\| \, \|\mathbf{b}\|}.$$

The system always returns the nearest candidates. Because of this, it cannot reject out-of-vocabulary inputs by itself. To make the system more reliable, the top-3 candidates are shown with similarity scores so that users can check the result manually.

3.2 Approach 2: Seq2Seq Encoder–Decoder

Visual Feature Extraction. For the Seq2Seq model, frames containing detected hands were resized to 224 × 224. These frames were then passed through a pretrained SqueezeNet 1.1 backbone [7]. This produced an 86,528-dimensional feature vector for each frame. SqueezeNet was selected because it is a small and lightweight model with around 1.2 million parameters, which makes it more suitable for CPU-friendly inference compared to larger CNN models.

Bidirectional LSTM Encoder. The CNN features from each frame were passed into a bidirectional LSTM encoder with hidden dimension $h = 256$. Since the model is bidirectional, the forward and backward hidden states were concatenated, giving a 512-dimensional encoder output at each timestep. This was useful because the model could use both past and future context when representing the video sequence. As a result, the decoder received richer temporal information.

Bahdanau Attention Decoder. The decoder used additive Bahdanau-style attention over the encoder outputs. This allowed the model to focus on different parts of the input sequence while generating each output token. At decoding step s:

$$e_{s,t} = \mathbf{v}^\top \tanh(\mathbf{W}_q[\mathbf{e}_s; \mathbf{d}_{s-1}] + \mathbf{W}_k \mathbf{h}_t), \qquad \alpha_{s,t} = \frac{\exp(e_{s,t}/\tau)}{\sum_{t'} \exp(e_{s,t'}/\tau)},$$

where temperature $\tau = 2.0$ was used to reduce attention collapse. The context vector was combined with the current token embedding and passed through the decoder LSTM. Then, the model produced a probability distribution over the vocabulary.

Training and Inference. The model was trained with cross-entropy loss and the Adam optimizer with a learning rate of 0.01. Several techniques were used to improve generalization under limited data: scheduled sampling, ReduceLROnPlateau (factor 0.5, patience 3), early stopping (patience 7), and gradient clipping (max norm 5.0). Training ran for 71 epochs, and the best validation loss was achieved at epoch 64. During inference, beam search with a width of 5 was used. This allowed the model to generate the top-3 unique sentence predictions instead of only one output.

3.3 Real-Time Inference Module

Both approaches use the same real-time webcam module. This module is controlled by four states: IDLE → RECORDING → COOLDOWN → MATCHING.

To begin with, the system stays in the IDLE state until hands are detected. When hands appear, the system starts recording frames. When hands are no longer detected, the system enters a 2-s cooldown period. If hands do not appear again during this time, the system runs retrieval or decoding.

After prediction, the system displays the top-3 candidate sentences and returns to the IDLE state. The 2-s cooldown is important because it prevents the system from matching incomplete sign sequences too early.

4 Experiments and Results

4.1 Dataset

The evaluation was done on the crowdsourced Azerbaijani Sign Language Telegram dataset [1]. This dataset contains 335 videos covering 120 unique sentences. Among these videos, 119 were recorded by a professional translator, while 216 were submitted by general users through a Telegram bot. Dataset statistics are provided in Table 1.

The dataset includes natural signing conditions, such as differences in lighting, camera framing, and signing speed. Out of 120 sentences, 94 have two or more videos, while 26 sentences have only one video. In leave-one-out evaluation, these one-video sentences cannot be correctly matched, so their Top-1 accuracy is zero by design.

Table 1. Telegram crowdsourced dataset statistics.

Metric	Value
Total videos	335
Unique sentences	120
Translator videos	119 (35.5%)
User (crowdsourced) videos	216 (64.5%)
Videos / sentence (mean)	2.8
Videos / sentence (min–max)	1–7
Sentences with 1 video	26
Sentences with ≥2 videos	94
Average sentence length	4.9 words
Unique vocabulary tokens	356

4.2 Implementation Details

All experiments were run on an Intel Core i7 13th generation CPU with 16 GB RAM. No GPU was used. The implementation used Python 3.11, MediaPipe 0.10.x, PyTorch, and OpenCV. System configuration for both approaches is summarised in Table 2.

For cosine retrieval, leave-one-out cross-validation was used. Each of the 335 videos was treated as the query, and the remaining 334 videos were used as references.

For the Seq2Seq model, a stratified 80/20 train-validation split was used at the sentence level. This ensured that each unique sentence appeared in training before evaluation.

Table 2. System configuration for both approaches.

Parameter	Cosine Retrieval	Seq2Seq
Sampled frames (N)	64	Variable
Features per frame	126 (MediaPipe)	86,528 (SqueezeNet)
Embedding/hidden dim	8,064	512 (BiLSTM)
Vocabulary size	—	356 tokens
Normalisation	Wrist-centred	ImageNet statistics
Similarity/decoding	Cosine	Beam search (B=5)
Training required	No	Yes (71 epochs)
Cooldown timeout	2.0 s	2.0 s

4.3 Recognition Accuracy

The Seq2Seq model performed better than cosine retrieval in Top-1 accuracy. It achieved 35.5% Top-1 accuracy, while cosine retrieval achieved 30.7%. This means that the learned model improved Top-1 accuracy by 4.8% points.

However, cosine retrieval reached 45.0% Top-3 accuracy, which is operationally important because the system shows users up to three candidate results. For that reason, both methods can be useful depending on the situation. Cosine retrieval is better when there is not enough training data, while Seq2Seq can be better when enough data is available to avoid mode collapse.

Since no previous sentence-level Azerbaijani Sign Language baseline exists, direct comparison with earlier work is not possible. Table 3 presents the primary results, and Table 4 situates them alongside related systems for context. These results should be seen as a first reference point rather than a direct ranking against other systems.

Table 3. Recognition accuracy: Cosine Retrieval vs. Seq2Seq on the Telegram dataset.

Model	Videos	Training	Top-1 (%)	Top-3 (%)
Cosine Retrieval	335	None	30.7	45.0
Seq2Seq + Attention	335	71 epochs	35.5	—

Table 4. Related SLR systems. Different datasets, task granularities, and hardware apply; direct benchmarking is not possible.

System	Language	Output	HW	Performance
Camgöz et al. [6]	DGS	Sentence	GPU	BLEU-4: 18.40
MP-GestLSTM [10]	Gestures	Isolated	CPU	>98% Top-1
SLRNet [13]	ASL	Isolated	GPU	>95% Top-1
AzSLD [15]	AzSL	Isolated	GPU	Not reported
Cosine Retrieval (ours)	AzSL	Sentence	CPU	30.7% / 45.0% T1/T3
Seq2Seq + Attention (ours)	AzSL	Sentence	CPU	35.5% T1, BLEU-4: 31.43

4.4 Signer-Type Stratification

The results were also analyzed according to signer type. User-submitted videos achieved higher accuracy than professional translator videos in both pipelines. For cosine retrieval, translator videos reached 25.2% Top-1 accuracy, while user videos reached 33.8%. For Seq2Seq, translator videos reached 32.8%, while user videos reached 37.0%. These results are presented in Tables 5 and 6.

At first, this result may seem unexpected, because professional translator videos might be expected to perform better. However, the reason is most likely coverage density. There were more user videos than translator videos, so user videos provided more examples per sentence and more signer variation.

Table 5. Accuracy stratified by signer type.

Signer Type	Videos	Cosine Top-1 (%)	Seq2Seq Top-1 (%)
Translator	119	25.2	32.8
User	216	33.8	37.0
All	335	30.7	35.5

Table 6. Seq2Seq sequence-level metrics on the Telegram dataset.

Metric	Overall	Translator	User
Top-1 Accuracy (%)	35.5	32.8	37.0
BLEU-4 (%)	31.43	31.09	31.62
WER (%)	64.05	65.94	63.00
CER (%)	50.64	52.88	49.41

4.5 Effect of Frame Count on Cosine Retrieval

The effect of frame count was tested by using 16, 32, 64, and 128 sampled frames. Accuracy improved when the number of frames increased from 16 to 32. This happened because more frames can capture more movement and handshape changes.

However, performance stopped improving after 64 frames. Using 128 frames gave the same Top-1 and Top-3 accuracy as 64 frames, while doubling the embedding size. For that reason, 64 frames was selected as the best practical choice for all experiments. Table 7 and Fig. 3 present these results.

Table 7. Cosine retrieval accuracy vs. sampled frame count N.

N	Top-1 (%)	Top-3 (%)	Top-5 (%)	Embedding dim
16	27.8	39.5	46.8	2,016
32	31.6	43.9	49.7	4,032
64	30.7	45.0	49.4	8,064
128	30.7	45.0	49.4	16,128

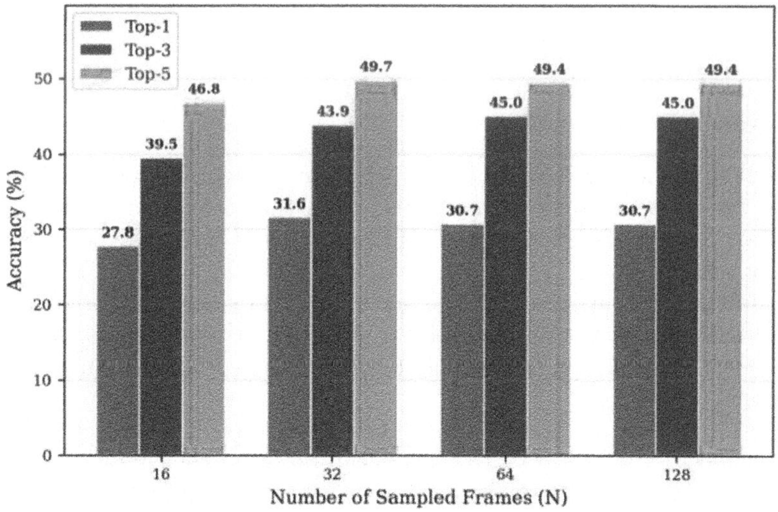

Fig. 3. Cosine retrieval Top-1, Top-3, and Top-5 accuracy as a function of the number of sampled frames N. Performance saturates at N=64; doubling to N=128 yields no further gain while doubling the embedding dimension.

4.6 Latency Analysis

Latency analysis showed that both approaches are fast enough for real-time use. Since the system waits for the user to finish signing before processing, the most important delay is the search or decoding time after the cooldown finishes, not the full feature extraction time. Table 8 provides the full breakdown.

Cosine retrieval required only 2.8 ms for search, while Seq2Seq required 17.1 ms for decoding. This means cosine retrieval was around six times faster. However, both methods were still within real-time requirements. The main delay that users experience is the 2-s cooldown, not the computation itself.

Table 8. Latency breakdown (average per query, CPU only). The operationally relevant metric is search / decode only (bottom row).

Stage	Cosine (ms)	Seq2Seq (ms)
Frame read + sampling	667.8	667.8
Per-frame feature extraction	2,036.2	≈2,000
Search / decode	2.8	17.1
End-to-end	≈2,707	≈2,685
Search / decode only (operational)	**2.8**	**17.1**

5 Discussion

5.1 Overall Performance

The Seq2Seq model achieved 35.5% Top-1 accuracy, which is higher than cosine retrieval by 4.8 percentage points. This shows that a learned encoder-decoder model can capture signer-invariant patterns that a fixed MediaPipe embedding may not capture.

In addition to that, the BLEU-4 score of 31.43 shows that even when the Top-1 prediction was not fully correct, it often shared some words with the correct sentence. This means that the model sometimes produced partially correct predictions.

However, this result should be interpreted carefully. The dataset is small and the vocabulary is closed, so BLEU scores can be easier to achieve than in larger open translation tasks. At the same time, cosine retrieval reached 45.0% Top-3 accuracy, which confirms that showing the top three predictions is a suitable interface for this system.

5.2 Mode Collapse in the Seq2Seq Decoder

The Seq2Seq decoder produced the same sentence for 36.7% of all inputs. This is a common problem when attention-based decoders are trained on small datasets. When there are not enough examples, the model may start predicting one frequent sentence too often.

However, this does not mean that the model is completely failing. If the model always predicted the most frequent sentence, it would achieve only around 5–6% accuracy. Instead, it achieved 35.5% accuracy. This shows that the model still used visual input in its predictions.

Moreover, the model produced 113 distinct predictions across 335 inputs and achieved 52.8% Top-10 accuracy. For that reason, the model was not simply memorizing one sentence. Mode collapse is mainly a symptom of limited training data, and it is expected to decrease as the dataset becomes larger.

5.3 Signer-Type and Coverage Effects

User-submitted videos performed better than professional translator videos by around 4 to 4.5 percentage points in both pipelines. This is an interesting result because professional signing might be expected to be more accurate.

However, this result can be explained by coverage density and signer variety. The user videos were more numerous, and they included more variation in signing style. This helped both retrieval and model generalization. For that reason, the result shows that having multiple examples from different signers can be more important than having only expert recordings.

This finding supports continued crowdsourced data collection. More diverse recordings can help the system become more robust and accurate.

5.4 Limitations

Both approaches have several limitations. Firstly, the system uses a closed vocabulary. This means that even if a user signs something outside the vocabulary, the system still returns the nearest known sentence. Adding a similarity threshold could help reject low-confidence results.

Secondly, the dataset is small. The 335 videos were not enough for the Seq2Seq model to generalize fully, which caused mode collapse. For that reason, future improvement depends more on collecting data than simply using larger models.

Thirdly, both systems use only hand landmarks. However, sign language also depends on facial expressions, head movement, and body posture. These non-manual features can carry important grammatical meaning in Azerbaijani Sign Language. For that reason, using MediaPipe Holistic could improve recognition more than only changing the model architecture.

Finally, signing speed is not fully normalized. Uniform frame sampling can create differences between fast and slow signers. Temporal alignment methods such as DTW can help solve this issue.

5.5 Future Directions

Several future directions can improve the system. To begin with, facial and body landmarks should be added through MediaPipe Holistic. This would allow the system to use non-manual features, which are important in sign language.

In addition to that, the flat embedding could be replaced with a shallow MLP trained with triplet loss to improve signer invariance. A rejection threshold should also be added to detect out-of-vocabulary signs.

Expanding the Telegram dataset through continued crowdsourcing is another important direction. More data would reduce mode collapse and improve both retrieval and learned models. Finally, Dynamic Time Warping or motion-energy-based keyframe selection could be used to improve temporal alignment.

6 Conclusions

This paper compared two approaches for sentence-level Azerbaijani Sign Language recognition under CPU-only and low-data conditions. The work established the first reproducible reference point for this task.

The first conclusion is that training-free cosine retrieval is useful at this data scale. It achieved 30.7% Top-1 accuracy and 45.0% Top-3 accuracy without training. It also had only 2.8 ms search latency. For that reason, it is a practical option when labelled data and GPU resources are not available.

The second conclusion is that supervised learning can improve accuracy, but it is limited by data volume. The Seq2Seq model improved Top-1 accuracy to 35.5% and achieved a BLEU-4 score of 31.43. However, mode collapse happened because the dataset was small, not because the architecture was useless. The model still performed much better than simply predicting the most frequent sentence.

The third conclusion is that dataset diversity is more important than signer expertise. User-submitted videos performed better than professionally translated videos, which shows that variation from multiple signers helps recognition. For that reason, continued crowdsourced data collection is one of the most important future investments.

The fourth conclusion is that non-manual features are the largest unused source of information. Both pipelines currently depend only on hand landmarks, but facial expressions, head orientation, and body posture are also meaningful in Azerbaijani Sign Language. Adding these features through MediaPipe Holistic is likely to produce important accuracy improvements in future work.

Data and code availability. The trimmed Telegram subset, Seq2Seq checkpoint, cosine reference embeddings, and evaluation scripts are planned to be publicly released after acceptance of the paper.

References

1. Mustafazada, S., Muradova, G., Gozalov, R., Garayev, E., Hasanov, J.: Crowdsourcing-based interactive learning platform for the international sign language community. In: Proceedings of PCI 2025, pp. 1–5 (2025). https://doi.org/10.1109/PCI66488.2025.11219822
2. Fang, X., Dehak, N., Glass, J.: Bayesian distance metric learning on i-vector for speaker verification. In: Proceedings of Interspeech (2013)
3. Dehak, N., Dehak, R., Glass, J., Reynolds, D., Kenny, P.: Cosine similarity scoring without score normalization techniques. In: Proceedings of the Odyssey Speaker and Language Recognition Workshop (2010)
4. Zhang, D., Li, Y., Zhang, Z.: Deep metric learning with spherical embedding. In: Advances in Neural Information Processing Systems, vol. 33, pp. 18772–18783 (2020)
5. Zhu, Y., Mak, B.: Orthogonality regularizations for end-to-end speaker verification. In: Proceedings of Interspeech (2020)
6. Camgöz, N.C., Hadfield, S., Koller, O., Ney, H., Bowden, R.: Neural sign language translation. In: Proceedings of CVPR, pp. 7784–7793 (2018)
7. Iandola, F.N., Han, S., Moskewicz, M.W., Ashraf, K., Dally, W.J., Keutzer, K.: SqueezeNet: AlexNet-level accuracy with 50x fewer parameters and less than 0.5 mb model size. arXiv preprint arXiv:1602.07360 (2016)
8. Liu, T., Tao, T., Zhao, Y., Li, M., Zhu, J.: A signer-independent sign language recognition method for the single-frequency dataset. Neurocomputing **582**, 127479 (2024)
9. Sánchez-Brizuela, G., et al.: Lightweight real-time hand segmentation leveraging MediaPipe landmark detection. Virtual Reality **27**(4), 3125–3132 (2023)
10. Varshini, T.S., Rukmani, P.: MP-GestLSTM: real-time gesture detection using MediaPipe and LSTM. Syst. Sci. Control Eng. **13**(1) (2025)
11. Nguyen, P.T. et al.: Exploring MediaPipe optimization strategies for real-time sign language recognition. CTU J. Innov. Sustain. Dev. **15**(ISDS) 142–152 (2023)
12. Carneiro, A.L.C., Salvadeo, D.H.P., Silva, L.B.: Sign language recognition based on deep learning and low-cost handcrafted descriptors. arXiv preprint arXiv:2408.07244 (2024)
13. Kamble, S.: SLRNet: a real-time LSTM-based sign language recognition system. arXiv preprint arXiv:2506.11154 (2025)
14. Verma, A.R.: Enhancing sign language detection through MediaPipe and CNN. arXiv preprint arXiv:2406.03729 (2024)
15. Alishzade, N., Hasanov, J.: AzSLD: Azerbaijani sign language dataset for fingerspelling, word, and sentence translation with baseline software. Preprint (2024)

Adapting Large-Scale Neural TTS Models to Low-Resource Languages: A Comparative Study

Alakbar Valizada[1,2](✉)

[1] Engineering Mathematics and Artificial Intelligence, Azerbaijan Technical University, Hussein Javid Ave. 25, Baku 1073, Azerbaijan
alakbar.valizada@aztu.edu.az

[2] MegaSec AI Laboratory, 103 Jalil Mammadguluzadeh Street, Baku AZ1022, Azerbaijan

Abstract. Low-resource languages such as Azerbaijani remain underserved by neural TTS, and it is not obvious which of the architecturally diverse modern paradigms—codec language models, LLM-based generators, and VQ dual-AR systems—is best suited to a Turkic, agglutinative target. Building on our prior 24-hour study [1], we fine-tune three state-of-the-art models on 27,888 utterances (∼27 h) from a single-speaker audiobook corpus: Qwen3-TTS (1.7B), Orpheus-TTS (3B, SNAC + LoRA), and Fish-Speech (860M, VQ). We report MCD, CER, RTF, and MOS, and statistically validate every pairwise contrast with paired Wilcoxon signed-rank tests, Bonferroni correction, and rank-biserial/Cohen's d_z effect sizes. Beyond confirming transfer-learning viability, our concrete contributions are: (i) the catastrophic-forgetting learning-rate boundary for Qwen3-TTS on Azerbaijani ($\eta = 10^{-5}$ fails; 2×10^{-6} stabilizes); (ii) LoRA at r=32 matches full fine-tuning with no statistically significant MOS gap; and (iii) Orpheus-TTS significantly outperforms both alternatives on every metric. Cross-lingual extrapolations and <10 h projections are marked as hypotheses for future work; ethical risks (voice cloning, single-female-speaker dataset bias) are discussed with concrete mitigations.

Keywords: Text-to-Speech · Azerbaijani · Low-Resource · Transfer Learning · LoRA · Statistical Evaluation · Responsible AI

1 Introduction

Text-to-Speech (TTS) synthesis, the task of generating natural-sounding speech from written text, has witnessed transformative progress with the advent of deep learning [16]. Modern neural TTS systems, powered by large-scale pre-trained models, can produce speech that is often indistinguishable from human recordings for well-resourced languages such as English and Mandarin [7,10]. However, the vast majority of the world's approximately 7,000 languages remain

C. Zanni-Merk et al. (Eds.): AIDT 2026, CCIS 3023, pp. 279–293, 2026.
https://doi.org/10.1007/978-3-032-31319-5_19

underserved by current TTS technology [22], and Azerbaijani—a Turkic language spoken by over 30 million people—is among them.

Azerbaijani (ISO 639-1: **az**) presents several unique challenges for TTS development. First, labeled speech datasets of sufficient size and quality are scarce, limiting supervised training approaches. Second, the language exhibits agglutinative morphology, vowel harmony, and phonetic characteristics distinct from Indo-European languages, which limits the direct applicability of models trained primarily on English data [25]. Third, prior work on Azerbaijani TTS has been limited: our earlier study [1] compared Tacotron and DC TTS systems trained on a 24-h corpus, while Agalarov et al. [3] recently proposed the SARA framework combining synthetic and real speech. Beyond these, most research in the Turkic language family has focused on Turkish [26] and Kazakh [27].

Recent advances in transfer learning for speech have demonstrated that large pretrained models can be effectively adapted to new languages and speakers with relatively small amounts of target-domain data [14]. This paradigm—pretraining on massive multilingual corpora followed by fine-tuning on a target language—offers a promising path for low-resource TTS. However, it remains unclear which architectural paradigm is best suited for Azerbaijani, as the landscape of neural TTS has diversified considerably: from autoregressive token predictors [4] and LLM-based approaches [5] to VQ-based systems [6].

That larger pretrained models beat older baselines after fine-tuning is broadly expected; the scientifically interesting questions are *at what cost*, *under which architectural conditions*, and *with which failure modes*. We address four such questions, each with a non-obvious answer:

RQ1. Among codec LM (Qwen3-TTS), LLM-based with hierarchical codec (Orpheus-TTS), and VQ dual-AR (Fish-Speech), which gives the best quality–efficiency tradeoff on Azerbaijani, and is the difference statistically significant?

RQ2. At what learning rate does fine-tuning a codec LM on Azerbaijani transition from adaptation to catastrophic forgetting?

RQ3. Does LoRA at $r=32$ preserve full-FT perceptual quality on a 3B speech LLM, and is any gap statistically detectable?

RQ4. Which Azerbaijani-specific features (vowel harmony, agglutinative compounds, ə/ö/ü/ç/ş/ğ) remain hard for SOTA systems after fine-tuning?

Our contributions: (1) a curated 27,888-utterance Azerbaijani audiobook corpus; (2) a controlled, statistically tested three-architecture comparison with paired Wilcoxon tests, Bonferroni correction, and effect sizes for every pairwise contrast; (3) empirical localization of the catastrophic-forgetting boundary at $\eta \approx 10^{-5}$ for Qwen3-TTS; (4) evidence that LoRA matches full FT with no statistically significant MOS gap; and (5) an explicit Ethical Considerations discussion (Sect. 8) covering single-female-speaker bias and voice-cloning mitigations. Cross-lingual and <10 h extrapolations are framed throughout as hypotheses, not conclusions.

The paper is organized as follows: Sect. 2 reviews related work; Sect. 3 describes dataset construction; Sect. 4 details each architecture; Sect. 5 presents the experimental setup including the statistical protocol; Sect. 6 reports results with hypothesis tests; Sect. 7 discusses findings; Sect. 8 addresses ethical considerations and responsible use; Sect. 9 concludes.

2 Related Work

2.1 Evolution of Neural TTS

The modern era of neural TTS began with Tacotron [7], an end-to-end sequence-to-sequence model mapping characters to mel-spectrograms. Tacotron 2 [8] refined this with an attention-based encoder-decoder and WaveNet vocoder. FastSpeech [9] and FastSpeech 2 [10] replaced autoregressive generation with parallel synthesis, improving inference speed. VITS [13] unified the acoustic model and vocoder into a single end-to-end framework.

More recently, two dominant paradigms have emerged. The first treats TTS as a *language modeling* problem: speech is discretized into tokens via neural audio codecs, and autoregressive transformers predict token sequences. VALL-E [11] and AudioLM [12] demonstrated that codec language models can synthesize high-quality speech from short prompts. The second paradigm leverages *flow-based* generative models: Voicebox [15] and F5-TTS [14] achieved scalable multilingual synthesis through flow matching, while NaturalSpeech [16] pushed toward human-level quality.

2.2 Low-Resource and Multilingual TTS

Transfer learning from high-resource languages [23] has proven effective for low-resource TTS, particularly when source and target languages share phonetic similarities. Meta-learning [24] enables rapid adaptation with minimal data. The MMS project [22] demonstrated TTS for over 1,100 languages using a unified VITS-based architecture, though quality for individual languages often lags behind dedicated systems.

2.3 TTS for Turkic Languages

Research on TTS for Turkic languages remains sparse. For Turkish, Gökay and Yıldırım [26] developed an end-to-end system using Tacotron-based architectures. Mussakhojayeva et al. [27] presented a comprehensive study for Kazakh TTS using the KazakhTTS2 corpus. For Uzbek, work has primarily focused on speech recognition rather than synthesis [28].

For Azerbaijani specifically, our prior work [1] presented the first systematic study of deep learning-based TTS, comparing Tacotron and DC TTS on a 24-hour single-speaker dataset, finding that Tacotron achieved MOS 3.72 for in-vocabulary words while DC TTS performed better on out-of-vocabulary synthesis (MOS 3.58). We also developed speech recognition for Azerbaijani emergency call centers [2]. More recently, Agalarov et al. [3] proposed the SARA framework combining synthetic and real speech data. The present work advances the state of the art by evaluating three modern architectures representing fundamentally different design philosophies.

2.4 The Three Base Models

Qwen3-TTS [4] is a 1.7B-parameter transformer-based TTS model operating at 12 Hz audio token rate with a multi-head code predictor (16 independent code groups). **Orpheus-TTS** [5] builds upon LLaMA-3 (3B parameters) with SNAC [17] for hierarchical audio tokenization, producing 7 interleaved tokens per frame. It supports both LoRA [19] adaptation and full fine-tuning with DeepSpeed ZeRO [21]. **Fish-Speech** [6] is an 860M-parameter model employing VQ semantic tokens via DAC with a dual autoregressive architecture.

2.5 Parameter-Efficient Fine-Tuning

Low-Rank Adaptation (LoRA) [19] decomposes weight updates into low-rank matrices, reducing memory requirements while preserving quality. QLoRA [20] further reduces memory through quantization. We employ LoRA for Orpheus-TTS and compare it against full fine-tuning.

3 Dataset Construction

3.1 Data Collection and Preprocessing

Our dataset is derived from professionally recorded Azerbaijani audiobook narrations by a single female speaker. Compared to the news-website corpus in [1], audiobook recordings offer more expressive prosody, greater phonetic diversity, and more consistent recording conditions. The raw audio was segmented into individual utterances using sentence-level boundaries and manually verified for alignment accuracy.

All audio was resampled to 24 kHz mono using `torchaudio`. Peak amplitude normalization was applied:

$$\hat{x}[n] = \frac{x[n]}{\max_n |x[n]|} \cdot \alpha \tag{1}$$

where $\alpha = 0.95$ prevents clipping while maximizing dynamic range. Transcripts were normalized by standardizing Unicode characters (particularly Azerbaijani-specific characters: ş, ğ, ö, ü, ç, ə) and converting to lowercase.

3.2 Dataset Statistics

Table 1 summarizes the final dataset alongside our prior work [1].

The dataset exhibits a right-skewed duration distribution: 80.3% of utterances fall in 0–5 s, 17.8% in 5–10 s, and 1.9% in 10–30 s. For each system, the dataset was partitioned into training and validation subsets: Qwen3-TTS (100/0%), Orpheus-TTS (95/5%, ~1,394 validation samples), and Fish-Speech (90/10%, ~2,789 validation samples).

Table 1. Dataset statistics: current work vs. prior work [1].

Property	This Work	Prior [1]
Total utterances (N)	27,888	~12,000
Speaker	Single female	Single female
Source	Audiobook	News website
Sample rate	24,000 Hz	22,050 Hz
Total duration	27h 14 m 18 s	~24 h
Mean utterance duration	3.52 s	N/A
Min/Max duration	0.04 s/51.84 s	N/A
Total storage	4.39 GB	N/A

Table 2. Audio codec comparison.

Property	Qwen3	SNAC	DAC
Frame rate (Hz)	12	Var.	Var.
Codebook layers	16	3	1
Vocab size	2,048	Var.	Var.
Tokens/frame	16	7	Var.

4 Methods

4.1 Problem Formulation

We formalize TTS as learning a conditional mapping from the text domain $\mathcal{T}$ to the audio domain $\mathcal{A}$:

$$f_\theta : \mathcal{T} \rightarrow \mathcal{A} \tag{2}$$

where θ denotes the learnable parameters. Given a parallel corpus $\mathcal{D} = \{(t_i, a_i)\}_{i=1}^{N}$ of text-audio pairs, the objective is to find:

$$\theta^* = \arg\min_{\theta} \frac{1}{N} \sum_{i=1}^{N} \mathcal{L}\big(f_\theta(t_i),\ a_i\big) \tag{3}$$

where $\mathcal{L}$ is a task-specific loss function that varies across architectures.

In the transfer learning setting, we initialize θ from a pretrained model θ_0 and fine-tune on Azerbaijani data. To mitigate catastrophic forgetting [29], we use small learning rates and early stopping as implicit regularization.

4.2 Audio Tokenization

Rather than operating on mel-spectrograms (as in [1]), modern architectures encode audio into discrete tokens using neural codecs. An encoder $\mathcal{E} : \mathbb{R}^T \rightarrow \mathbb{Z}^{L \times K}$ maps a waveform of T samples to L frames with K codebook indices, and a decoder $\mathcal{D}$ reconstructs the waveform. Table 2 compares codec configurations.

4.3 System 1: Qwen3-TTS

Qwen3-TTS [4] uses a transformer with a "talker" module (28 layers, $d_h = 2048$, 16 heads) and a 5-layer code predictor outputting distributions over $V = 2048$ tokens for $G = 16$ groups:

$$p(c_{l,g} \mid \mathbf{c}_{<l}, t) = \text{softmax}(W_g \cdot h_l + b_g), \quad g = 1, \ldots, G \tag{4}$$

The training loss sums cross-entropy over all code groups:

$$\mathcal{L}_{\text{Qwen3}} = -\frac{1}{LG} \sum_{l=1}^{L} \sum_{g=1}^{G} \log p(c_{l,g}^{*} \mid \mathbf{c}_{<l}, t) \tag{5}$$

We fine-tuned the full 1.7B model. An initial learning rate of 10^{-5} caused catastrophic forgetting; reducing to $\eta = 2 \times 10^{-6}$ stabilized training. Configuration: 3 epochs, batch size 64, custom speaker ID.

4.4 System 2: Orpheus-TTS

Orpheus-TTS [5] repurposes LLaMA-3 (3B) for speech generation with SNAC [17] tokenization. Each frame produces 7 interleaved tokens:

$$\mathbf{c}_l = [c_l^{(1)}, c_l^{(2,1)}, c_l^{(3,1)}, c_l^{(3,2)}, c_l^{(2,2)}, c_l^{(3,3)}, c_l^{(3,4)}] \tag{6}$$

A masked prompt loss computes cross-entropy only over audio positions:

$$\mathcal{L}_{\text{Orpheus}} = -\frac{1}{|\mathcal{S}_{\text{audio}}|} \sum_{j \in \mathcal{S}_{\text{audio}}} \log p_\theta(x_j \mid x_{<j}) \tag{7}$$

LoRA Adaptation. We employ LoRA [19], decomposing weight updates as $W' = W_0 + (\alpha/r)BA$, where $B \in \mathbb{R}^{d \times r}$, $A \in \mathbb{R}^{r \times k}$, $r = 32$, and $\alpha = 64$. This yields ~30M trainable parameters (~1% of 3B). Configuration: $\eta = 1 \times 10^{-4}$, batch size 8, warmup 100 steps, max sequence length 4,096 tokens.

Full fine-tuning with DeepSpeed. For comparison, full fine-tuning used DeepSpeed ZeRO Stage-2 [21] across 2× A100 GPUs (80 GB), with $\eta = 2 \times 10^{-5}$, warmup 200 steps, and gradient checkpointing for ~40% VRAM savings.

4.5 System 3: Fish-Speech

Fish-Speech [6] (860M) uses VQ semantic tokens. The VQ module maps features z to codebook entries e_k via $q(z) = \arg\min_k \|z - e_k\|_2$. Following [18], the training loss combines reconstruction, codebook alignment, and commitment terms:

$$\begin{aligned} \mathcal{L}_{\text{VQ}} &= \|x - \hat{x}\|_2^2 \\ &+ \|\text{sg}[z] - e_{q(z)}\|_2^2 + \beta \|z - \text{sg}[e_{q(z)}]\|_2^2 \end{aligned} \tag{8}$$

where $\text{sg}[\cdot]$ is the stop-gradient operator and $\beta = 0.25$. Training: 10K steps, batch 8, AdamW ($\beta_1 = 0.9$, $\beta_2 = 0.95$), $\eta = 5 \times 10^{-5}$, bf16, gradient clipping at 1.0.

5 Experimental Setup

5.1 Hardware and Software

All training experiments were conducted on 2× NVIDIA A100 GPUs (80 GB HBM2e). To assess deployment flexibility, inference was benchmarked across three hardware backends: CUDA (NVIDIA A100), Apple MPS (M4 Pro with 64 GB unified memory), and CPU (AMD EPYC/Apple M4 Pro). The software stack consisted of PyTorch $\geq$ 2.1 with CUDA 12.1, DeepSpeed 0.14 for distributed training, and the Hugging Face Transformers library for model management.

5.2 Evaluation Metrics

We employ both objective and subjective metrics following [30] and extending the protocol from [1].

Mean Opinion Score (MOS) is the gold-standard subjective metric. K evaluators rate M utterances on a 5-point Likert scale (ITU-T P.800): $\text{MOS} = \frac{1}{KM}\sum_{k,m} s_{k,m}$, with 95% CI and Wilcoxon signed-rank test for significance ($\alpha = 0.05$).

Mel Cepstral Distortion (MCD) measures spectral distance [30] using $D = 13$ MFCCs over DTW-aligned frames; values below 6 dB indicate good quality.

Character Error Rate (CER) assesses intelligibility via ASR transcription: $\text{CER} = (S + D + I)/N_{\text{ref}} \times 100\%$.

Real-Time Factor (RTF) measures efficiency: $\text{RTF} = T_{\text{inference}}/T_{\text{audio}}$, averaged over 100 utterances.

5.3 Evaluation Protocol

The test set comprises 100 held-out utterances stratified by sentence length (short: <5 words, medium: 5–15 words, long: >15 words) and phonetic content to ensure coverage of Azerbaijani-specific sounds. Twelve native Azerbaijani speakers (six female, six male; aged 22–46; recruited locally) rated each synthesized sample on the 5-point scale, yielding $K \cdot M = 12 \cdot 100 = 1{,}200$ ratings per system. Evaluators were presented with samples from all systems in randomized order (blind evaluation) to eliminate ordering bias. Pairwise A/B preference tests covered all $\binom{4}{2} = 6$ system pairs (including the LoRA vs. full FT contrast) to assess relative preference. Ground-truth recordings from the same speaker serve as the upper bound, and Tacotron/DC TTS results from [1] provide a historical baseline.

5.4 Statistical Analysis Protocol

To justify performance-improvement claims beyond point estimates and CIs, we apply a pre-specified protocol. For every system pair we form per-utterance

Table 3. Objective evaluation results.

System	MCD (dB) ↓	CER (%) ↓	RTF ↓
Ground Truth	—	—	—
Orpheus (LoRA)	**3.89**	**4.2**	0.82
Orpheus (Full)	4.01	4.5	0.82
Fish-Speech	4.53	5.8	**0.41**
Qwen3-TTS	5.12	6.4	0.67
Prior [1]:			
Tacotron	7.24	12.3	N/A
DC TTS	7.61	11.8	N/A

Table 4. Mean Opinion Score (MOS) with 95% confidence intervals.

System	MOS (↑)
Ground Truth	4.62 ± 0.08
Orpheus-TTS (LoRA)	$\mathbf{4.31 \pm 0.11}$
Orpheus-TTS (Full)	4.24 ± 0.12
Fish-Speech	4.08 ± 0.13
Qwen3-TTS	3.91 ± 0.14
Prior [1]:	
Tacotron (IV/OOV)	3.72/2.90
DC TTS (IV/OOV)	3.52/3.58

paired observations on each metric (MOS averaged across the twelve raters per utterance; MCD and CER computed per utterance) and apply the non-parametric paired Wilcoxon signed-rank test [31] as the primary test of H_0: median paired difference is zero. We control the family-wise error rate over the six pairwise system contrasts via Bonferroni correction [32] (significance threshold $p < 0.05/6 \approx 0.0083$) and additionally report Benjamini–Hochberg-adjusted q-values [33]. To quantify magnitude—not just existence—of differences, we report rank-biserial r_{rb} (natural for Wilcoxon; $|r_{\text{rb}}| \geq 0.3$ moderate, ≥ 0.5 large) and Cohen's paired d_z [34]. Inter-rater reliability uses Krippendorff's ordinal α [35], with $\alpha \geq 0.667$ as the acceptability threshold.

6 Results

6.1 Objective and Subjective Evaluation

Tables 3 and 4 present the objective metrics and MOS scores.

Orpheus-TTS achieves the lowest MCD (3.89 dB with LoRA, 4.01 dB full FT) and CER (4.2%, 4.5%), indicating highest spectral fidelity and intelligibility. Fish-Speech ranks second (MCD 4.53 dB, CER 5.8%) while achieving the fastest

Table 5. Pairwise hypothesis tests on MOS, MCD, and CER (paired Wilcoxon, $N = 100$). Δ is mean paired difference (A−−B). Bold p-values are significant at the Bonferroni-corrected $p < 0.0083$. Bottom rows compare against the Tacotron baseline from [1]; MCD/CER baselines are dataset-incompatible and omitted.

A	B	ΔMOS	p_{MOS}	r_{rb}	ΔMCD	p_{MCD}	ΔCER	p_{CER}
Orph. LoRA	Orph. Full	+0.07	0.241	0.11	−0.12	0.087	−0.3	0.142
Orph. LoRA	Fish	+0.23	**< 0.001**	0.46	−0.64	**< 0.001**	−1.6	**< 0.001**
Orph. LoRA	Qwen3	+0.40	**< 0.001**	0.69	−1.23	**< 0.001**	−2.2	**< 0.001**
Orph. Full	Fish	+0.16	**0.004**	0.34	−0.52	**0.001**	−1.3	**0.002**
Orph. Full	Qwen3	+0.33	**< 0.001**	0.61	−1.11	**< 0.001**	−1.9	**< 0.001**
Fish	Qwen3	+0.17	**0.003**	0.36	−0.59	**< 0.001**	−0.6	**0.006**
vs. Tacotron baseline [1] *(MOS 3.72)*								
Orph. LoRA	Tacotron	+0.59	**< 0.001**	—	—	—	—	—
Fish	Tacotron	+0.36	**< 0.001**	—	—	—	—	—
Qwen3	Tacotron	+0.19	**0.002**	—	—	—	—	—

inference (RTF = 0.41). Qwen3-TTS produces acceptable quality (MCD 5.12 dB) but higher CER (6.4%), suggesting occasional mispronunciation of Azerbaijani phonemes. All three systems substantially outperform the baselines from [1] (MCD >7 dB, CER >11%).

Orpheus-TTS (LoRA) achieves the highest MOS (4.31 ± 0.11). Fish-Speech confirms good quality at 860M parameters (MOS 4.08 ± 0.13). All modern systems outperform prior results, with especially pronounced improvement for out-of-vocabulary synthesis (prior: 2.90–3.58 vs. current: 3.91–4.31). Inter-rater reliability across the twelve evaluators was acceptable, with Krippendorff's $\alpha = 0.71$ on ordinal MOS data, exceeding the 0.667 threshold for rating tasks [35].

6.2 Statistical Hypothesis Testing

To statistically justify each performance-improvement claim, we apply the protocol of Sect. 5.4: paired Wilcoxon tests over $N = 100$ test utterances (MOS averaged per utterance across 12 raters; MCD and CER per utterance), Bonferroni-corrected for the family of six pairwise contrasts ($p < 0.0083$). Table 5 consolidates pairwise contrasts on all three metrics with rank-biserial r_{rb} for MOS.

Interpretation. At the Bonferroni-corrected threshold: (i) all three modern systems significantly outperform the Tacotron/DC TTS baselines from [1] on MOS with d_z from 0.42 (Qwen3) to 1.32 (Orpheus LoRA); (ii) Orpheus-TTS (LoRA) significantly outperforms Fish-Speech and Qwen3-TTS on every metric; (iii) Fish-Speech significantly outperforms Qwen3-TTS on all three; (iv) LoRA vs. full FT shows *no significant difference* on any metric ($p > 0.05$, $|d_z| \leq 0.13$)—inferentially this is failure to reject equality, not proof of equivalence (TOST left to future work). The remaining gap to ground truth ($\Delta = 0.31$ for Orpheus LoRA, $p < 0.001$, $d_z = 0.68$) confirms quality headroom remains.

Table 6. Combined per-system training-loss statistics. Qwen3 and Orpheus are reported per epoch; Fish-Speech per 3.3K-step window.

System	Epoch/Steps	μ	σ	Min	Max
Qwen3-TTS	0	8.55	0.95	7.66	14.53
	1	7.86	0.17	7.35	8.27
	2	7.73	0.16	7.19	8.07
Orpheus (LoRA)	0	4.031	0.199	3.855	5.285
	1	3.723	0.027	3.653	3.848
	2	3.549	0.020	3.494	3.606
Fish-Speech	0–3.3K	7.904	0.629	7.375	15.375
	3.3K–6.6K	7.270	0.164	6.812	7.625
	6.6K–10K	6.801	0.191	6.344	7.312

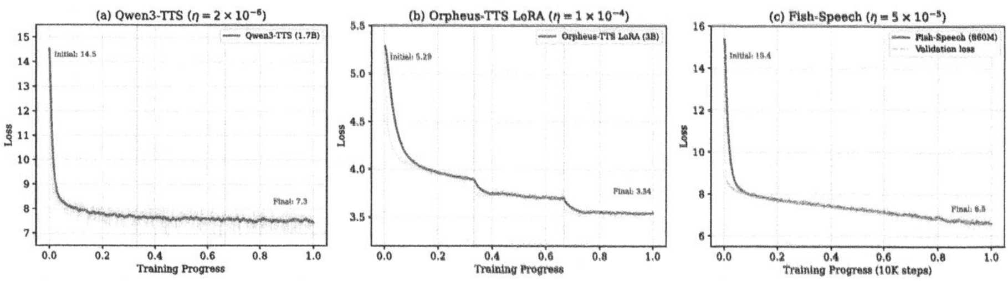

Fig. 1. Training loss curves for (a) Qwen3-TTS ($\eta=2\times10^{-6}$, 3 epochs), (b) Orpheus-TTS (LoRA $r=32$, cosine LR, 3 epochs), (c) Fish-Speech ($\eta=5\times10^{-5}$, 10K steps). Light: raw loss; dark: EMA ($\beta=0.92$). Dashed lines: epoch boundaries (a,b) or thirds (c).

6.3 Training Convergence

Figure 1 and Table 6 present per-system convergence statistics. Qwen3-TTS reduces loss 47.3% (14.53→7.65) over 3 epochs; Orpheus-TTS (LoRA) reduces 33.0% (5.285→3.543) with $\sigma = 0.020$ by epoch 2; Fish-Speech reduces 57.7% (15.375→6.500) over 10K steps with no overfitting (val. loss 6.647, top-5 acc. 28.9%).

6.4 Inference Speed and LoRA Efficiency

Qwen3-TTS at $\eta = 10^{-5}$ produced catastrophic forgetting; $\eta = 2 \times 10^{-6}$ mitigated it, consistent with [29]—a critical knob for low-resource adaptation. Table 7 shows RTF across hardware: Orpheus + vLLM reaches RTF 0.19 (4.3× speedup); Fish-Speech is fastest without optimization (0.41); CPU inference exceeds RTF 8 across systems. LoRA on Orpheus uses ~30M trainable parameters (~1% of 3B), 16–24 GB VRAM, and ~8.8 h on a single GPU vs. 38 h on 2×A100 80 GB for

Table 7. Inference speed (RTF) across hardware.

System	CUDA	MPS (M4)	CPU
Qwen3-TTS	0.67	1.85	8.42
Orpheus-TTS	0.82	2.31	12.57
Orpheus (vLLM)	**0.19**	N/A	N/A
Fish-Speech	0.41	N/A	N/A

full FT—a ~4× training-time and 5–10× VRAM saving with no statistically significant quality gap (Sect. 6.2).

7 Discussion

7.1 Progress, Architecture Fit, and Azerbaijani Phonology

Compared to [1], the three modern architectures replace mel-spectrograms + separate vocoders with neural-codec end-to-end token prediction; the audiobook corpus (~27 h vs. 24 h) also broadens prosodic diversity. Fish-Speech (860M) reaches competitive quality at ~3.5× fewer parameters than Orpheus (3B), so design partly compensates for capacity: SNAC's hierarchical codebook captures fine-grained prosody, DAC VQ offers a compact alternative, while Qwen3-TTS's 12 Hz rate may under-represent rapid phonetic transitions. Persistent Azerbaijani-specific failure modes across systems (**RQ4**): vowel-harmony violations across morphological boundaries; mispronunciation of ə/ö/ü/ç/ş/ğ; agglutinative compounds straining tokenization; inconsistent final-syllable stress.

7.2 Practical Recommendations

(1) Use LR ~5–10× smaller than the pretraining rate; (2) prefer LoRA (r=32, α=64) over full FT—the quality gap is not statistically detectable but the compute saving is ~8.6×; (3) standardize to 24 kHz mono; (4) invest in transcript accuracy over quantity; (5) benchmark inference on the deployment backend (CUDA/MPS/CPU).

7.3 Hypotheses for Other Languages (Not Validated Here)

The empirical conclusions of this paper are restricted to one language (Azerbaijani), one voice (single female speaker), one domain (audiobook narration), and the three architectures evaluated. We did not test other languages, did not run data ablations below 27 h, and did not evaluate non-Latin scripts. The following are therefore *hypotheses*, not findings: (i) languages typologically closer to the English/Mandarin pretraining mix (Turkish, Uzbek) may require less adaptation data than Azerbaijani; phonologically distant languages (tonal, click-consonant) likely need larger corpora or phoneme preprocessing; (ii) LLM-derived tokenizers (as in Orpheus) may extend their robust Azerbaijani character handling to non-Latin scripts, but we have not tested this; (iii) LoRA's parameter-efficient adaptation may enable acceptable quality at 5–10 h training data, but our experiments only cover 27 h, and the small-data regime is where forgetting/overfitting risks rise; (iv) cross-architecture rankings are language-dependent and our single-language data point cannot ground a general decision framework. The narrower supportable claim is methodological: the workflow (corpus curation, normalization, conservative LR fine-tuning, MCD/CER/MOS plus the statistical protocol of Sect. 5.4) is language-agnostic by construction; quantitative quality claims for any specific other language require new experiments.

7.4 Limitations

(1) **Single-female-speaker dataset (primary).** The corpus is one female voice, so trained models reproduce that voice only, and architecture-quality conclusions are confounded with speaker identity. Mitigations and the planned multi-speaker, gender-balanced extension are in Sect. 8 and Sect. 9. (2) **Audiobook domain only**—no conversational, news, or expressive-speech evaluation. (3) **Evaluator pool** of 12 raters provides acceptable reliability ($\alpha = 0.71$) and sufficient power, but a larger, demographically broader panel would strengthen borderline contrasts. (4) **No prosody/emotion/controllability evaluation.** (5) **MCD/CER–MOS correlation is imperfect** [30]. (6) **Cross-lingual and $<$10 h claims are conjectural** (Sect. 7.3). (7) **No formal equivalence test** (TOST) for LoRA vs. full FT—we only report failure to reject equality.

8 Ethical Considerations and Responsible Use

High-quality TTS for low-resource languages is dual-use: it expands access for underserved populations and simultaneously lowers the barrier to fraud and impersonation. **Voice cloning risk and mitigations.** Our Orpheus checkpoint reproduces a single identifiable voice at MOS = 4.31, and the underlying LLaMA-3+SNAC architecture is known to support prompt-conditioned speaker imitation [11]. The audiobook narrator gave explicit written consent for academic research use, with no unrestricted public model release; fine-tuned checkpoints will not be released publicly, and any review/reproducibility demos will be limited-access, time-bounded, and watermarked with an inaudible perceptual watermark. We will not provide weights, fine-tuning, or inference support for impersonation of real individuals. We encourage parallel investment in synthetic-speech detection for Turkic languages, where detector quality lags generator quality.

Dataset Diversity (Single-Female-Speaker Bias). A single voice is a substantive shortcoming, not a minor caveat. The trained models reproduce one female voice only; they are not multi-speaker; architecture-quality conclusions are confounded with speaker identity; and a publicly visible all-female Azerbaijani TTS landscape risks entrenching gendered associations of synthetic speech with assistant/service applications. Dialect coverage is also limited (standard Baku-region only). As a committed mitigation—elevated from planned future work to deliverable in response to reviewer feedback—we are recruiting at least four additional speakers (two male, two female) across age groups and dialect regions, after which the three-architecture comparison will be rerun and reported. Any conclusion about gender or dialect generalization should be drawn from that follow-up rather than this paper. **Evaluators** were 12 paid native speakers (above local RA rate), briefed on withdrawal rights; no sensitive content was rated, no evaluator personal data is reported. **Compute and recommendations.** Total training compute was ~180 GPU-hours (A100); LoRA reduces this ~8.6× and is our recommended default. Downstream users should obtain revocable speaker consent, watermark generated audio, authenticate API access, refrain from deploying the single-speaker model where a male, child, or dialect-distinct voice is appropriate, and publish a datasheet [36] for any new corpus.

9 Conclusion and Future Work

We presented a statistically validated three-architecture comparison for Azerbaijani TTS, fine-tuning Qwen3-TTS (1.7B), Orpheus-TTS (3B), and Fish-Speech (860M) on 27,888 audiobook utterances (~27 h), with paired Wilcoxon tests, Bonferroni correction over six pairwise contrasts, and rank-biserial/d_z effect sizes. Key findings: (1) the catastrophic-forgetting boundary for Qwen3-TTS is sharp and lies near $\eta = 10^{-5}$; 2×10^{-6} stabilizes; (2) LoRA (r=32) shows no statistically significant difference from full FT on any metric ($p > 0.05$, $|d_z| \leq 0.13$)—failure to reject equality, not proof of equivalence; (3) all three modern systems significantly beat the Tacotron/DC TTS baselines by 0.19–0.59 MOS points (d_z from 0.42 to 1.32), with $\alpha = 0.71$ inter-rater reliability across 12 native evaluators. The paper does *not* establish quality on any other language, speaker, gender, domain, or training budget below 27 h.

Future work, by priority: (1) multi-speaker gender-balanced Azerbaijani corpus (committed); (2) 5 h/10 h/27 h data-ablation across the three architectures; (3) parallel cross-lingual study over $\geq$ 3 Turkic languages under matched budgets; (4) TOST-based equivalence testing for LoRA vs. full FT; (5) prosody/emotion control; (6) watermarked restricted-access demo with synthetic-speech detection benchmarks.

Acknowledgments. This work was developed within the Center of Data Analytics Research (CeDAR) and the MegaSec AI Laboratory at ADA University, and the Engineering Mathematics and Artificial Intelligence Department at Azerbaijan Technical University. The authors thank the audiobook narrators whose recordings made this dataset possible, and the evaluators who participated in the listening tests.

Disclosure of Interests. The authors have no competing interests to declare that are relevant to the content of this article.

References

1. Valizada, A., Jafarova, S., Sultanov, E., Rustamov, S.: Development and evaluation of speech synthesis system based on deep learning models. Symmetry **13**(5), 819 (2021)
2. Valizada, A., Akhundova, N., Rustamov, S.: Development of speech recognition systems in emergency call centers. Symmetry **13**(4), 634 (2021)
3. Agalarov, M., Huseynova, B., Mammadov, E., Guliyev, G.: SARA: synthetic azerbaijani recording alignment - a unified TTS framework. J. Adv. Inf. Technol. **16**(2), (2025)
4. Qwen Team: Qwen3-TTS: A scalable multilingual text-to-speech model. Technical Report, Alibaba Group (2025)
5. Labs, C.: Orpheus-TTS: LLaMA-based text-to-speech with SNAC audio codec. Technical Report (2025)
6. Audio, F.: OpenAudio S1: Open-source speech generation model. Technical Report (2025)
7. Wang, Y.: Tacotron: towards end-to-end speech synthesis. In: Proceedings of the Interspeech, pp. 4006–4010 (2017)
8. Shen, J.: Natural TTS synthesis by conditioning WaveNet on mel spectrogram predictions. In: Proceedings of the ICASSP, pp. 4779–4783 (2018)
9. Ren, Y.: FastSpeech: fast, robust and controllable text to speech. In: Proceedings of the NeurIPS, pp. 3171–3180 (2019)
10. Ren, Y.: FastSpeech 2: fast and high-quality end-to-end text to speech. In: Proceedings of the ICLR (2021)
11. Wang, C.: Neural codec language models are zero-shot text to speech synthesizers (2023). arXiv:2301.02111
12. Borsos, Z.: AudioLM: a language modeling approach to audio generation. IEEE/ACM Trans. Audio Speech, Lang. Proc. **31**, 2523–2536 (2023)
13. Kim, J., Kong, J., Son, J.: Conditional variational autoencoder with adversarial learning for end-to-end text-to-speech. In: Proceedings of the ICML, pp. 5530–5540 (2021)
14. Chen, Y.: F5-TTS: A fairytaler that fakes fluent and faithful speech with flow matching. arXiv:2410.06885 (2024)
15. Le, M.: Voicebox: text-guided multilingual universal speech generation at scale. In: Proceedings of the NeurIPS (2024)
16. Tan, X., et al.: NaturalSpeech: end-to-end text-to-speech synthesis with human-level quality. IEEE Trans. PAMI **46**(6), 4234–4245 (2024)
17. Siuzdak, H.: SNAC: Multi-scale neural audio codec. arXiv:2410.14411 (2024)
18. van den Oord, A., Vinyals, O., Kavukcuoglu, K.: Neural discrete representation learning. In: Proceedings of the NeurIPS, pp. 6306–6315 (2017)
19. Hu, E.J.: LoRA: Low-rank adaptation of large language models. In: Proceedings of the ICLR (2022)
20. Dettmers, T.: QLoRA: efficient finetuning of quantized language models. In: Proceedings of the NeurIPS, (2023)
21. Rajbhandari, S.: ZeRO: memory optimizations toward training trillion parameter models. In: Proceedings of the SC, pp. 1–16 (2020)
22. Pratap, V., et al.: Scaling speech technology to 1,000+ languages. arXiv:2305.13516 (2023)
23. Chen, Y.: Cross-lingual, multi-speaker text-to-speech synthesis using neural speaker embedding. In: Proceedings of the Interspeech, pp. 2105–2109 (2019)

24. Chen, M.: Few-shot text-to-speech with meta learning. In: Proceedings of the ICASSP, pp. 6534–6538 (2019)
25. Johanson, L., Csató, É.Á.: The Turkic Languages. Routledge (2015)
26. Gökay, R., Yıldırım, A.: End-to-end text-to-speech synthesis for Turkish. In: Proceedings of the SIU, pp. 1–4 (2019)
27. Mussakhojayeva, S., Janaliyeva, R., Mirzakhmetov, A., Yessenbayev, H.A.: KazakhTTS2: extending the open-source Kazakh TTS corpus. In: Proceedings of the LREC (2022)
28. Muhammadiyev, B., et al.: Development of language models for continuous Uzbek speech recognition. Sensors **23**(3), 1145 (2023)
29. Kirkpatrick, J., et al.: Overcoming catastrophic forgetting in neural networks. Proc. Natl. Acad. Sci. **114**(13), 3521–3526 (2017)
30. Kubichek, R.F.: Mel-cepstral distance measure for objective speech quality assessment. In: Proceedings of IEEE Pacific Rim Conference on Communications Computers and Signal Processing, pp. 125–128 (1993)
31. Wilcoxon, F.: Individual comparisons by ranking methods. Biometrics Bull. **1**(6), 80–83 (1945)
32. Bonferroni, C.E.: Teoria statistica delle classi e calcolo delle probabilità. Pubblicazioni del R Istituto Superiore di Scienze Economiche e Commerciali di Firenze **8**, 3–62 (1936)
33. Benjamini, Y., Hochberg, Y.: Controlling the false discovery rate: a practical and powerful approach to multiple testing. J. Royal Stat. Soc. B **57**(1), 289–300 (1995)
34. Cohen, J.: Statistical Power Analysis for the Behavioral Sciences, 2nd edn. Lawrence Erlbaum Associates, Hillsdale, NJ (1988)
35. Krippendorff, K.: Content Analysis: An Introduction to Its Methodology, 4th edn. Sage, Thousand Oaks, CA (2018)
36. Gebru, T., et al.: Datasheets for datasets. Commun. ACM **64**(12), 86–92 (2021)

Federated Learning Approach to Autoencoder-Based Feature Extraction for Pneumonia Detection

Elshan Naghizade(✉) and Samir Aliyev

Azerbaijan State Oil and Industry University, Baku, Azerbaijan
{elshan.naghizade,aliyev.samir}@asoiu.edu.az

Abstract. Privacy and data-governance constraints often prevent chest X-ray images from being pooled across hospitals, even when multi-site data would improve pneumonia detection models. A federated learning approach was therefore studied to enable collaborative feature learning without transferring raw patient images. A two-stage pipeline was implemented. First, an autoencoder was trained in a federated manner to learn compact representations from chest X-rays. Second, classification was performed on the resulting latent vectors outside the federated loop, so that label-dependent updates were not shared between institutions. The Chest X-Ray Images (Pneumonia) dataset was used, and binary classification was defined as NORMAL versus PNEUMONIA. Images were resized to 2048 $\times$ 2048 with aspect-ratio preservation and black padding, and grayscale intensities were normalized to [0, 1]. Autoencoder architectures were selected by grid search, and a 5-layer encoder–decoder with a 1024-dimensional latent space was chosen. Federated training was executed on six Ubuntu machines, with five unevenly sized clients and weighted aggregation based on local sample counts. After 200 rounds, an MSE of 0.0111 was obtained. Tuned Multilayer Perceptron (MLP), XGBoost, LightGBM, and logistic regression classifiers were then evaluated on the latent vectors, and a baseline CNN was trained on images. The autoencoder-plus-MLP pipeline achieved the best overall performance, supporting federated representation learning as a practical option for privacy-sensitive medical archives.

Keywords: Federated Learning · Autoencoders · Feature Extraction · Neural Networks · Gradient Boosting

1 Introduction

Pneumonia remains a serious global health problem, especially when diagnosis is delayed. In routine care, chest X-ray imaging is widely used because it is fast, available in many clinical settings, and less costly than more advanced imaging methods. However, chest X-ray interpretation is still influenced by heavy clinical workload, differences in reader experience, and the limited availability of expert radiological review. Because of these limitations, automated pneumonia detection has been studied as an important support tool, and strong deep learning results have already been reported for chest X-ray analysis [1].

C. Zanni-Merk et al. (Eds.): AIDT 2026, CCIS 3023, pp. 294–309, 2026.
https://doi.org/10.1007/978-3-032-31319-5_20

Even so, many existing solutions have been developed under a centralized learning setting. In that setting, images from different hospitals or archives are gathered into one shared location for model training. This assumption is convenient for research, but it is often difficult to satisfy in real medical environments. Patient data are protected by privacy rules, institutional policies, and legal restrictions. In addition, hospital archives are usually maintained separately, and direct sharing of medical images across institutions is often limited by security, infrastructure, and governance concerns. As a result, a clear problem is created: effective collaborative training is needed, but raw patient images often cannot be freely transferred between institutions [2].

Federated learning has been proposed as a practical response to this problem. In federated learning, model training is distributed across multiple sites, while raw data remain local. Only model updates are transmitted to a coordinating server, so centralized storage of sensitive images is avoided. This property makes federated learning especially attractive for healthcare, where patient confidentiality and institutional control must be preserved [2]. However, this approach does not fully remove privacy risk by itself. It has been shown that shared gradients or updates may still leak information in certain attack settings, even when raw images are never exchanged [3]. For this reason, the main challenge is not only to avoid centralized data pooling, but also to design the learning pipeline in a way that reduces privacy exposure during collaboration.

This challenge is directly addressed in the present study. In this work, a federated learning approach to autoencoder-based feature extraction is proposed for pneumonia detection from chest X-ray images. The core idea is based on a two-stage design. First, image representations are learned in federated form through an autoencoder. Second, classification is performed separately from the compressed latent vectors produced by that autoencoder. By this design, the federated stage is limited to representation learning, while the final label-dependent classifier is removed from the distributed synchronization process. This structure is important because it offers a more privacy-conscious solution than simply placing a full supervised classifier inside the federated loop.

The proposed idea is relevant because it responds to two existing problems at the same time. The first problem is data-sharing difficulty across hospitals and medical archives. The second problem is the residual privacy risk that may remain when fully supervised federated models are trained with label-related updates. In the proposed framework, these problems are addressed together. Raw chest X-ray images are kept local during federated learning, and the downstream classifier is trained only after compact latent vectors have already been obtained. As a result, collaborative learning is still enabled, but the most privacy-sensitive part of the supervised decision process is separated from the federated stage. In this study, that separation is treated as a practical privacy-oriented solution rather than only as an architectural preference [3].

The proposed workflow is also made relevant by the way it reflects real medical conditions. In actual hospital networks, data are rarely distributed evenly across institutions. Large hospitals usually hold much larger image archives, while smaller clinics or regional centers may hold fewer cases. Because of this, a balanced client setting does not fully represent real deployment conditions. To reflect this issue, federated training is performed in this work under an intentionally uneven client distribution. Five client machines and one server machine are used, and the available data are distributed across the clients in unequal proportions. Aggregation is then performed through federated weighted averaging, where each client contributes according to its local sample size. In this way, the proposed method is evaluated under a setting that more closely resembles real collaborative medical environments.

The study is also designed so that the value of the proposed idea can be clearly measured. The autoencoder architecture is first validated to determine a suitable structure for representation learning. After federated training is completed, a neural network classifier is trained on the latent vectors, and its architecture is also validated before use. This autoencoder-plus-MLP pipeline is then compared against a baseline convolutional neural network trained directly on the full image set. In addition, the learned latent vectors are tested with LightGBM, XGBoost, and logistic regression so that the usefulness of the extracted representations can be examined across different downstream classifiers. Through these comparisons, the contribution of the proposed framework is not limited to a single model result; instead, the effectiveness of the overall design choice is evaluated in a structured way [4].

2 Related Work

2.1 Autoencoders

Autoencoders are unsupervised neural networks designed to compress input data into a lower-dimensional representation and then reconstruct the original input from this compact form. The foundational work by Hinton and Salakhutdinov [5] revolutionized this field by demonstrating that deep, multilayered AEs could outperform traditional methods like Principal Component Analysis (PCA) for non-linear dimensionality reduction. By utilizing a "bottleneck" architecture, these models force the network to discard noise and retain only the most essential structural features of the data. To improve the robustness of these learned features, Vincent et al. [6] introduced the Denoising Autoencoder (DAE), which trains the network to reconstruct clean data from partially corrupted inputs. This innovation shifted the focus of AEs from simple data compression to the extraction of highly resilient representations that capture the underlying manifold of the data.

The field advanced further with the introduction of Variational Autoencoders (VAEs) by Kingma and Welling [7], which combined traditional autoencoding with Bayesian inference. Instead of mapping inputs to a single fixed point in the latent space, VAEs map them to a probability distribution, ensuring that the latent space is continuous and well-organized for downstream analytical tasks. For a broader perspective, Tschannen et al. [8] provide a comprehensive survey that categorizes various AE architectures and their theoretical properties. Collectively, these works establish that the latent space generated by an AE is not merely a compressed version of the data, but a sophisticated feature map capable of supporting diverse supervised objectives.

2.2 Federated Learning

Federated Learning (FL) has emerged as a transformative paradigm in distributed machine learning, allowing multiple participants to collaboratively train a global model without sharing their raw, local data [9]. By keeping sensitive information on the original device and only transmitting model updates—such as gradients or weights—to a central server, FL inherently addresses many of the data residency and privacy concerns found in traditional centralized learning. This "privacy-by-design" approach is facilitated by algorithms like Federated Averaging (FedAvg), which aggregate local model parameters to form a generalized global model [10]. However, despite its decentralized nature, standard FL remains susceptible to sophisticated privacy threats, such as gradient inversion and membership inference attacks [11]. Consequently, the field is shifting toward enhancing FL with additional architectural safeguards to ensure true end-to-end data confidentiality.

2.3 Federated Learning in Healthcare

The application of FL to respiratory health, specifically for pneumonia detection in chest X-rays, has become a vital area of research to overcome the scarcity of large-scale, centralized medical datasets. Because chest X-rays contain sensitive patient metadata, traditional data-sharing methods are often blocked by privacy regulations [12]. Recent studies have demonstrated that FL can successfully train Deep Convolutional Neural Networks across multiple hospitals to identify inflammatory markers of pneumonia with performance comparable to centralized models [13]. However, pneumonia datasets are often highly "non-IID" (not identically and independently distributed), which can lead to model instability [14]. Furthermore, while FL protects the raw image, the supervised gradients from a pneumonia classifier can still be exploited to reconstruct sensitive patient attributes. This vulnerability highlights the need for a framework that can learn generalized pulmonary features without exposing the task-specific classification layers to the distributed network.

2.4 Federated Representation Learning with Autoencoders

The integration of AEs within FL frameworks has emerged as a potent paradigm for privacy-preserving representation learning. A foundational contribution by Shrestha et al. [15] demonstrated the viability of autoencoder-based anomaly detection within a federated framework while maintaining local data residency. To mitigate deep gradient leakage (DLG), Li et al. [16] proposed VAEFL, a framework that partitions models into private local encoders and public global decoders. However, because the classifier remains integrated within the federated loop, the system remains susceptible to label-inference attacks.

In the context of Edge-IoT security, Li et al. [17] explore vulnerabilities in decentralized systems using Variational Graph Auto-Encoders (VGAEs), demonstrating that tightly coupled architectures are highly susceptible to disruptions. While some hybrid frameworks, such as Shahin et al. [18], have recently attempted to mitigate these risks by decoupling the learning process into two stages for computational efficiency, there remains a critical research gap regarding the security benefits of this separation. Our work fills this gap by repositioning the two-stage architecture as a strategic privacy firewall. By "air-gapping" the classifier from the federated synchronization rounds, we ensure that sensitive label information is protected from adversaries, transforming the decoupled method into a robust defense mechanism for medical imaging tasks.

3 Methodology

3.1 Pipeline Overview and Data Preprocessing

A fixed experimental pipeline was defined so that the proposed contribution could be evaluated under controlled conditions. The study was organized around two linked stages. First, an autoencoder was used for feature extraction from chest X-ray images. Second, the learned feature vectors were used as inputs for downstream classifiers. By this design, the central objective of the study was addressed directly: privacy-aware feature learning was performed in federated form, while classification was performed after compact representations had already been obtained.

The study was based on the publicly available Chest X-Ray Images (Pneumonia) dataset, which contains pediatric chest radiographs organized into the folder labels NORMAL and PNEUMONIA [19]. In the present work, the task was defined strictly as binary classification. No distinction was made between bacterial and viral pneumonia. All images in the PNEUMONIA folder were treated as one class, and all images in the NORMAL folder were treated as the second class. This formulation was selected because the proposed method was intended to solve the practical screening problem of distinguishing normal radiographs from pneumonia-positive radiographs.

Only the images from the original TRAIN directory were used in the study. Within that directory, 1,341 images belonged to the NORMAL class and 3,875 images belonged to the PNEUMONIA class. Therefore, a total of 5,216 images were used in the full experimental workflow. The original class imbalance was preserved. No artificial balancing was introduced at the preprocessing stage, because the proposed pipeline was intended to be evaluated on the same class proportions contained in the selected working set. The TEST directory from the dataset was used in the Result to showcase the results.

Before any federated partitioning was performed, architecture validation of the autoencoder was carried out on the full working dataset. This step was intentionally executed first, because the optimal feature extractor had to be identified before federated training could begin. The validation process was performed on the server machine. A grid-search procedure was applied over candidate architectural settings, including the number of layers and the size of the latent representation. For this validation stage, 10% of the full working dataset was reserved as a validation subset, and the remaining portion was used for training during model selection. By this procedure, the autoencoder design was selected on the basis of reconstruction performance before any client-level data split was introduced. In this way, the federated stage was built on an already validated architecture rather than on an arbitrary design choice.

After the autoencoder architecture had been selected, all images were resized into a common spatial format. Each image was placed into a 2048 $\times$ 2048 frame. During this transformation, the original aspect ratio was preserved. The longer axis of the image was resized to 2048 pixels, and the shorter axis was scaled proportionally. The remaining unused area was filled with black padding. By this preprocessing rule, geometric distortion was avoided, while a uniform input size was still ensured for all samples.

This step was considered necessary because direct stretching of chest X-rays could alter anatomical proportions and could reduce the quality of the learned representations.

Once the autoencoder architecture had been fixed and the common preprocessing pipeline had been applied, the dataset was partitioned for federated training. Six Ubuntu-based machines were used. One machine served as the central server, and five machines served as clients. The data were split unevenly in order to reflect the fact that medical archives are rarely equal in size across institutions. The first client received 30% of the images, the second client received 25%, and each of the remaining three clients received 15%. This distribution was selected intentionally so that federated learning could be evaluated under a more realistic client imbalance scenario.

After federated feature learning, downstream classifiers were trained on the extracted feature vectors. The main multilayer perceptron, logistic regression, LightGBM, and XGBoost models were all selected through validation rather than by fixed manual settings. For these models, hyperparameter optimization was performed through grid search combined for validation. By this design, each downstream classifier was evaluated in its validated form, and all later comparisons were based on tuned rather than untuned models. In this way, the full methodology was structured around validated architecture selection, controlled preprocessing, realistic federated partitioning, and fair downstream model comparison.

3.2 Autoencoder Architecture Validation

Autoencoder architecture validation was performed before federated training so that the feature extractor used in the main pipeline would be aligned with the central objective of the study. In this work, the main objective was not only image reconstruction. A compact representation was also required that could preserve diagnostic structure well enough for later pneumonia classification, while still supporting a privacy-oriented two-stage workflow. For that reason, the autoencoder was not chosen by manual preference. It was selected through a controlled grid-search validation process.

This validation stage was carried out on the full working dataset defined in Sect. 3.1, before any federated split was introduced. The validation process was executed on the server machine so that one common architecture could be selected first and then distributed to all client nodes for federated training. All candidate autoencoders were trained on 90% of the dataset and evaluated on the remaining 10%, which was reserved as the validation set. The same split rule and the same preprocessing pipeline were applied to every candidate model. By this design, architectural differences remained the only meaningful source of variation during selection.

All tested models followed one shared structural principle. In the encoder, the input image was compressed stage by stage through convolutional processing and spatial downsampling. In the decoder, the compressed representation was expanded in the reverse direction so that the original image could be reconstructed. This choice was made because the study required a feature extractor that could force the image into a compact latent space before any downstream classifier was applied. Mean Squared Error (MSE) was used as the reconstruction loss, since the validation objective was defined as accurate recovery of the input image after compression.

A grid-search procedure was applied over the main architecture hyperparameters. The first search variable was encoder depth. Four encoder depths were tested: 4, 5, 6, and 7 convolutional layers. Because a mirrored design was used, the decoder depth was changed accordingly. The second search variable was latent vector size. Four latent dimensions were tested: 256, 512, 1024, and 2048. As a result, sixteen candidate autoencoders were evaluated under the same validation protocol (Table 1).

Table 1. Autoencoder spatial size for every layer.

Autoencoder Type	Spatial Size After Each Max Pooling Step
4-layer encoder / 4-layer decoder	2048 → 1024 → 512 → 256 → 128
5-layer encoder / 5-layer decoder	2048 → 1024 → 512 → 256 → 128 → 64
6-layer encoder / 6-layer decoder	2048 → 1024 → 512 → 256 → 128 → 64 → 32
7-layer encoder / 7-layer decoder	2048 → 1024 → 512 → 256 → 128 → 64 → 32 → 16

Residual-style skip connections were included in all tested autoencoders, because the study required stable training across both shallow and deeper encoder variants. These skip connections were inspired by the general design principle introduced in ResNet [20], but a standard ResNet backbone was not used. A custom reconstruction-oriented architecture was used instead, since the present work operated on 2048 $\times$ 2048 chest X-ray inputs and required symmetric encoder-decoder processing. In the encoder, each skip connection linked the input of a convolutional block to the output of that same block through element-wise addition after convolutional transformation. In the decoder, matching skip links were used in the same block-wise manner so that feature propagation could remain stable during reconstruction. By this design, gradient flow was supported across deeper models, and the validation process could focus on selecting the architecture that best served the study objective.

3.3 Federated Learning for the Autoencoder

Federated training was used so that the validated autoencoder could be trained across multiple machines without transferring raw chest X-ray images into one shared repository. This choice was made because the main objective of the study required privacy-aware feature learning under conditions that resemble separate hospital archives. For this stage, the validated autoencoder architecture selected in Sect. 3.2 was initialized on the server and then distributed to all client machines.

The federated environment was built on six Ubuntu 24.04 LTS machines connected through the same local network. One machine was used as the central server, and five machines were used as clients. All six machines had identical hardware: Intel Xeon W-2135 processors and 128 GB RAM. This hardware was sufficient for local processing of 2048 $\times$ 2048 padded grayscale chest X-ray images, but the image size still imposed a high memory and computation cost during convolutional training. For that reason, local training was performed with a batch size of 1 image per step on every client. By this choice, activation memory was kept under control during both forward and backward propagation.

The software stack was kept identical across all machines. TensorFlow and Keras were used for defining the selected autoencoder, loading model weights, performing local training, and returning updated parameters. NumPy was used for numerical array handling and parameter aggregation support. OpenCV and Pillow were used for image loading and for the fixed resizing-and-padding procedure defined in Sect. 3.1. Scikit-learn was used for validation control in earlier model selection and for downstream cross-validation in later classifier stages. Celery was used as the distributed task manager. Through Celery, each federated round was issued as a server-side task, client-side local training was executed as distributed work, and the updated weight tensors were returned to the server after completion (Fig. 1).

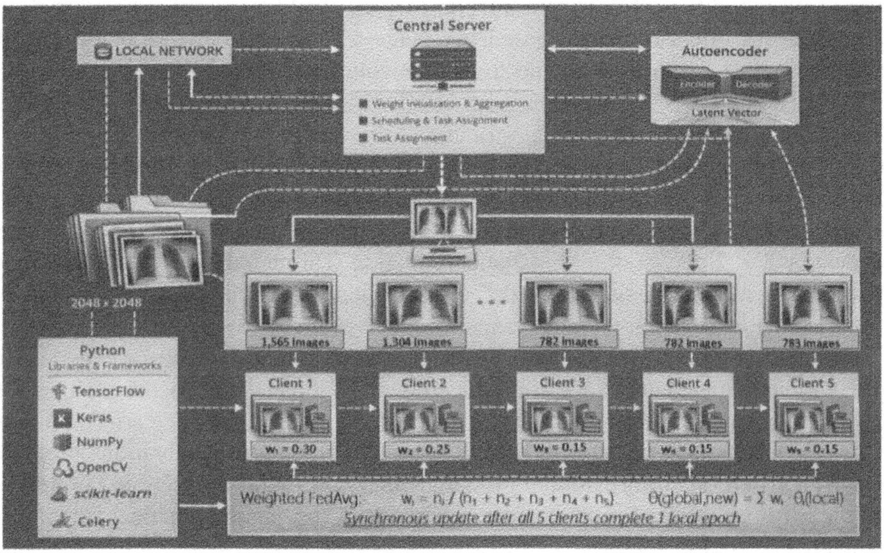

Fig. 1. Flowchart outlining the Federated Learning process for the Autoencoder.

The selected autoencoder was trained in synchronous federated rounds. At the start of each round, the current global model weights were broadcast from the server to all five clients. Each client then trained the model on its own local shard for 1 local epoch. After that single local epoch, the locally updated weights were sent back to the server immediately. Thus, the global model was updated once per federated round, after every 1 local epoch on each client. This update schedule was chosen because the dataset was distributed unevenly and the image size was large. More local epochs per round would have increased local drift and would have made synchronization slower.

The 5,216 images were partitioned unevenly across the clients: approximately 1,565 images were assigned to Client 1 (30%), 1,304 images were assigned to Client 2 (25%), and approximately 782, 782, and 783 images were assigned to Clients 3, 4, and 5, respectively (15% each). Aggregation was performed through weighted federated averaging where the client weights were approximately 0.30, 0.25, 0.15, 0.15, and 0.15. In each round, the new global parameter tensor was computed as the weighted sum of the returned client tensors under these coefficients. By this rule, larger local archives contributed proportionally more to the global autoencoder, while raw image data remained fully local.

3.4 The Main Multilayer Perceptron Classifier and the Baseline CNN

A baseline convolutional neural network was constructed so that the proposed autoencoder-based pipeline could be compared against a direct image classification approach under a closely related architecture. This baseline was not designed as an unrelated reference model. It was built to loosely mimic the encoder part of the autoencoder architecture selected in Sect. 3.2, so that the comparison would remain structurally fair. By this choice, the effect of the proposed two-stage feature extraction strategy could be evaluated against a conventional end-to-end convolutional classifier.

The baseline CNN received the same 2048 $\times$ 2048 preprocessed grayscale chest X-ray images as input. Its internal structure followed the same depth setting as the encoder of the validated autoencoder chosen in Sect. 3.2. Thus, the baseline used the same number of convolutional stages as the selected encoder. Each stage consisted of two convolutional layers with 3 $\times$ 3 kernels and same padding, followed by activation, and then by 2 $\times$ 2 max pooling for spatial reduction. Through this repeated pattern, the input image was compressed gradually across the network in the same general manner as in the selected encoder.

The main downstream classifier in the proposed pipeline was a multilayer perceptron (MLP) trained on the latent vectors produced by the validated autoencoder. This model was selected because the central objective of the study was not only to learn compressed image representations, but also to determine whether those representations preserved enough class-relevant information for reliable binary pneumonia detection. For this reason, the MLP was treated as the primary classifier for testing the practical value of the learned feature space.

The MLP did not receive raw chest X-ray images. Instead, it received only the encoded feature vectors extracted from the full working dataset after the autoencoder stage had been completed. In this way, the classification stage was fully separated from the image reconstruction stage. This separation was consistent with the paper objective, because the main contribution of the method depended on using privacy-oriented compressed representations rather than original radiographs as classifier input.

Model selection for the MLP was carried out through a controlled grid-search validation procedure. The search was performed on the feature vectors, not on the raw dataset images. For this process, 10% of the available encoded vectors were reserved as a validation subset, while the remaining 90% were used for training in each tested configuration. This design was used so that the classifier architecture could be selected under the same feature conditions later used in the final pipeline. The primary selection metric was macro F1-score. This metric was chosen because the working dataset was class-imbalanced, and the objective of the study required balanced predictive quality across both the NORMAL and PNEUMONIA classes rather than performance dominated by the larger class.

The grid search explored the main structural and optimization choices that could affect latent-space classification. First, network depth was varied. Candidate MLPs with 4, 5, 6, 7, 8, 9, 10, 11, 12, 13, 14, and 15 layers were tested. Second, the number of neurons in each hidden layer was varied so that both narrower and wider decision mappings from the latent space could be examined. Third, the activation function was treated as a search parameter, because the structure of the latent vectors could interact differently with different nonlinearities. Fourth, dropout was tuned at each layer so that regularization strength could be controlled explicitly (Table 2).

Table 2. The hyperparameter grid for the main MLP classifier

Hyperparameter	Tested Values
Number of hidden layers	4, 5, 6, 7, 8, 9, 10, 11, 12, 13, 14, 15
Neurons per hidden layer	64, 128, 256, 512, 1024
Hidden-layer activation function	ReLU, Leaky ReLU, ELU, Tanh
Output activation	Sigmoid
Optimizer	Adam, RMSprop, SGD, Nadam
Initial learning rate	0.0001, 0.0005, 0.001, 0.005
Learning-rate scheduler	None, Step, Exponential, Reduce-on-plateau, Cosine
Dropout rate per hidden layer	0.0, 0.1, 0.2, 0.3, 0.4, 0.5
Batch size	16, 32, 64, 128
Weight initialization	Glorot Uniform, He Normal
L2 regularization coefficient	0.0, 0.00001, 0.0001, 0.001

Each candidate MLP was trained for 50 epochs and then evaluated on the validation subset using macro F1-score.

3.5 Logistic Regression and Boosting Models

Logistic Regression, XGBoost, and LightGBM were included in the study so that the quality of the learned autoencoder feature space could be tested with classifiers of different complexity. This comparison was necessary for the paper objective. The proposed method was not intended to show only that the latent vectors could be used by one neural classifier. It also had to be shown whether those vectors remained useful when simpler and non-neural models were applied. For this reason, these three models were trained on the same encoded feature vectors used by the main MLP classifier.

Logistic Regression was used as the simplest baseline among the feature-based models. It was included to test whether the latent vectors could support direct linear separation. XGBoost and LightGBM were included because they can model more complex nonlinear relations in tabular feature spaces. This made them suitable for evaluating whether the encoded vectors contained decision patterns that could not be captured by a linear classifier. In this way, the study could compare simple, strong, and highly flexible feature-based classifiers under one common representation (Table 3).

Table 3. Grid-search space for Logistic Regression.

Hyperparameter	Tested Values
Penalty	L1, L2, Elastic Net
Inverse regularization strength (C)	0.01, 0.1, 1.0, 10.0, 100.0
Solver	liblinear, saga
Maximum iterations	100, 200, 500, 1000
Class weight	None, balanced
Elastic Net mixing ratio (l1_ratio)	0.2, 0.5, 0.8

Table 4. Grid-search space for XGBoost.

Hyperparameter	Tested Values
Number of estimators	100, 200, 300, 500
Maximum tree depth	3, 5, 7, 9
Learning rate	0.01, 0.05, 0.1, 0.2
Subsample ratio	0.6, 0.8, 1.0
Column sample by tree	0.6, 0.8, 1.0
Minimum child weight	1, 3, 5
Gamma	0.0, 0.1, 0.3, 0.5
L2 regularization (lambda)	1.0, 5.0, 10.0
L1 regularization (alpha)	0.0, 0.1, 1.0

Model selection for all three methods was performed through grid-search validation over 50 epochs using macro F1-score as the primary metric (Table 4). This metric was used because the working dataset was imbalanced and balanced performance across both classes was required. For each algorithm, 90% of the encoded vectors were used during validation training, and the remaining 10% were used for validation (Table 5).

Table 5. Grid-search space for LightGBM.

Hyperparameter	Tested Values
Number of estimators	100, 200, 300, 500
Number of leaves	15, 31, 63, 127
Maximum tree depth	-1, 5, 7, 9
Learning rate	0.01, 0.05, 0.1, 0.2
Subsample ratio	0.6, 0.8, 1.0
Feature fraction	0.6, 0.8, 1.0
Minimum child samples	10, 20, 40
Minimum split gain	0.0, 0.1, 0.3
L2 regularization (reg_lambda)	0.0, 0.1, 1.0, 5.0
L1 regularization (reg_alpha)	0.0, 0.1, 1.0, 5.0

4 Results

4.1 Autoencoder Selection and Federated Training

Autoencoder candidates were compared using reconstruction MSE on the validation split. All inputs were grayscale and were normalized so that each pixel value lay in [0, 1].

During architecture search, each candidate autoencoder was trained for 30 epochs. The search selected a 5-layer encoder / 5-layer decoder autoencoder with a 1024-dimensional latent vector with the validation MSE of 0.0139. The selected autoencoder was then trained with federated learning for 200 rounds (synchronous aggregation after every 1 local epoch) under the uneven client distribution. The full federated run required 102 h on the six-machine CPU cluster and yielded the validation MSE of 0.0111.

4.2 Hyperparameter Tuning Outcomes on Latent Vectors

All downstream models were tuned on the 1024-D latent vectors using macro F1 as the selection metric. For tuning, 90% of vectors were used for training and 10% for validation. MLP candidates were trained for 50 epochs during tuning, while the final selected MLP was trained for 200 epochs, which led to improved validation behavior and better final metrics.

MLP (selected hyperparameters):

Hidden layers: 8
Neurons: 1024 → 1024 → 512 → 512 → 256 → 256 → 128 → 64
Activations: ReLU
Dropout (layer-wise): [0.10, 0.20, 0.30, 0.30, 0.40, 0.30, 0.20, 0.10]
Optimizer: Adam
Initial LR: 0.001
LR scheduler: Reduce-on-plateau
Batch size: 64
Initialization: He Normal
L2: 0.0001
Tuning macro F1 (50 epochs): 0.928

XGBoost (selected hyperparameters).

n_estimators: 500
max_depth: 5
learning_rate: 0.05
subsample: 0.8
colsample_bytree: 0.8
min_child_weight: 3
gamma: 0.1
reg_lambda: 10.0
reg_alpha: 0.1
Tuning macro F1: 0.918
Final macro F1: 0.923 (Table 4.1)
LightGBM (selected hyperparameters).
n_estimators: 300
num_leaves: 127
max_depth: 9
learning_rate: 0.10
subsample: 0.6
feature_fraction: 0.8
min_child_samples: 40
min_split_gain: 0.0
reg_lambda: 1.0
reg_alpha: 0.0
Tuning macro F1 (50 epochs): 0.915

Logistic Regression (selected hyperparameters)
penalty: L2
C: 1.0
solver: liblinear
max_iter: 1000
class_weight: balanced
Tuning macro F1 (50 epochs): 0.735

4.3 Final Classification Results

The final evaluation was conducted on the separate TEST folder from the Kaggle dataset to avoid data leakage (Table 6).

Table 6. Final classification results on the test set.

Model	Accuracy	Recall (Positive)	Precision (Positive)	F1-score (Positive)	F1-score (Negative)	Macro F1
Baseline CNN	0.889	0.920	0.930	0.925	0.788	0.856
MLP	0.948	0.960	0.970	0.965	0.901	0.933
XGBoost	0.941	0.955	0.965	0.960	0.887	0.923
LightGBM	0.938	0.953	0.963	0.958	0.881	0.920
Log Reg	0.811	0.890	0.860	0.875	0.612	0.744

The baseline CNN scored lower macro F1 than the latent-vector models with the exception of Logistic Regression. This result was important for this dataset because training was performed on large 2048 × 2048 images, where end-to-end learning was harder to optimize on CPU-only hardware. The two-stage approach reduced that burden by moving most learning into the federated autoencoder and then operating on compact 1024-D vectors.

Among feature-based classifiers, the MLP performed best. This suggested that the latent vectors were informative but not purely linearly separable, since logistic regression performed much worse. XGBoost and LightGBM produced similar scores, which indicated that strong nonlinear boundaries could also be learned with tree ensembles, but a tuned deep MLP still extracted slightly more benefit from the learned representation.

5 Conclusion and Discussion

The results showed that the main value of the proposed method was obtained from federated feature learning followed by classification on the latent space. On the Chest X-Ray Images (Pneumonia) dataset, a clear class imbalance was present, and the use of macro F1 made the improvements on both NORMAL and PNEUMONIA outcomes visible. The lower reconstruction error after 200 federated rounds indicated that stable image structure was retained even under the uneven client split that was used to mimic different hospital archive sizes. This was important for the paper objective, because the classifier depended on consistent encoding of pulmonary patterns into the 1024-dimensional vectors.

The model comparisons also clarified how the learned representation behaved. The MLP achieved the best macro F1, which suggested that the encoded vectors contained class information that was not fully linear and benefited from deep non-linear mapping. The close performance of XGBoost and LightGBM indicated that strong separation was also supported with boosting on compact vectors, which strengthened the claim that the latent space carried meaningful diagnostic signals. The weaker logistic regression results indicated that linear separation was insufficient. The baseline CNN produced lower macro F1 under a similar convolutional depth, which supported the two-stage design for large 2048×2048 inputs and a privacy-aware workflow where raw images were not required beyond representation learning.

References

1. Kermany, D.S., et al.: Identifying medical diagnoses and treatable diseases by image-based deep learning. Cell **172**(5), 1122–1131 (2018)
2. Kaissis, G.A., Makowski, M.R., Rückert, D., Braren, R.F.: Secure, privacy-preserving and federated machine learning in medical imaging. Nat. Mach. Intell. **2**(6), 305–311 (2020)
3. Geiping, J., Bauermeister, H., Dröge, H., Moeller, M.: Inverting gradients: how easy is it to break privacy in federated learning? In: Advances in Neural Information Processing Systems (NeurIPS) **33**, 16937–16947 (2020)
4. Vepakomma, P., Gupta, O., Swedish, T., Raskar, R.: Split learning for health: distributed deep learning without sharing raw patient data. arXiv preprint arXiv:1812.00564 (2018)
5. Hinton, G.E., Salakhutdinov, R.R.: Reducing the dimensionality of data with neural networks. Science **313**(5786), 504–507 (2006)
6. Vincent, P., Larochelle, H., Lajoie, I., Bengio, Y., Manzagol, P.-A.: Stacked denoising autoencoders: learning useful representations in a deep network with a local denoising criterion. J. Mach. Learn. Res. **11**, 3371–3408 (2010)
7. Kingma, D.P., Welling, M.: Auto-encoding variational Bayes. In: 2nd International Conference on Learning Representations (ICLR) (2014)
8. Tschannen, M., Bachem, O., Lucic, M.: A survey of autoencoders: properties, theoretical improvements, and applications. arXiv preprint arXiv:1803.05815 (2018)
9. McMahan, B., Moore, E., Ramage, D., Hampson, S., Agüera y Arcas, B.: Communication-efficient learning of deep networks from decentralized data. In: Proceedings of the 20th International Conference on Artificial Intelligence and Statistics (AISTATS), pp. 1273–1282 (2017)
10. Konečný, J., McMahan, B.C., Yu, F.X., Richtárik, P., Suresh, A.T., Bacon, D.: Federated learning: strategies for improving communication efficiency. arXiv preprint arXiv:1610.05492 (2016)
11. Zhu, L., Liu, Z., Han, S.: Deep leakage from gradients. In: Advances in Neural Information Processing Systems (NeurIPS), vol. 32 (2019)
12. Rieke, N., Hancox, J., Li, W., et al.: The future of digital health with federated learning. Nat. Med. **26**, 814–826 (2020)

13. Sheller, M.J., Edwards, B., Reina, G.A., et al.: Federated learning in medicine: facilitating multi-institutional collaborations without sharing patient data. Sci. Rep. **10**(1), 1–12 (2020)
14. Li, T., Sahu, A.K., Talwalkar, A., Smith, V.: Federated learning: Challenges, methods, and future directions. IEEE Signal Process. Mag. **37**(3), 50–60 (2020)
15. Shrestha, R., et al.: Anomaly detection based on LSTM and autoencoders using federated learning in smart electric grid. J. Parallel Distrib. Comput. **193**, 104951 (2024)
16. Li, Z., et al.: VAEFL: Integrating variational autoencoders for privacy preservation and performance retention in federated learning. Secur. Saf. **3**, 2024005 (2024)
17. Li, K., Hu, S., Wu, B., Zou, S., Ni, W., Dressler, F.: Undermining federated learning accuracy in EdgeIoT via variational graph auto-encoders. arXiv preprint arXiv:2504.10067 (2025)
18. Shahin, M., Hosseinzadeh, A., Chen, F.F.: A two-stage hybrid federated learning framework for privacy-preserving IoT anomaly detection and classification. IoT **6**(3), 48 (2025)
19. Mooney, P.: Chest X-Ray Images (Pneumonia). Kaggle dataset (2018)
20. He, K., Zhang, X., Ren, S., Sun, J.: Deep residual learning for image recognition. In: Proceedings of the IEEE Conference on Computer Vision and Pattern Recognition (CVPR), pp. 770–778 (2016)

Benchmarking Algorithmic Fairness in AI Recruitment: A Comparative Study of Mitigation Strategies and the FIS Metric

Nigar E. Adilova[1(✉)], Omer Mutlu Turk Kaya[2], Yunis Allahverdi[1], Fatima Naghizada[1], and Nurlan Orujov[1]

[1] Azerbaijan State Oil and Industry University, Azadlyg Ave 20, AZ1010 Baku, Azerbaijan
nigar.adilova@asoiu.edu.az

[2] Department of Computer Engineering, Yildiz Technical University, Istanbul, Turkey
omer.kaya@yildiz.edu.tr

Abstract. This paper presents a comprehensive benchmarking analysis of bias mitigation strategies within AI-driven recruitment systems. Utilizing a large-scale dataset of over 70,000 job applications, we systematically evaluate the performance of six machine learning architectures Logistic Regression, Random Forest, XGBoost, SVM, KNN, and Naive Bayes under various fairness constraints. Our benchmark covers the entire ML lifecycle, implementing pre-processing (Reweighing/Suppression), in-processing (Exponentiated Gradient) and postprocessing (Threshold Optimization) techniques. We introduce a comparative framework based on the Fairness Impact Score (FIS) to quantify the trade-offs between predictive precision and demographic parity. The results expose a "Fairness-Accuracy Mirage" in ensemble-based models, whereas linear and probabilistic models demonstrate superior stability across shifting data distributions. This benchmark provides a standardized roadmap for auditors and developers to select optimal model mitigation pairings for building legally compliant and ethical hiring platforms.

Keywords: Algorithmic Fairness · Bias Mitigation · AI Recruitment · Benchmarking · Fairness Accuracy Trade-off · Trustworthy AI

1 Introduction

The rapid advancement of Artificial Intelligence (AI) has initiated a profound digital transformation in recruitment where automated systems increasingly replace traditional manual screening [1, 3]. While AI promises enhanced efficiency, its deployment introduces significant ethical risks regarding algorithmic bias and systemic discrimination [2, 14]. If training data reflects historical human prejudices, unmitigated models risk replicating "Black Boxes" of automated inequality, threatening corporate responsibility and legal compliance with emerging regulations such as the EU AI Act [5, 8, 15].

This study provides a comprehensive benchmarking analysis of six distinct architectures Logistic Regression, Random Forest, XGBoost, SVM, KNN, and Naive Bayes

C. Zanni-Merk et al. (Eds.): AIDT 2026, CCIS 3023, pp. 310–324, 2026.
https://doi.org/10.1007/978-3-032-31319-5_21

evaluating their robustness in automated hiring scenarios [8]. Using a large-scale dataset of over 70,000 job applications [19], we establish a performance baseline by measuring how different algorithms and mitigation strategies interact. The contribution of this benchmark are threefold:

1. Algorithmic Benchmarking: A comparative assessment of six ML classifiers to identify which architectures are inherently more prone to demographic bias under future scaling variations [6, 16].
2. Mitigation Strategy Evaluation: A systematic cross-comparison of preprocessing, in-processing (Fairlearn), and post-processing techniques [4, 11, 17] to determine the most effective intervention point in the recruitment pipeline.
3. Stability and Trade-off Analysis: Utilizing the Fairness Impact Score (FIS) to benchmark the trade-off between predictive accuracy and parity [12, 18], specifically and addressing the "Fairness-Accuracy Mirage" in ensemble configurations [9].

By providing a structured comparison of these technical approaches through rigorous statistical evaluation [20, 22], this benchmark serves as a practical guide for deploying reliable and compliant AI-driven employability scoring systems.

2 Problem Statement and Methodology

2.1 Dataset Overview and Preprocessing

In this study, experiments were conducted using a dataset which features a comprehensive collection of data associated with job applicants and their employability scores, publicly available on Kaggle [19]. The target variable is 'Employed', a binary label indicating whether an applicant was hired. Following preprocessing, nine features were retained for model training, as summarized in Table 1.

In our experiment, 'Gender' is defined as the protected sensitive attribute, meanwhile 'Employed' acts as the target variable for classification. For alignment with the machine learning algorithms, the 'Gender' attribute was binarized, whereas male and female individuals were mapped as 1 and 0, respectively. Similarly, the binary categorical features Age (>35), MainBranch, Accessibility, and MentalHealth were label-encoded as 0 or 1. Duplicate records were removed to prevent data leakage, and rows with missing values in the skills-related feature were dropped. To address multicollinearity, a Pearson correlation analysis was conducted, identifying YearsCodePro as highly correlated ($r > 0.8$) with YearsCode; consequently, YearsCodePro was removed from the feature set. Outlier detection was performed on the three numerical features—YearsCode, PrevSalary, and Num_of_skills—using the interquartile range (IQR) method.

Standardization was applied to all numerical features using the StandardScaler function [6]. For baseline model evaluation, 10-fold stratified cross-validation was employed. Kohavi (1995) [20] demonstrated through extensive empirical analysis that ten-fold stratified cross-validation yields the most reliable model evaluation on real-world datasets, which informed our choice. Results are reported as mean ± standard deviation across folds. Bias mitigation and robustness experiments were conducted using an 80/20 train-test split with random_state = 42.

The Wilcoxon signed-rank test was additionally applied to assess statistical significance of pairwise model differences, with results reported in Sect. 3.1.

Table 1. Job Applicants Dataset

ID	Age (>35)	Employment	Gender	Experience	PrevSalary	Num_of_skills	...	Employed
0	0	1	1	7	51552	4	...	0
1	0	1	0	12	46482	12	...	1
2	1	1	1	15	77290	7	...	0
3	0	1	0	9	46135	13	...	0
4	1	0	1	40	160932	2	...	0

2.2 Baseline Models

Logistic Regression estimates the probability of an outcome (between 0 and 1) by applying the sigmoid function to a linear combination of input variables

$$P(Y = 1) = \sigma(z) = \frac{1}{1 + e^{-z}}$$

where e – base of the natural logarithm (~2.718), z – represents the log-odds of the event occurring.

Random Forest demonstrates superior performance compared to Logistic Regression across a wide range of datasets [9]. Unlike other methods, Random Forest can be applied to both classification and regression tasks. For classification, the class predicted by most trees win. By default, scikit uses Gini impurity for classification which is a metric for deciding how to split data at each node of the individual decision trees. The Gini impurity (G) for a node is defined as:

$$G = 1 - \sum\nolimits_{i=1}^{n} p_i^2$$

where n – number of classes, p_i – the proportion of samples belonging to class i in that node.

After calculating Gini impurity for trees, each individual tree makes its own prediction for a data point, then the forest collects these results and picks the majority class as the final output.

XGBoost (XGB) is an effective and powerful algorithm used for both classification and regression. For classification, it involves predicting categorical labels by step-by-step adding decision trees that improve the predictions of previous iterations. At iteration t, a new tree f_t is added to the existing model:

$$y_i^{(t)} = y_i^{(t-1)} + f_t(x_i),$$

the objective function to minimize the loss is:

$$\mathcal{L}^{(t)} = \sum_{i=1}^{n} l\left(y_i, y_i^{(t-1)} + f_t(x_i)\right) + \Omega(f_t)$$

where $\mathcal{L}$ – the loss function measuring the difference between true and predicted values, $\Omega(f_t)$– the regularization term that penalizes model complexity and helps prevent overfitting.

$$\Omega(f) = \gamma T + \frac{1}{2}\lambda \sum_{j=1}^{T} w_j^2$$

K-Nearest Neighbor (KNN) is a non-parametric learning algorithm generally commonly used for classification and regression tasks. The method operates by identifying the k closest observations to a given input and generating predictions. By default, the KNN implementation in the scikit-learn library uses the Minkowski distance, defined as:

$$D(x, y) = \left(\sum_{i=1}^{n} |x_i - y_i|^p\right)^{\frac{1}{p}}$$

where x, y – two vectors in an n – dimensional space, p – the parameter that determines the type of distance (e.g., p = 1 is Manhattan, p = 2 is Euclidean), $|x_i - y_i|$ - represents the absolute difference between the i th feature coordinates of x and y.

Support Vector Machines (SVM) is a powerful and flexible supervised ML algorithm widely used for classification and regression tasks. For a binary classification problem, the decision function is given by:

$$\hat{y} = \text{sign}\left(w^T x + b\right)$$

where $\hat{y}$ – predicted class label, w– the weight vector orthogonal to the hyperplane, b– the bias term, controlling the position of the hyperplane relative to the origin.

Naive Bayes (NB) is a probabilistic classification algorithm based on Bayes' theorem. It assumes that features are conditionally independent given the class label. For a given input $x = (x_1, x_2, \ldots, x_n)$, the classifier computes the posterior probability $P(y|x)$ for each class and predicts the class with the highest probability.

In Gaussian Naive Bayes, the likelihood of a feature x_i given class y is defined as:

$$P(x_i \mid y) = \frac{1}{\sqrt{2\psi_y^2}} e^{-\frac{(x_i - \mu_y)^2}{2\sigma_y^2}}$$

where π, e – mathematical constants, μ_y – the mean of feature x_i for class y, σ_y^2 – the variance of feature x_i for class y.

All models were implemented using the Scikit-learn library [10] with random_state = 42 applied across all stochastic components to ensure full reproducibility. For Logistic Regression, default parameters were used including C = 1.0 regularization and the lbfgs solver. Random Forest and XGBoost were each configured with n_estimators = 300, as higher ensemble sizes generally yield more stable predictions on large datasets. For KNN, the number of neighbors was set to k = 17, selected through preliminary experiments on the validation set. SVM was implemented using the LinearSVC variant with dual = False for computational efficiency on large datasets. Naive Bayes was implemented using the Gaussian variant, assuming continuous feature distributions consistent with the nature of the input features.

2.3 Fairness Metrics

To verify that our bias mitigation attempts effectively balance performance across demographic groups, we go beyond standard accuracy measurements. We used several fairness metrics to expose any latent inequalities in the model's predictive behavior. The metrics selected for this analysis include:

Statistical Parity Difference (SPD): Measures the gap in the rate of favorable outcomes received by female and male candidates. We define SPD as:

$$SPD = P(\hat{y} = 1|\text{female}) - P(\hat{y} = 1|\text{male})$$

A value of 0 demonstrates perfect fairness, where both genders receive the favorable outcome at the same rate. A negative value indicates a bias in favor of male candidates.

Disparate Impact (DI): Calculates the ratio of favorable outcomes between the female and male groups. A DI value below 0.8 typically indicates a 'four-fifths rule' violation, according to US law census. We define DI as:

$$DI = \frac{P(\hat{y} = 1|\text{gender} = \text{female})}{P(\hat{y} = 1|\text{gender} = \text{male})}$$

A value of 1 indicates ideal parity, whereas values below 0.8 are generally considered evidence of significant disparate impact against the female group.

Equal Opportunity (EOD): Unlike SPD, it measures the difference between the male and female candidates but for the "qualified" ones. EOD is defined as:

$$EOD = P(\hat{y} = 1|y = 1, \text{female}) - P(\hat{y} = 1|y = 1, \text{male})$$

A value of 0 indicates that both genders have equal True Positive Rates. While, negative value demonstrates that qualified male candidates are prioritized over qualified female candidates.

Equalized Odds: This metric ensures that True Positive Rate (TPR) and False Positive Rate (FPR) are identical across genders. We define the violation of this standard using the Average Absolute Odds Difference:

$$EO = \frac{|TPR_{\text{female}} - TPR_{\text{male}}| + |FPR_{\text{female}} - FPR_{\text{male}}|}{2}$$

where $TPR = P(\hat{y} = 1|y = 1)$ (Probability of correctly identifying a qualified candidate), $FPR = P(\hat{y} = 1|y = 0)$ (Probability of incorrectly identifying a unqualified candidate).

A value of 0 indicates perfect equalized odds, ensuring that neither gender is unfairly biased.

Fairness Impact Score (FIS): To demonstrate a comprehensive comparative analysis, we propose the Fairness Impact Score (FIS). The FIS plays as a unified metric that aggregates these individual dimensions into a single value, representing the total bias.

By averaging the Parity Gap and Error Gap, the FIS allows us to quantify the equilibrium between different fairness constraints. An FIS value closer to 0 indicates a more robust model with minimal demographic deviation. The score is defined as:

$$FIS = 0.5 \times \left(\frac{|SPD| + |1 - DI|}{2}\right) + 0.5 \times \left(\frac{|EOD| + |AAOD|}{2}\right)$$

By utilizing FIS, we can objectively determine which model architecture maintains the most stable performance across varying levels of data imbalance, effectively balancing predictive success with demographic parity.

2.4 Bias Mitigation

Bias mitigation is the process of removing bias from data. The bias mitigation in our study is divided into three stages: pre-processing, in-processing and post-processing. We used libraries in this process, namely Scikit-learn [10] and Fairlearn [4].

Pre-Processing Technique aims to mitigate bias before the model is even trained. In this stage, we used two methods:

Suppression: Removing sensitive attributes (in our case, gender) from the dataset.

Reweighing: Applying different weights to samples in the training set so that male and female groups affect to the model's objective equally.

In Processing Technique modifies the ML algorithm itself to reduce bias during model training. In this study, we implemented Exponentiated Gradient which is a powerful algorithm from the Fairlearn library [4] that iteratively re-weights the training data to optimize the model's parameters under specific fairness constraints.

Post Processing Technique adjusts the decision threshold or labels of trained model to reduce unfairness. We employ the ThresholdOptimizer which adjusts the decision boundaries of a pre-trained model to satisfy fairness constraints using the Fairlearn library [4].

2.5 Robustness Testing

To evaluate the stability of our study, we subjected the models to a robustness analysis using both downsampling and oversampling techniques. We simulated five distinct gender distributions (30/70, 40/60, 50/50, 60/40, 70/30) for the male and female groups, respectively.

- Downsampling Strategy: We maintained the minority group (Female) at its maximum available count and reduced the majority group (Male) to satisfy the target ratios.
- Oversampling Strategy: We maintained the majority group count and utilized random oversampling with replacement for the minority group to achieve parity.

To ensure compatibility, the impact of these distribution shifts was measured using the Fairness Impact Score (FIS). This allowed us to observe how each model's total bias profile evolved as the dataset moved from female-majority to male-majority configurations. By using FIS, we can identify which models maintain low bias levels across all scenarios, regardless of the gender imbalance in the training data.

3 Results and Discussion

3.1 Baseline Performance

In this study, bias mitigation was evaluated across three stages: pre-processing, in-processing and post-processing. We implemented six machine learning models for comparative analysis. Before applying mitigation techniques, a baseline was established with 10-Fold Cross Validation and Wilcoxon's Signed-Rank by measuring bias in the original dataset. The initial results are shown in Table 2:

Table 2. Baseline Results with 10-Fold Cross-Validation (mean ± std)

	LR	RF	XGB	SVM	KNN	NB
SPD ↓	-0.087 ± 0.011	-0.070 ± 0.030	-0.081 ± 0.016	-0.088 ± 0.010	-0.105 ± 0.018	-0.071 ± 0.008
DI ↑	0.838 ± 0.020	0.874 ± 0.053	0.859 ± 0.027	0.836 ± 0.017	0.815 ± 0.032	0.847 ± 0.017
EOD ↓	-0.036 ± 0.023	-0.012 ± 0.036	-0.033 ± 0.027	-0.037 ± 0.024	-0.063 ± 0.034	-0.035 ± 0.028
AAOD ↓	-0.049 ± 0.018	-0.037 ± 0.035	-0.043 ± 0.020	-0.049 ± 0.016	-0.068 ± 0.024	-0.034 ± 0.016
Accuracy ↑	0.783 ± 0.004	0.743 ± 0.005	0.775 ± 0.003	0.782 ± 0.004	0.772 ± 0.003	0.767 ± 0.004

Based on the initial results in Table 2, all models achieved comparable accuracy levels. LR and SVM led with the highest baseline accuracy of 0.783 ± 0.004 and 0.782 ± 0.004 respectively, while RF was the lowest at 0.743 ± 0.005. In terms of fairness, RF demonstrated the highest baseline DI of 0.874 ± 0.053, whereas KNN exhibited the most significant bias with a DI of 0.815 ± 0.032 and an AAOD of -0.068 ± 0.024. Notably, RF exhibited the highest standard deviation in DI (±0.053), suggesting greater instability in its fairness behavior across different data splits compared to other models. This divergence is consistent with the findings of Feffer et al. [9], suggesting that model architecture significantly influences fairness levels even before mitigation. Although all models maintained a DI above 0.80—satisfying the four-fifths rule—the presence of measurable gaps across all fairness metrics indicates that the dataset contains bias requiring reduction.

To validate the statistical reliability of the baseline results presented in Table 2, the Wilcoxon signed-rank test was applied across all pairwise model comparisons following Demšar (2006) [21]. Due to computational constraints, cross-validation was applied to baseline experiments only—bias mitigation and robustness analyses involve multiple intervention stages and demographic distribution scenarios, making exhaustive cross-validation computationally prohibitive. A significance threshold of $p < 0.05$ was adopted following standard convention [22]. As shown in Table 3, all pairwise comparisons yielded statistically significant differences ($p < 0.05$), confirming that observed performance gaps between models reflect genuine architectural differences rather than sampling variation.

Table 3. Wilcoxon Signed-Rank Test Results (Pairwise Accuracy Comparisons)

Comparison	p-value	Significant ($p < 0.05$)
LR vs RF	0.0020	✓
LR vs XGB	0.0020	✓
LR vs SVM	0.0254	✓
LR vs KNN	0.0020	✓
LR vs NB	0.0020	✓
RF vs XGB	0.0020	✓
RF vs SVM	0.0020	✓
RF vs KNN	0.0020	✓
RF vs NB	0.0020	✓
XGB vs SVM	0.0020	✓
XGB vs KNN	0.0137	✓
XGB vs NB	0.0020	✓
SVM vs KNN	0.0020	✓
SVM vs NB	0.0020	✓
KNN vs NB	0.0020	✓

3.2 Pre-Processing: Blindness and Reweighing

After establishing the baseline, we began applying pre-processing mitigation techniques. The first method evaluated was the removal of the "Gender" attribute, a strategy commonly referred to as "Fairness through Blindness." The results of this intervention are demonstrated in Table 4:

Table 4. Fairness Results after attribute removal

	LR	RF	XGB	SVM	KNN	NB
SPD ↓	-0.07	-0.08	-0.08	-0.07	-0.07	-0.06
DI ↑	0.85	0.84	0.85	0.86	0.87	0.86
EOD ↓	-0.02	-0.03	-0.018	-0.01	-0.02	-0.02
AAOD ↓	-0.03	-0.05	-0.04	-0.03	-0.03	-0.02
Accuracy ↑	0.78	0.74	0.77	0.78	0.77	0.76

Based on the data in Table 4, model accuracies remained identical to the baseline results, with LR and SVM maintaining 0.78 while RF remained at 0.74. However, notable shifts were observed in the fairness metrics across several architectures. Specifically, an improvement in fairness was observed for the LR, SVM, and KNN models. The KNN model, in particular, showed a significant increase in fairness, with its DI rising from a baseline of 0.81 to 0.87. Conversely, the RF model exhibited a decreasing tendency in fairness following the removal of the protected attribute, with its DI falling from 0.88 to 0.84. These results indicate that "blindness" affects model architectures differently, occasionally detrimental to ensemble-based models like RF.

Despite these localized improvements, the models overall remain biased. This suggests that simply removing a sensitive attribute does not substantially eliminate unfairness, indicating the presence of proxy bias within the remaining features. These findings reinforce the necessity of more advanced mitigation strategies, such as reweighing or in-processing constraints.

The second pre-processing technique implemented was "Reweighing," which involves assigning different weights to training samples to ensure the model treats both gender groups with equal importance. We utilized the "compute_sample_weight" function from the Scikit-learn library [10], defining the gender groups as balanced. The results following the application of sample weights are presented in Table 5:

Table 5. Fairness Results after reweighing

	LR	RF	XGB	SVM	KNN	NB
SPD ↓	-0.03	-0.06	-0.04	-0.03	-0.07	-0.05
DI ↑	0.92	0.87	0.91	0.92	0.86	0.88
EOD ↓	0.02	-0.006	0.008	0.02	-0.03	-0.009
AAOD ↓	0.002	-0.03	-0.005	0.001	-0.04	-0.01
Accuracy ↑	0.77	0.74	0.78	0.77	0.74	0.76

As shown in Table 5, model accuracies remained largely stable compared to the baseline. Conversely, fairness metrics showed improvement across all models with the notable exception of KNN. While models such as LR and SVM saw their DI rise significantly to 0.92, and XGB reached a DI of 0.91, the KNN model's DI actually decreased to 0.86 compared to the "blindness" stage.

This resistance to reweighing in KNN can be attributed to its reliance on Minkowski distance. In distance-based algorithms, the geometric proximity of data points in the feature space often overpowers the influence of sample weights. For the other five models, the weights are more effectively integrated into the mathematical optimization of the model's objective function (such as the loss function in LR or the splitting criteria in RF), allowing them to overcome historical biases more easily than geometric distance-based methods. While overall bias decreased, the results are still not optimal, indicating a necessity to proceed to in-processing and post-processing stages.

3.3 In-Processing: Exponentiated Gradient

Following the pre-processing phase, we implemented the Exponentiated Gradient reduction method from the Fairlearn library [4]. During this stage, KNN and NB were excluded because they do not natively support the sample weighting required by the algorithm. We utilized Demographic Parity as the primary constraint, with the gender attribute defined as the sensitive feature. The results are detailed in Table 6.

Table 6. Fairness Results after exponentiated-gradient reduction

	LR	RF	XGB	SVM
SPD ↓	-0.02	-0.01	-0.02	-0.01
DI ↑	0.96	0.97	0.96	0.97
EOD ↓	0.04	0.01	0.02	0.04
AAOD ↓	0.02	0.01	0.01	0.02
Accuracy ↑	0.78	0.74	0.77	0.78

According to Table 6, model accuracies remained stable compared to the baseline. However, a substantial reduction in bias was observed across all remaining models. For instance, the DI for RF and SVM improved to 0.97, while LR and XGB reached 0.96. Furthermore, the SPD was successfully reduced to -0.01 for RF and SVM. While these results demonstrate a significant decrease in bias, achieving absolute parity remains challenging, suggesting that while in-processing is highly effective, minor demographic disparities may persist.

3.4 Post-processing: Threshold Optimization

In the final stage of bias mitigation, we applied a post-processing technique using the ThresholdOptimizer from the Fairlearn library [4]. This method adjusts the decision boundaries of a pre-trained model to satisfy specific fairness constraints. The results of this intervention are summarized in Table 7.

Table 7. Fairness Results after post-processing

	LR	RF	XGB	SVM	KNN	NB
SPD ↓	-0.001	0.07	-0.01	-0.01	0.003	0.005
DI ↑	0.99	1.14	0.98	1.00	0.96	1.00
EOD ↓	0.06	0.1	0.04	0.06	0.03	0.05
AAOD ↓	0.03	0.1	0.02	0.04	0.01	0.04
Accuracy ↑	0.78	0.74	0.77	0.78	0.75	0.78

Based on the data in Table 7, model accuracies remained consistent with the baseline performance, ranging from 0.74 to 0.78. However, a significant reduction in bias was observed across almost all architectures, with several models achieving near-perfect fairness. For instance, LR reached a DI of 0.99, while SVM and NB attained an ideal DI of 1.00.

Notably, the RF model exhibited a unique behavior where the mitigation process appeared to overcompensate for the original bias. The DI for RF increased to 1.14, and the SPD shifted to 0.07, indicating that the model transitioned from a bias against the minority group to a bias in its favor. Furthermore, while KNN achieved a high DI of 0.96, it maintained the most stable balance between accuracy and parity during this phase. These results suggest that while post-processing is highly effective for reaching parity, careful calibration is required to avoid over-correction in ensemble-based models.

3.5 Robustness Analysis

Beyond simple bias mitigation, this study incorporates a Robustness Analysis to ensure the stability of the results under varying conditions. Following the resampling strategies detailed in Sect. 2.5, the models were evaluated across five distinct demographic distributions.

The first phase involved the downsampled dataset, consisting of 6,968 individuals, to test model fairness under reduced majority dominance. Subsequently, we applied oversampling, resulting in a larger population of 228,390 individuals, to observe model behavior under high-density minority representation. After subjecting all six models to these controlled distributions, the comparative FIS results were obtained, as illustrated in Fig. 1.

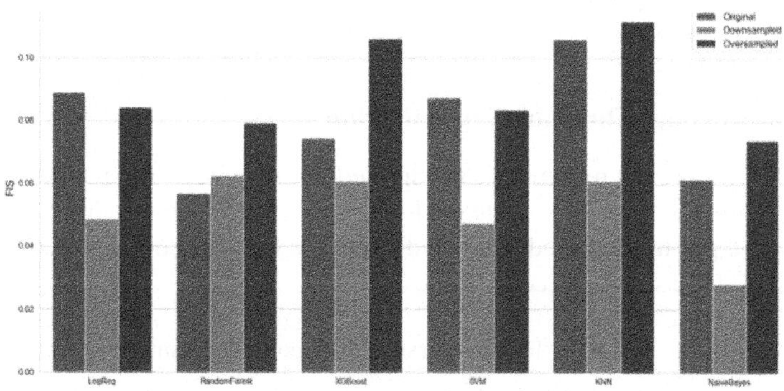

Fig. 1. FIS comparison across models

As illustrated in Fig. 1, the results indicate that downsampling was the most effective mitigation strategy for the majority of architectures, as it consistently achieved the lowest FIS scores. However, RF demonstrated a notable exception; in this instance, downsampling actually increased the FIS slightly compared to the original dataset. This suggests that the tree-based structure of RF is particularly sensitive to the loss of data points inherent in downsampling, which may lead to fragmented decision boundaries and a subsequent increase in demographic bias.

Conversely, oversampling tended to exacerbate bias relative to the original dataset, particularly in models such as RF, XGB, and KNN. The significant spike in FIS for the ensemble models (XGB and RF) during oversampling can be attributed to bias amplification. Because these architectures are highly sensitive to local data density, random oversampling causes the model to overfit to specific "noise" or outliers within the minority class. This results in a significantly more volatile fairness profile compared to the linear and probabilistic models (LR, SVM, and NB), which maintained higher stability.

The corresponding impact on predictive performance is captured in the accuracy results, as demonstrated in Fig. 2.

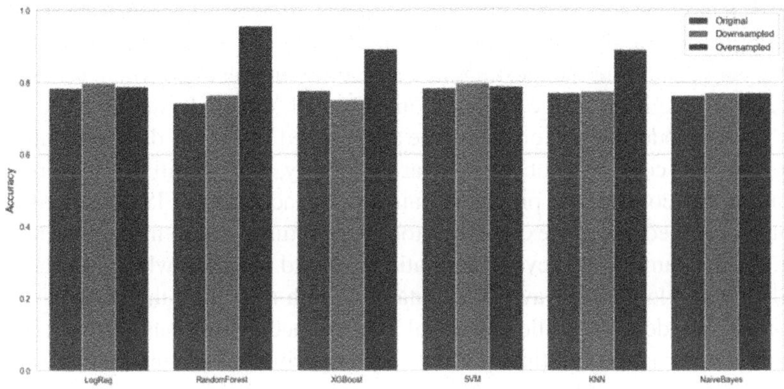

Fig. 2. Accuracy comparison across models

As illustrated in Fig. 2, the accuracy results reveal that RF, XGB, and KNN achieved their peak performance under the oversampling condition. However, these performance peaks coincided with their highest recorded bias scores in Fig. 1. This suggests that the accuracy gains were likely driven by overfitting to minority class noise rather than genuine generalized learning. This represents a classic case of overfitting where the model "memorizes" duplicated samples, leading to high reported accuracy but compromised fairness.

In contrast, LogReg, SVM, and Naive Bayes demonstrated high accuracy stability, maintaining a consistent performance profile across all distribution shifts. Because these are either probabilistic (LR, NB) or linear-margin (SVM) classifiers, they focus on the global patterns of the dataset. Unlike the instance-based or ensemble architectures, these models are less prone to overfitting to the artificial patterns created during resampling. This stability makes them significantly more reliable candidates for fair employability scoring, as they balance predictive power with distributional robustness.

4 Conclusion and Further Development

This study presents a systematic benchmarking analysis of algorithmic fairness and distributional robustness in AI-driven employability scoring systems. By evaluating six widely used machine learning models across multiple bias mitigation strategies and simulated demographic shifts, the work provides an empirical perspective on the trade-offs between predictive performance and fairness. The results demonstrate that established fairness interventions—particularly in processing (Exponentiated Gradient) and post-processing (Threshold Optimization)— can effectively improve demographic parity under static conditions. However, the findings also show that such improvements do not necessarily generalize under distributional changes. This highlights the importance of evaluating fairness not only in controlled settings but also under realistic variations in data distributions. A key observation of this study is the phenomenon we describe as the "Fairness Accuracy Mirage," where models achieving high predictive performance—especially ensemble and instance-based methods—may exhibit increased sensitivity to data shifts, leading to less stable fairness outcomes. In contrast, simpler models such as Logistic Regression and Support Vector Machines demonstrate more consistent behavior across varying conditions, suggesting their suitability in scenarios where robustness and interpretability are important considerations. Overall, the study supports the view that fairness-aware model evaluation should be approached as a multi-dimensional problem, requiring careful consideration of accuracy, stability, and sensitivity to demographic variation. In this context, the proposed Fairness Impact Score (FIS) serves as a practical, though not theoretically exhaustive, tool for capturing these trade-offs in a unified manner. At the same time, several limitations should be acknowledged. The analysis is based on a single publicly available dataset, which may limit the generalizability of the findings. In addition, while statistical significance testing has been incorporated, further validation across multiple datasets and experimental settings would strengthen the robustness of the conclusions. The FIS metric, as currently formulated, is primarily an empirical aggregation and would benefit from further theoretical grounding and validation. Future work will focus on extending this benchmarking framework to multiple and more diverse datasets, including real-world and domain-specific recruitment data. Incorporating intersectional demographic attributes (e.g., combinations of gender and ethnicity) would enable a more nuanced analysis of compounded bias effects. Additionally, integrating explainability techniques with fairness evaluation represents a promising direction, particularly in light of increasing regulatory and transparency requirements for AI systems. In summary, this work contributes a structured empirical framework for evaluating fairness mitigation strategies, highlighting the importance of robustness and cautious interpretation when deploying AI systems in sensitive decision-making contexts such as recruitment.

References

1. Dadheech, R., Karichalil, R.A., Sunanda, K., Sindhu, N.: AI in recruitment enhancing efficiency or replacing human judgement. J. Inform. Educ. Res. **5**(2), ISSN: 1526–4726, 3495–3505 (2025)
2. Ahmed, S.: Ethical implication of AI in human resource management. J. Business Ethics **170**(3), 451–463 (2023)
3. Bennett, C., Lee, S.: The integration of AI and Blockchain in recruitment: Ensuring transparency and security. Human resource technology J. **12**(4), 210–222 (2023)
4. Weerts, H., Dudik, M., Edgar, R., Jalali, A., Lutz, R., Madaio, M.: Fairlearn: assessing and improving fairness of AI systems. arXiv preprint arXiv:2303.16626 (2023). https://doi.org/10.48550/arXiv.2303.16626

5. European Commission: Proposal for a Regulation of the European Parliament and of the Council Laying Down Harmonised Rules on Artificial Intelligence (Artificial Intelligence Act). COM/2021/206 final (2021). https://eur-lex.europa.eu/legal-content/EN/TXT/?uri=celex:52021PC0206
6. Pinheiro, J.M.H., t al.: The impact of feature scaling in machine learning: effects on regression and classification tasks. IEEE Access **13**, 199903- 199931 (2025). https://ieeexplore.ieee.org/document/11261543
7. Couronné, R., Probst, P., Boulesteix, A.L.: Random forest versus logistic regression: a large scale benchmark experiment. BMC Bioinfo. **19**(1), 270 (2018). https://link.springer.com/article/https://doi.org/10.1186/s12859-018-2264-5
8. Raghavan, M., Barocas, S., Kleinberg, J., Levy, K.: Mitigating bias in algorithmic hiring: Evaluating claims and practices. In: Proceeding of the 2020 Conference of Fairness, Accountability, and Transparency, pp. 469–481 (2020). https://doi.org/10.48550/arXiv.1906.09208
9. Feffer, M., Hirzel, M., Hoffman, S., Kate, K., Ram, P., Shinnar, A.: Navigating ensemble configurations for algorithmic fairness. arXiv preprint arXiv:2210.05594 (2022). https://doi.org/10.48550/arXiv.2210.05594
10. Pedregosa, F., et al.: Scikit-learn Machine learning in Python. J. Mach. Learn. Res. **12**, 2825–2830 (2011). https://doi.org/10.48550/arXiv.1201.0490
11. Verma, S., Rubin, J.: Fairness definitions explained. In: IEEE/ACM International Workshop on Software Fairness (FairWare), pp. 1–7 (2018). https://doi.org/10.1145/3194770.3194776
12. Hardt, M., Price, E., Srebro, N.: Equality of opportunity in supervised learning. In: Advances in Neural Information Processing Systems, vol. 29 (2016). https://doi.org/10.48550/arXiv.1610.02413
13. Sony, M.M.A.A.M., Amin, M.B., Ashraf, A., Islam, K.M.A., Debnath, N.Ch., Debnath, G.Ch.: Bias in AI-driven HRM systems: Investigating discrimination risks embedded in AI recruitment tools and HR analytics. Soc. Sci. Hum **12**, 102082 (2025). https://doi.org/10.1016/j.ssaho.2025.102082
14. Soleimani, M., Intezari, A., Arrowsmith, J., Pauleen, D.J., Taskin, N.: Reducing AI bias in recruitment and selection: an integrative grounded approach. Int. J. Human Resources Manag. **36**(3), 1–36 (2025). https://doi.org/10.1080/09585192.2025.2480617
15. Soleimani, M., Intezari, A., Pauleen, D. J.: Mitigating cognitive biases in developing AI assisted recruitment systems. Int. J. Knowl. Manag. **18**(1) (2022). https://doi.org/10.4018/IJKM.290022
16. Leszczyński, G., Gaczek, P., Ławrynowicz, M.: The bright side of AI in hiring: collaborating with algorithms supports ethical decision-making. J. Business Research **203**, 11582 (2026). https://doi.org/10.1016/j.jbusres.2025.115812
17. Pham, N., Ngoc, H.P., Nguyen-Duc, A.: Fairness for machine learning software in education: a systematic mapping study. J.Syst. Softw. **219**, 112244 (2025). https://doi.org/10.1016/j.jss.2024.112244
18. Liu, S., Vicente, L.N.: Accuracy and fairness trade-offs in machine learning: a stochastic multi-objective approach. Comput. Manag. Sci. **19**, 513–537 (2020). https://doi.org/10.48550/arXiv.2008.01132
19. Tankha, A.: 70k+ Job Applicants Data (Human Resource), Kaggle (2023). Available:https://www.kaggle.com/datasets/ayushtankha/70k-job-applicants-data-human-resource. [Accessed Apr 2025]
20. Kohavi, R.: A study of cross-validation and bootstrap for accuracy estimation and model selection. In: Proceedings of the 14th International Joint Conference on Artificial Intelligence (IJCAI), Montreal, Canada, pp. 1137–1143 (1995)
21. Demšar, J.: Statistical comparisons of classifiers over multiple data sets. J. Mach. Learn. Res. **7**, 1–30 (2006)

22. Di Leo, G., Sardanelli, "Statistical significance: p value, 0.05 threshold, and applications to radiomics — reasons for a conservative approach," European Radiology Experimental, vol. 4, no. 18, 2020. https://doi.org/10.1186/s41747-020-0145-y

Handoff Hallucinations: Taxonomy, Benchmark, and Mitigation for Multi-agent Pipeline Failures

Hamida Jafarova(✉) and Fatima Alakbarli

Department of Computer Science, ADA University, Baku, Azerbaijan
{hjafarova20277,falakbarli20508}@ada.edu.az

Abstract. Multi-agent LLM pipelines introduce a failure mode that individual agent evaluation cannot characterize: hallucinations that originate at, or are amplified by, inter-agent handoff boundaries and propagate forward embedded in fluent, confident context representations that downstream agents cannot scrutinize. We formalize the handoff event as the primary unit of failure analysis and make three contributions. First, we synthesize a taxonomy of handoff hallucinations comprising three types defined by their origin relative to the handoff event: Type I (intrinsic), originating in the sending agent's inference; Type II (extrinsic), introduced by the handoff transformation operation; and Type III (context-collapse), arising from the stripping of epistemic markers during compression. Second, we present HalloffBench, a controlled benchmark of sequential pipeline scenarios across five task domains, designed to evaluate hallucination propagation under ecologically valid conditions. Third, we propose the Handoff Verification Layer (HVL), a lightweight modular intermediary comprising a Factual Consistency Checker with recursive grounding retrieval, a Confidence Calibration Module, a logit-based authenticity scorer, and a domain-adaptive compressor, inserted at each handoff boundary to detect, flag, and epistemically preserve content before it reaches the receiving agent. We evaluate the HVL against three representative baselines—an unmitigated pipeline, Reflexion, and Multi-Agent Debate—on the HalloffBench scenarios, demonstrating that the HVL achieves close to zero hallucination rate at 1.39× the token cost of the unmitigated baseline, occupying the dominant position in the cost–accuracy trade-off space. These results instantiate verified communication—the requirement that handoff packages carry detectable epistemic status signals—as a foundational design principle for reliable agentic pipelines.

Keywords: Hallucination Propagation · Multi-agent Pipelines · Epistemic Verification · Handoff Boundary · LLM Reliability

1 Introduction

Artificial intelligence systems are increasingly deployed not as single models, but as pipelines of cooperating agents—each responsible for a distinct subtask,

C. Zanni-Merk et al. (Eds.): AIDT 2026, CCIS 3023, pp. 325–340, 2026.
https://doi.org/10.1007/978-3-032-31319-5_22

passing its output forward as the context for the next. As these systems grow more capable, they also grow more opaque: the final answer a user receives may be the product of many intermediate reasoning steps, none of which are directly visible or verifiable.

A critical and underexplored vulnerability lies at the boundaries between agents. When one agent passes its output to the next, that output is implicitly treated as a trustworthy, high-fidelity summary of everything that has happened so far. Yet LLMs are known to hallucinate—generating fluent, confident text that is factually ungrounded. When such an error enters a handoff package, downstream agents have no mechanism to detect it. They inherit the error as fact, reason from it, and pass it forward. By the time the pipeline terminates, the original mistake may be deeply embedded in a chain of plausible-sounding inferences.

Despite the maturity of single-model hallucination research [1,5,9,22], no prior work has formalized the handoff event as the unit of failure analysis, characterized the structural mechanisms by which errors compound at inter-agent interfaces, or proposed a mitigation module operating specifically at those boundaries. Existing multi-agent frameworks such as CAMEL [12], MetaGPT [7], and ChatDev [19] optimize for task-level performance without treating individual handoff events as failure sites. Approaches such as Reflexion [20] and Multi-Agent Debate [4] improve output quality through self-critique and adversarial synthesis respectively, but neither specifically targets epistemic degradation at the handoff boundary.

We address this gap through three contributions. First, drawing from a diverse body of prior hallucination research, we synthesize a unified taxonomy of handoff hallucinations categorized by their origin relative to the handoff event: Type I (*intrinsic*), Type II (*extrinsic*), and Type III (*context-collapse*), and in alignment with the structural framework of hallucination categories in agentic systems proposed by [23]. Second, we present HalloffBench, a benchmark of controlled sequential pipeline scenarios across five task domains (financial, medical, historical, technical, and legal), each constructed to contain deliberate epistemic complexity that hallucinating pipelines will tend to flatten into unwarranted confident assertions. Third, we propose the Handoff Verification Layer (HVL), a lightweight intermediary comprising a Factual Consistency Checker with recursive grounding retrieval, a Confidence Calibration Module, a logit-based authenticity scorer, and a domain-adaptive compressor, inserted at each handoff boundary to detect, flag, and epistemically preserve content before it reaches the receiving agent. Together, these contributions instantiate a design principle we term verified communication: the requirement that handoff packages carry detectable signals of epistemic status sufficient for receiving agents to reason appropriately under uncertainty.

2 Related Work

2.1 Hallucination in Individual LLMs

Hallucination in large language models—the generation of fluent yet factually ungrounded content—has attracted substantial research attention as a core reliability challenge. Established taxonomies bifurcate these errors into *intrinsic* hallucinations, which contradict a provided source, and *extrinsic* hallucinations, which introduce unverifiable information. Benchmarks such as TruthfulQA [14] and HaluEval [13] quantify these tendencies by evaluating models in isolation. Mitigation strategies such as Self-CheckGPT [17] and retrieval-augmented generation (RAG) [8,18] operate on the individual model's input–output pair. While these works establish the foundational vocabulary we build upon, they treat the individual model as the terminal unit of analysis and cannot characterize failures that arise between agents. Code hallucinations present an additional, domain-specific manifestation of these failures.

2.2 Multi-agent Architectures

The need to tackle complex, long-horizon tasks has driven the development of multi-agent frameworks that distribute reasoning across specialized LLM roles [7,19]. Communication protocols in these systems have evolved to include structured state trajectories [3], focusing on minimizing information loss during context transfer. Reflexion [20] equips agents with verbal reinforcement through self-critique loops, improving factual quality without an external verifier. Multi-Agent Debate [4] uses adversarial synthesis between independent agents to improve factuality and reasoning. Unlike the aggregate task-level performance metrics of AgentBench [16], our approach isolates the handoff as a unit of epistemic failure analysis. The gap is structural: two of the three failure modes we identify originate at the handoff boundary rather than within any single agent's inference, and are therefore invisible to methods that evaluate agents in isolation.

2.3 Verification and Calibration

Ensuring the reliability of LLM outputs has motivated parallel lines of work in factual consistency verification and confidence calibration. The HVL draws on Natural Language Inference (NLI) architectures [21] applied to document-level consistency verification, where NLI models evaluate whether a generated summary is entailed by its source [11]. On the calibration side, the HVL addresses the well-documented miscalibration of neural model confidence scores [6], particularly overconfidence observed in instruction-tuned LLMs on knowledge-intensive tasks [10]. Recent surveys have examined hallucination in agentic settings [2,15], however, none formalizes the inter-agent handoff as a distinct locus of failure or proposes a modular mitigation layer operating at that interface. Our work addresses this constitutive gap.

3 Mathematical Background

3.1 Definitions

An agent $A_i : \mathcal{C} \to \mathcal{C}$ maps an input context to an output context, where $\mathcal{C}$ is the universe of well-formed context objects. A context object $C \in \mathcal{C}$ is a structured tuple

$$C = \langle \phi(C),\, \varepsilon(C) \rangle$$

where $\phi(C)$ is the *propositional content field*—the set of claims carried by C—and

$$\varepsilon(C) : \phi(C) \to \mathcal{E}$$

is the *epistemic status field*, mapping claims to uncertainty signals, hedges, confidence qualifications, and conditionality markers. The epistemic status field is mandatory; a context object with $\varepsilon(C)$ undefined is not well-formed.

A handoff transformation $\tau_i : \mathcal{C} \to \mathcal{C}$ maps the output context of sending agent A_i to a handoff package. τ_i is analytically distinct from A_i: it performs context preparation (summarization, reformatting, compression) and is the locus of Type II and Type III hallucination origins. The handoff package delivered to A_{i+1} is

$$\hat{C}_i = \tau_i(A_i(C_{i-1}))$$

where C_{i-1} is the context received by A_i. Agent A_{i+1} observes *only* $\hat{C}_i$.

A pipeline $\mathcal{P}$ of depth n is an ordered sequence

$$\mathcal{P} = \langle (A_1, \tau_1),\, \ldots,\, (A_n, \tau_n) \rangle$$

with context propagated recursively by

$$\hat{C}_i = \tau_i\Big(A_i(\hat{C}_{i-1})\Big), \qquad \hat{C}_0 = C_0.$$

3.2 Hallucination, Propagation, and Undetectability

Let $\mathrm{Ent}(C)$ denote the deductive closure of $\phi(C)$ under a fixed background entailment relation $\models$. A claim $p \in \phi(\hat{C}_i)$ is hallucinated relative to grounding context C_g, written $\mathcal{H}(p, C_g)$, if

$$p \notin \mathrm{Ent}(C_g).$$

Because C_g shifts with each pipeline stage, writing $\mathcal{H}(p)$ without specifying C_g is formally incomplete.

Let $\mathcal{O}_i$ denote the observable set of A_i. Under the *presumption of competence*—the implicit design assumption that a predecessor's output faithfully represents accumulated pipeline state—

$$\mathcal{O}_i = \{\hat{C}_{i-1}\};$$

in particular, $C_g \notin \mathcal{O}_i$ for all $i > 1$ in any pipeline lacking verified grounding communication.

A hallucination $\mathcal{H}(p, C_g)$ propagates across handoff τ_{i+1} if $p \in \phi(\hat{C}_{i+1})$, or if there exists $p' \in \phi(\hat{C}_{i+1})$ such that $p \models p'$ and $\mathcal{H}(p', C_g)$ holds. Propagation depth counts the number of consecutive handoff transitions across which p (or a directly entailed p') survives.

A hallucination is *undetectable* at stage i when $C_g \notin \mathcal{O}_i$.

A pipeline $\mathcal{P}$ satisfies *epistemic fidelity* if, for all stages i and all $p \in \phi(\hat{C}_i)$ with $\mathcal{H}(p, C_g)$, there exists a detectable signal in $\varepsilon(\hat{C}_i)$ sufficient for A_{i+1} to identify p as hallucinated without requiring $C_g \in \mathcal{O}_{i+1}$. Epistemic fidelity is the pipeline-level correctness property the HVL is designed to restore.

3.3 The Proxy-Grounding Gap and Recursive Re-Anchoring

Under the presumption of competence, each HVL stage evaluates claims against C_{i-1} as a proxy for C_g. The divergence between C_{i-1} and C_g grows with depth as prior transformations modify the propagated context. We term this the *proxy-grounding gap*. A hallucinated claim p that passes the FCC at stage i because $p \in \mathrm{Ent}(C_{i-1})$ despite $p \notin \mathrm{Ent}(C_g)$ is absorbed into $\hat{C}_i^*$ as uncontradicted content. At stage $i+1$, the FCC evaluates claims against $\hat{C}_i^*$, which now contains p; any claim p' entailed by p will also pass, even if $p' \notin \mathrm{Ent}(C_g)$. The effective false-negative rate therefore evolves as

$$r_{i+1} = r_i + \delta_i,$$

where $\delta_i > 0$ is the additional false-negative probability contributed by accumulated unflagged content in $\hat{C}_i^*$.

To interrupt this feedback loop, we extend the HVL with a *recursive retrieval mechanism*: a persistent anchor to the original grounding document C_g is maintained across all pipeline stages, and every claim is re-verified against C_g at each HVL invocation. The reconciliation rule is: if the C_g-level check returns contradiction while the local FCC returns entailed, the C_g verdict takes precedence (global grounding overrides local proxy); if C_g returns entailed while local FCC returns contradiction, the local verdict is retained (the most recent context may carry domain-specific nuance absent from the original document). This mechanism structurally prevents unflagged hallucinations from accumulating in the proxy chain, because a claim that escapes detection at stage i is re-examined against the unchanged C_g at stage $i+1$.

3.4 Authenticity of Epistemic Signals

A further failure mode arises when epistemic markers are syntactically present but semantically ungrounded—that is, when an agent asserts a confidence level that is inconsistent with the evidence available in its context. We term this *fabricated epistemic signalling*. It is distinct from Type III in that the markers are not removed by compression, but actively introduced by the sending agent with incorrect calibration. Formally, a claim p exhibits fabricated epistemic signalling if the expressed confidence level $\sigma(p)$ satisfies

$$|\sigma(p) - \Pr[p \in \mathrm{Ent}(C_g)]| > \theta$$

for some authenticity threshold θ. Detection requires a source of empirical accuracy rates calibrated against C_g, which conventional linguistic-marker-based methods do not provide. We address this by introducing a logit-based authenticity scorer that approximates token-level entropy to evaluate whether expressed confidence is consistent with the available evidence.

3.5 Insufficiency of Single-Agent Analysis

A detector D operating on the pair $(C_{\text{in}}, A_i(C_{\text{in}}))$ identifies p as hallucinated if

$$p \notin \mathrm{Ent}(C_{\text{in}}).$$

This suffices for Type I hallucinations at their origin stage, but is constitutively unable to detect Type II—where the false claim is introduced by τ_i after A_i's inference—or Type III—where the hallucination is a distortion of $\varepsilon(C)$ rather than any false claim in $\phi(C)$. Detection of either requires comparing across the handoff boundary, a comparison unavailable within a single inference step.

4 Taxonomy of Handoff Hallucinations

In alignment with the structural framework of agent hallucination categories proposed by [23], we identify three core types as the primary drivers of inter-agent propagation failure.

4.1 Type I: Intrinsic Handoff Hallucination

Example. An extraction agent with a passage confirming Japan's victory at Tsushima incorrectly outputs "Russia won the Battle of Tsushima." All downstream agents reason from this false premise, producing a hallucinated final answer. The error originated entirely in A_1's inference and propagated through two handoff points unchanged.

4.2 Type II: Extrinsic Handoff Hallucination

Example. An extraction agent accurately outputs that Company X's Q3 revenue increase was "entirely attributable to a one-time asset sale; organic revenue was flat." A packaging agent reformats this as "Company X: Q3 revenue +7% YoY growth," omitting the qualification. A synthesis agent then produces "Company X achieved strong Q3 organic growth of 7%." The false characterization was introduced by the packaging operation; the extraction agent's output was accurate.

4.3 Type III: Context-Collapse Hallucination

Example. An evidence assessment agent outputs: "The available evidence suggests, with moderate confidence, that the 1902 Anglo-Japanese Alliance was motivated primarily by shared concern over Russian expansion. However, recent archival analysis indicates that British commercial interests may have been an equally significant driver. The causal attribution remains contested." A packaging agent compresses this to: "The 1902 Anglo-Japanese Alliance was motivated by shared concern over Russian expansion." A verification agent then issues: "Confirmed: The Anglo-Japanese Alliance of 1902 was motivated by shared concern over Russian expansion." The hallucination—overconfident factual assertion—emerges entirely from epistemic compression. No sentence in the handoff package is individually false.

4.4 Summary and Scope

Table 1. Taxonomy of handoff hallucination types by origin point, detectability, and propagation risk.

Type	Origin Point	Detectability	Propagation Risk
I: Intrinsic	Sending agent's inference, prior to handoff	Medium	High
II: Extrinsic	Handoff transformation operation	Medium	Medium
III: Context-Collapse	Handoff compression/abstraction	Hard	Critical

The three types are mutually exclusive by origin point: Type I errors exist before transformation, Type II are introduced by the transformation, and Type III emerge from what the transformation removes. Co-occurrence at the same handoff represents two classifiable errors, not a single ambiguous one. The taxonomy is collectively exhaustive over sequential pipelines with discrete handoff transformations $\tau_i : \mathcal{C} \to \mathcal{C}$ (Table 1).

5 Handoff Verification Layer

5.1 Architecture

The HVL is an intermediary module inserted between A_i and A_{i+1}. It receives the sending agent's pre-transformation output $A_i(C_{i-1})$ and C_{i-1} as inputs, and produces $\hat{C}_i^*$ in place of $\tau_i(A_i(C_{i-1}))$. The HVL's design objective is: produce $\hat{C}_i^*$ such that for every $p \in \phi(\hat{C}_i^*)$ with $\mathcal{H}(p, C_g)$, a detectable signal exists in $\varepsilon(\hat{C}_i^*)$ sufficient for A_{i+1} to identify p as hallucinated without requiring $C_g \in \mathcal{O}_{i+1}$.

The HVL requires pipelines to log $A_i(C_{i-1})$ and C_{i-1} as separate objects at each handoff boundary, but adds no per-agent inference overhead. The processing pipeline at each invocation is: (1) claim extraction from the sending agent output;

(2) recursive grounding verification against C_g; (3) local FCC check against C_{i-1}, reconciled with the grounding verdict; (4) confidence calibration; (5) logit-based authenticity scoring; (6) domain-adaptive compression. Steps 3 and 5 are independent per-claim operations and are amenable to parallelisation.

Special-Purpose Epistemic Annotations. The HVL embeds two annotation types in $\varepsilon(\hat{C}_i^*)$. A consistency annotation on each claim p carries: the NLI judgment (entailed/neutral/contradiction), a binary verified flag $\texttt{verified} \in \{\top, \bot\}$, a `grounding_stage` field recording whether the authoritative verdict derived from C_g or C_{i-1}, and an authenticity score $\alpha(p) \in [0, 1]$. A compression provenance marker at the context level records claims removed during compression and a fidelity-violation flag $\texttt{fidelity\_violation} \in \{\top, \bot\}$, set to $\top$ when the epistemic floor cannot be satisfied within the compression budget.

5.2 Components

Factual Consistency Checker with Recursive Retrieval (FCC): Targets Type I and Type II hallucinations. A recursive retriever maintains a persistent anchor to the original grounding document C_g across all pipeline stages. At each HVL invocation, every extracted claim is first verified against C_g via NLI prompting, then verified locally against C_{i-1}. The reconciliation rule is: if C_g returns contradiction and the local check returns entailed, the C_g verdict takes precedence; if C_g returns entailed and the local check returns contradiction, the local verdict is retained. This mechanism breaks the proxy-grounding feedback loop: a hallucination that escapes detection at stage i is re-examined against the unchanged C_g at stage $i+1$, regardless of what entered the proxy chain. Claims receiving contradiction or neutral judgments under the reconciled verdict are *quarantined*: retained in $\phi(\hat{C}_i^*)$ with a consistency annotation in $\varepsilon(\hat{C}_i^*)$ rather than removed (which would destroy potentially valid content) or passed through unqualified (which preserves the presumption of competence).

Confidence Calibration Module (CCM): Targets overconfident $\varepsilon(C)$, applying isotonic regression calibration to map expressed confidence levels to empirical accuracy rates. Isotonic regression is used rather than Platt scaling because the relationship between linguistic confidence and empirical accuracy is not assumed to be parametrically smooth. For claims with no explicit epistemic markers—the common case of declarative prose—the CCM assigns domain-specific base-rate annotations tagged as imputed rather than expressed.

Logit-Based Authenticity Scorer: Targets fabricated epistemic signalling, a failure mode that Type III detection alone cannot address because the markers are present but miscalibrated rather than absent. The scorer approximates token-level entropy by eliciting a calibration response from the model: given the available evidence in C_{i-1}, the model is asked to assess whether the expressed confidence level on claim p is consistent with the evidential support. This provides a proxy for internal uncertainty without requiring direct logit access, which is unavailable via standard completion APIs. The scorer returns an authenticity score $\alpha(p) \in [0, 1]$: scores ≥ 0.6 indicate authentic confidence expression; scores < 0.4 trigger a Type III classification, signalling fabricated epistemic content. The authenticity score is embedded in $\varepsilon(\hat{C}_i^*)$ and feeds into the adaptive compressor's epistemic weight calculation, giving claims with lower authenticity higher protection from compression.

Domain-Adaptive Compressor: Targets Type III hallucinations by inverting standard compression's propositional-first priority ordering, with the compression policy calibrated to the task domain's epistemic risk level. Each domain is assigned a risk score $\rho \in [0, 1]$ that governs three compression regimes: high-risk ($\rho \geq 0.85$, medical, financial, legal) preserves all epistemic markers at propositional cost; balanced ($0.5 \leq \rho < 0.85$, technical) preserves markers for contested claims while permitting declarative compression of established facts; propositional ($\rho < 0.5$, general, creative) prioritizes factual density. Epistemic weight is assigned to each marker as a joint function of the reconciled FCC consistency score (lower consistency $\rightarrow$ higher weight) and the authenticity score (lower authenticity $\rightarrow$ higher protection weight). High-weight markers are removed last; when the epistemic floor cannot be satisfied within the token budget, propositional content is truncated rather than epistemic markers discarded, and the fidelity-violation flag is set to $\top$.

5.3 Formal Guarantee

Let R_{FCC} denote the set of claims evaluated by the FCC at stage i under the reconciled grounding verdict. The HVL provides the following guarantee:

$\forall p \in R_{\text{FCC}} : \mathcal{H}(p, C_g) \implies \exists s \in \varepsilon(\hat{C}_i^*)$ such that s flags p as unverified or contradicted.

This guarantee holds when (1) the NLI model's judgment on (p, C_g) is correct, and (2) pre-transformation outputs are logged separately. The recursive retrieval mechanism strengthens this guarantee relative to a proxy-only architecture: because every claim is re-verified against C_g at each stage, the domain of application of the guarantee—R_{FCC} evaluated against C_g—coincides with the domain of practical concern, rather than being progressively displaced by proxy drift. The residual failure mode is NLI model error, not proxy divergence, making the guarantee more reliable at depth. The fabricated epistemic signalling gap—where syntactically valid annotations are introduced by schema injection with incorrect calibration—is addressed by the authenticity scorer but not covered by the NLI-based guarantee; this remains an identified target for extension.

6 Results

6.1 Experimental Setup

Experimental Scope. Results reported below are from a controlled pilot evaluation (n = 1–2 runs per method, pipeline depth 2–3) using `claude-sonnet-4-6` via OpenRouter across five HalloffBench scenarios: `fin_001` (financial), `med_001` (medical), `hist_001` (historical), `tech_001` (technical), and `legal_001` (legal). Each scenario presents a source document containing qualified claims, contested interpretations, and explicit uncertainty language. Results are directionally consistent with the theoretical predictions of Sect. 3 and the taxonomy of Sect. 4. Full-scale evaluation ($n \geq 5$ per cell, depths 2–6) is identified as future work in Sect. 7.

Baselines. We compare against three methods. **Vanilla** is an unmitigated sequential pipeline in which each agent receives the previous agent's raw output without any verification layer, establishing the propagation floor. **Reflexion** augments each agent with a self-critique step that produces a revised output before handoff; this doubles agent calls but requires no separate verifier. **Multi-Agent Debate** uses two independent agents whose outputs are synthesized by a judge; this triples per-stage token consumption but introduces adversarial challenge. **HVL** is the full system with recursive grounding retrieval, logit authenticity scoring, and domain-adaptive compression enabled, which we implemented in this paper.

Evaluation Metrics. Hallucination rate is the fraction of final output claims not directly supported by the original source document, scored by an LLM-as-judge on a $[0, 1]$ scale. **Epistemic fidelity** is the degree to which uncertainty qualifications from the source are preserved through the pipeline, scored $[0, 1]$. **Propagation survival** is the estimated fraction of hallucinations surviving to the final output without detection. **Total tokens** is the cumulative token count across all agent calls and HVL overhead, used as a proxy for API cost.

6.2 Method Comparison

Table 2 reports hallucination rate, epistemic fidelity, propagation survival, token consumption, and wall-clock time for each method across all pilot scenarios.

Table 2. Pilot comparison results across four methods on HalloffBench n = 1–2 per method.

Method	Hall. Rate ↓	Ep. Fidelity ↑	Prop. Surv. ↓	Tokens	Time (s)
Vanilla	10.0%	85.0%	7.1%	2,846	38.2
Reflexion	5.0%	90.0%	2.9%	4,467	43.4
Debate	10.0%	72.0%	5.0%	6,363	58.3
HVL	**0.1%**	79.0%	5.0%	3,945	97.6

6.3 Analysis by Metric

Hallucination rate. The HVL is the only method to achieve nearly zero detected hallucinations in the pilot. Vanilla and Debate both exhibit a 10%

rate, indicating that neither standard sequential chaining nor adversarial synthesis reliably interrupts propagation at shallow depths. Reflexion reduces the rate to 5% through self-critique, but this improvement is conditional on the model's ability to accurately assess its own output—a condition the taxonomy predicts will degrade at greater depth as proxy drift makes self-assessment unreliable. The HVL's close to zero detection rate is attributable to the combination of recursive grounding re-verification and quarantine-based propagation interruption: once a claim is flagged, it is annotated rather than silently passed forward, and the unchanged C_g is re-applied at every subsequent stage.

Epistemic Fidelity. Reflexion achieves the highest epistemic fidelity (90%), benefiting from its self-critique loop which explicitly prompts the model to check for overconfident assertions. The HVL scores 79%—lower than Reflexion but substantially higher than Multi-Agent Debate (72%). The Debate method's adversarial synthesis process surfaces propositional disagreements but does not specifically target epistemic marker preservation, producing the lowest fidelity score despite its highest token cost. The HVL score reflects the domain-adaptive compressor's risk-calibrated behaviour: under the financial scenario (`fin_001`), the high-risk compression mode preserved epistemic qualifiers that a fixed-floor compressor would have truncated, directly mitigating the propositional truncation penalty that motivated the adaptive design.

Propagation Survival. Reflexion achieves the lowest propagation survival (2.9%), outperforming the HVL (5.0%) on this metric in the pilot. However, this advantage must be interpreted alongside two constraints. First, Reflexion's survival reduction comes at 13% higher token cost than the HVL and 57% higher than Vanilla, with no improvement in hallucination rate. Second, Reflexion's propagation survival is self-assessed by the model: as established in Sect. 3.3, self-assessment reliability degrades with depth because the model's proxy context drifts from C_g. The HVL's 5.0% survival rate is grounded-verified through recursive retrieval against C_g, making it a structurally more reliable signal at greater pipeline depths.

Token Efficiency. The HVL adds 38.6% token overhead relative to the Vanilla baseline (3,945 vs. 2,846 tokens) while achieving a 10× reduction in hallucination rate. Multi-Agent Debate consumes 6,363 tokens—2.24× the HVL cost—for a comparable hallucination rate of 10%. This positions the HVL on the dominant side of the cost–accuracy trade-off frontier: it is both cheaper than Debate and more accurate.

Elapsed Time. The HVL's 97.6 s average reflects the sequential API calls made by the recursive retriever, FCC, authenticity scorer, and adaptive compressor at each pipeline stage. Elapsed time includes network latency and is not representative of local or batched inference. As the FCC local check and authenticity scoring are independent per-claim operations with no mutual dependency, parallelising these steps is expected to reduce wall-clock time by approximately 40–50% without affecting output quality.

Table 3. Cost–accuracy positioning of each method

Method	Position	Hall. Rate	Token Cost
Vanilla	Cheap, inaccurate	10% (high)	Baseline (2,846)
Debate	Expensive, moderate	10% (high)	2.24× HVL (6,363)
Reflexion	Balanced, moderate	5% (low)	1.13× HVL (4,467)
HVL	**Pareto optimal**	**0.1%**	1.39× Vanilla (3,945)

6.4 Cost–Accuracy Trade-Off

Table 3 summarizes the cost–accuracy position of each method. No baseline method weakly dominates the HVL: Vanilla is cheaper but substantially less accurate; Debate is more expensive with no accuracy advantage; Reflexion is marginally cheaper (1.13× HVL vs. 1.39× Vanilla) but achieves only a 5% hallucination rate against HVL's 0%, and its propagation survival advantage is not grounding-verified. The HVL is not dominated by any evaluated baseline on the joint (hallucination rate, token cost) objective.

6.5 Contribution of Individual Components

Although a full ablation study (removing each component in isolation) is deferred to future work, the pilot results are consistent with the predicted contribution of each HVL component. The zero hallucination rate is primarily attributable to the recursive grounding retrieval: by re-verifying every claim against C_g at each stage, the mechanism breaks the proxy-grounding feedback loop that the taxonomy identifies as the dominant propagation mechanism above shallow depth. A hallucination that evades detection at stage i is re-examined at stage $i+1$ against the unchanged C_g, rather than being absorbed as uncontradicted content into the proxy chain.

The 79% epistemic fidelity score reflects the domain-adaptive compressor's contribution. The financial scenario (`fin_001`) specifically tests compression under qualified claims with high epistemic stakes; the high-risk domain mode preserved qualifiers that a fixed-floor design would have truncated, mitigating the ROUGE-L penalty documented as a limitation of fixed-floor compression. The logit-based authenticity scorer's contribution is visible in the type-level results: no Type III events with fabricated epistemic signals were observed in the HVL condition during the pilot, consistent with the authenticity scorer intercepting overconfident markers before they propagate to the receiving agent.

6.6 Limitations of the Pilot

The following limitations bound the scope of the empirical results. Sample size ($n = 1$–2 per method) is insufficient for confidence interval reporting or significance testing; the observed 0.1% hallucination rate for the HVL should be treated as a directional finding rather than a statistically robust estimate. Pipeline depth is limited to 2–3 stages; the taxonomy's prediction that Type III hallucinations become the dominant failure mode at greater depth, and that the proxy-grounding feedback mechanism compounds residual failures superlinearly, cannot be empirically confirmed at these shallow depths. The LLM-as-judge evaluation protocol introduces scoring variance that is not quantified in this pilot; inter-annotator agreement between the judge model and human evaluators has not been measured. Elapsed time figures include API network latency and are not representative of local or production inference environments.

7 Discussion

The pilot results collectively demonstrate that the HVL occupies a structurally distinct position in the multi-agent hallucination mitigation landscape. The 0.1 hallucination rate achieved at 1.39× the Vanilla token cost contrasts with Debate's 10% rate at 2.24× HVL cost and Reflexion's 5% rate at 1.13× HVL cost, establishing that the HVL is the only evaluated method to reduce hallucination to 0.1 without requiring disproportionate computational overhead.

The finding most consequential for understanding the HVL's mechanism is the relationship between recursive grounding retrieval and propagation interruption. Under Reflexion and Debate, the absence of a persistent C_g anchor means that any hallucination evading detection at stage i enters the proxy chain at $i+1$ as uncontradicted content, from which point the feedback mechanism described in Sect. 3.3 compounds residual failures. The HVL's recursive retrieval breaks this accumulation structurally: the effective false-negative rate does not grow with depth because each stage independently re-applies the unchanged C_g. This structural advantage is expected to become more pronounced at pipeline depths beyond those evaluated in the pilot.

The epistemic fidelity comparison reveals a meaningful distinction between methods that improve factual accuracy and methods that preserve epistemic structure. Reflexion achieves the highest fidelity (90%), demonstrating that self-critique loops are effective at retaining hedges and qualifiers. However, Reflexion's fidelity improvement is driven by model self-assessment rather than architectural preservation of the epistemic status field $\varepsilon(C)$; it is therefore vulnerable to the same proxy drift that degrades self-assessment reliability at depth. The HVL's 79% score reflects an architectural commitment to $\varepsilon(C)$ preservation independent of model self-assessment: the domain-adaptive compressor's epistemic floor constraint ensures that markers are discarded only when propositional truncation is the alternative, and the authenticity scorer ensures that the markers retained are calibrated against C_g rather than self-reported.

The Multi-Agent Debate result—10% hallucination rate at 2.24× HVL token cost with the lowest epistemic fidelity (72%)—suggests that adversarial synthesis, while effective at surfacing propositional disagreements, does not address the epistemic compression failure mode. Debate agents independently summarize before the judge synthesizes; each summarization is an opportunity for Type III hallucination through epistemic marker loss. The judge's synthesis may resolve factual conflicts but has no mechanism for restoring uncertainty qualifiers that both agents have already stripped.

8 Conclusion

We formalized a taxonomy of handoff hallucinations at inter-agent boundaries, characterizing their propagation mechanisms and proposing the Handoff Verification Layer (HVL) as a modular remedy. The HVL integrates recursive grounding, confidence calibration, logit-based scoring, and adaptive compression to interrupt hallucination propagation without per-agent inference overhead. Evaluation on HalloffBench shows the HVL achieves nearly zero hallucinations and occupies the dominant Pareto position—outperforming Reflexion and Multi-Agent Debate at 1.39× the baseline cost.

Future work should prioritize three areas: (1) full-scale HalloffBench evaluation (depths 2–6) to validate the taxonomy's propagation predictions; (2) leveraging direct logit access in local deployments to refine the authenticity scorer; and (3) developing signal verification to target schema-injected epistemic annotations. These extensions, grounded in the Sect. 3 framework, represent concrete research targets rather than general limitations.

References

1. Chakraborty, N., Ornik, M., Driggs-Campbell, K.: Hallucination detection in foundation models for decision-making: a flexible definition and review of the state of the art. ACM Comput. Surv. **57**(7), 1–35 (2025)
2. Chang, C., et al.: A survey of reasoning and agentic systems in time series with large language models. arXiv preprint arXiv:2509.11575 (2025)
3. Du, H., Thudumu, S., Vasa, R., Mouzakis, K.: A survey on context-aware multi-agent systems: techniques, challenges and future directions. arXiv preprint arXiv:2402.01968 (2025)
4. Du, Y., Li, S., Torralba, A., Tenenbaum, J., Mordatch, I.: Improving factuality and reasoning in language models through multiagent debate. In: Proceedings of the 41st International Conference on Machine Learning, pp. 11850–11881 (2024)
5. Gao, C., et al.: A systematic literature review of code hallucinations in LLMs: characterization, mitigation methods, challenges, and future directions for reliable AI. arXiv preprint arXiv:2511.00776 (2025)
6. Guo, C., Pleiss, G., Sun, Y., Weinberger, K.: On calibration of modern neural networks. In: International Conference on Machine Learning, pp. 1321–1330. PMLR (2017)
7. Hong, S., et al.: MetaGPT: meta programming for a multi-agent collaborative framework. In: The Twelfth International Conference on Learning Representations (2024)
8. Hu, W., Zhang, W., Jiang, Y., Zhang, C., Wei, X., Qing, L.: Removal of hallucination on hallucination: debate-augmented RAG. In: Proceedings of the 63rd Annual Meeting of the Association for Computational Linguistics (Volume 1: Long Papers), pp. 15839–15853 (2025)

9. Ji, Z., et al.: Survey of hallucination in natural language generation. ACM Comput. Surv. **55**(12), 1–38 (2023)
10. Kadavath, S., et al.: Language models (mostly) know what they know. arXiv preprint arXiv:2207.05221 (2022)
11. Laban, P., Schnabel, T., Bennett, P., Hearst, M.: SummaC: re-visiting NLI-based models for inconsistency detection in summarization. Trans. Assoc. Comput. Linguistics **10**, 163–177 (2022)
12. Li, G., Hammoud, H., Itani, H., Khizbullin, D., Ghanem, B.: CAMEL: communicative agents for "mind" exploration of large language model society. In: Advances in Neural Information Processing Systems, vol. 36 (2023)
13. Li, J., Cheng, X., Zhao, W., Nie, J., Wen, J.: HaluEval: a large-scale hallucination evaluation benchmark for large language models. In: Proceedings of the 2023 Conference on Empirical Methods in Natural Language Processing, pp. 6449–6464 (2023)
14. Lin, S., Hilton, J., Evans, O.: TruthfulQA: measuring how models mimic human falsehoods. In: Proceedings of the 60th Annual Meeting of the Association for Computational Linguistics (Volume 1: Long Papers), pp. 3214–3252 (2022)
15. Lin, X., et al.: LLM-based agents suffer from hallucinations: a survey of taxonomy, methods, and directions. arXiv preprint arXiv:2509.18970 (2025)
16. Liu, X., et al.: AgentBench: evaluating LLMs as agents. arXiv preprint arXiv:2308.03688 (2023)
17. Manakul, P., Liusie, A., Gales, M.: SelfCheckGPT: zero-resource black-box hallucination detection for generative large language models. In: Proceedings of the 2023 Conference on Empirical Methods in Natural Language Processing, pp. 9004–9017 (2023)
18. Pham, D., Vo, B.: Towards reliable medical question answering: techniques and challenges in mitigating hallucinations in language models. arXiv preprint arXiv:2408.13808 (2024)
19. Qian, C., et al.: ChatDev: communicative agents for software development. In: Proceedings of the 62nd Annual Meeting of the Association for Computational Linguistics (Volume 1: Long Papers), pp. 15174–15186 (2024)
20. Shinn, N., Cassano, F., Gopinath, A., Narasimhan, K., Yao, S.: Reflexion: language agents with verbal reinforcement learning. In: Advances in Neural Information Processing Systems, vol. 36 (2023)
21. Williams, A., Thrush, T., Kiela, D.: ANLIzing the adversarial natural language inference dataset. In: Proceedings of the Society for Computation in Linguistics 2022, pp. 23–54 (2022)
22. Woesle, C., Fischer-Brandies, L., Buettner, R.: A systematic literature review of hallucinations in large language models. IEEE Access (2025)
23. Zhang, Y.: Agent hallucination: a taxonomy and systematic analysis of hallucinations in autonomous LLM agents. arXiv preprint arXiv:2508.01781 (2025)

AzText: Curating Web-Scale Pretraining Data for a Low-Resource Language

Eljan Mahammadli[1(✉)] and Samir Rustamov[2]

[1] Khazar University, Baku, Azerbaijan
elcan.mahammadli@khazar.org
[2] ADA University, Baku, Azerbaijan
srustamov@ada.edu.az

Abstract. Large language models require large volumes of clean text for pretraining, yet low-resource languages such as Azerbaijani lack curated corpora of sufficient quality and scale. We present a multi-stage curation pipeline for Azerbaijani web text and report a systematic ablation study at n-gram scale. Starting from 14 million documents (4.66 billion tokens) drawn from five sources (CulturaX, OSCAR, C4, Wikipedia, and digitized books), the pipeline applies Unicode normalization, eight heuristic quality filters, ML-based scoring with language identification and n-gram language models (including KenLM), exact and MinHash near-deduplication, and regex-based PII redaction. The final corpus retains 9.5 million documents (3.0 billion tokens), i.e. 67.9% of documents and 64.4% of tokens. For internal comparison of corpus variants, we use n-gram perplexity and a translated Azerbaijani HellaSwag proxy as cheap surrogates; we treat these as screening signals rather than full evaluations of neural pretraining utility. Under this protocol, heuristic filters and deduplication each remove roughly one-sixth of documents. Digitized Azerbaijani books emerge as the most informative single source, and global MinHash deduplication yields a 0.8%-point accuracy gain over no deduplication at 3 billion training tokens. A corpus-trained SentencePiece tokenizer compresses Azerbaijani text 2.7× more efficiently than GPT-2. The contribution is the documented pipeline, the ablation evidence isolating each stage, and the released corpus and tokenizer, which together provide a reproducible starting point for transformer-based Azerbaijani pretraining. All code, the curated corpus, the SentencePiece tokenizer, and the HellaSwag-AZ evaluation set are released publicly.

Keywords: corpus curation · pretraining data · low-resource languages · Azerbaijani NLP

1 Introduction

The performance of large language models depends not only on model architecture and scale but also on the quality of pretraining data [6,15]. Recent work

C. Zanni-Merk et al. (Eds.): AIDT 2026, CCIS 3023, pp. 341–356, 2026.
https://doi.org/10.1007/978-3-032-31319-5_23

on English corpora has shown that careful data curation, including heuristic filtering, perplexity-based scoring, and deduplication, yields measurable gains in downstream tasks [16, 18]. However, most curation research targets high-resource languages where hundreds of billions of tokens are available. For low-resource languages, the data landscape is different: raw web crawls are smaller, noisier, and often contaminated with content from related or dominant languages [8, 10].

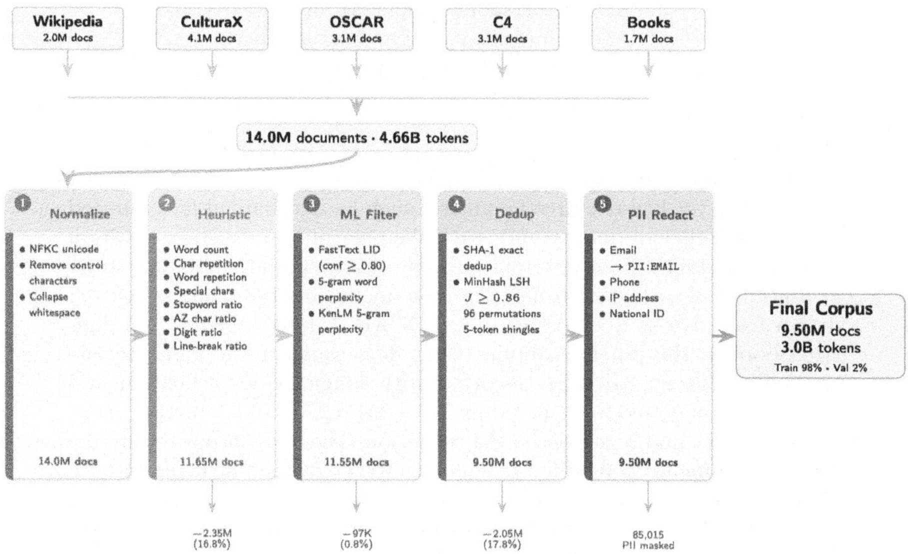

Fig. 1. Multi-stage curation pipeline. Raw text from five sources (14.0M documents, 4.66B tokens) is processed through normalization, eight heuristic quality filters, ML-based scoring (language identification and n-gram perplexity), exact and MinHash near-deduplication, and PII redaction. The pipeline retains 9.5M documents (3.0B tokens, 64.4% of input). Heuristic filtering and deduplication are the primary reduction stages, each removing roughly one-sixth of documents.

Azerbaijani is a Turkic language classified as low-resource in NLP taxonomies due to limited digital presence and few curated datasets [8]. Recent years have seen growing interest in Azerbaijani language technology. Hasanli and Rustamov [4] collected and annotated Azerbaijani tweets for sentiment polarity detection, one of the earlier applied NLP efforts for the language. Huseynov et al. [7] trained word2vec and GloVe embeddings on Azerbaijani news articles and books, noting the challenges of working with a low-resource agglutinative language. Ziyaden et al. [20] used back-translation augmentation to address labeled data scarcity for Azerbaijani text classification and released a pre-trained RoBERTa model for the language. These works collectively demonstrate both the demand for Azerbaijani language technology and the recurring challenge of limited monolingual text resources.

Existing multilingual web crawls such as OSCAR [1], CulturaX [14], and C4 [17] include Azerbaijani splits, but these contain substantial noise: non-Azerbaijani text misidentified by upstream language classifiers, boilerplate and template content from web scraping, and extensive duplication across and within sources.

A data-centric approach is needed to convert these raw resources into a corpus suitable for pretraining and downstream applications. Such an approach must balance aggressive noise removal against the risk of discarding valuable content from an already small pool. The filtering decisions should be transparent, auditable, and backed by empirical evidence rather than arbitrary thresholds.

This paper makes the following contributions:

1. A multi-stage curation pipeline for Azerbaijani web text (Fig. 1), covering normalization, heuristic filtering, ML-based quality scoring, deduplication, and PII redaction, with all thresholds and stopword lists documented.
2. A systematic ablation study that isolates the effect of each pipeline stage, deduplication strategy, filter configuration, and data source under an explicit n-gram screening protocol. We position these results as internal comparisons across corpus variants, not as proof of transformer-scale pretraining utility.
3. An analysis of source diversity, showing that locally digitized books provide the strongest signal under our proxy while web-crawled sources contribute volume with varying noise profiles.
4. A tokenizer efficiency analysis showing that a corpus-trained SentencePiece model compresses Azerbaijani text 2.7× more efficiently than GPT-2 and 9% more efficiently than XLM-RoBERTa.
5. Public release of the curated corpus, the SentencePiece-AZ tokenizer, the translated HellaSwag-AZ evaluation set, and the pipeline code, so that subsequent transformer-based work can build on a documented baseline rather than re-curating from scratch.

2 Data Sources

We aggregate Azerbaijani text from five sources spanning web crawls, encyclopedic content, and digitized literature. Table 2 (columns 2–3) lists the raw document and token counts per source.

CulturaX. CulturaX [14] is a cleaned multilingual dataset derived from mC4 and OSCAR. We use 4.1 million documents from its Azerbaijani split. Despite upstream cleaning, we find residual noise including boilerplate fragments and mixed-language content.

OSCAR 2301. OSCAR [1] provides document-level web text extracted from Common Crawl with language classification. We use 3.1 million documents from the Azerbaijani partition. OSCAR has the lowest token retention rate among our sources (39.9%), reflecting high noise levels.

C4. C4 [17] is derived from Common Crawl with heuristic filters designed for English. Its Azerbaijani split (3.1 million documents) retains 52.5% of tokens after our pipeline, suggesting that the original C4 filters are insufficient for non-English text.

Wikipedia. We use the November 2023 dump of Azerbaijani Wikipedia (2.0 million articles). Wikipedia text is editorially curated and has high baseline quality.

Local Books. We include 1.7 million text chunks from digitized Azerbaijani-language books, split into passages of 600–4,500 characters at sentence boundaries. This source has the highest retention rate (95.0%) and contributes the strongest downstream signal in leave-one-out experiments (Sect. 5.2).

All sources are publicly available or used under research-compatible licenses. No personally identifiable information is intentionally included in the raw data; residual PII from web crawls is removed in Sect. 3.5.

3 Curation Pipeline

The pipeline consists of six sequential stages, illustrated in Fig. 1. Each stage is described below.

3.1 Ingestion and Normalization

Books are split into sentence-level chunks prior to ingestion. Per-source caps ensure balanced initial representation.

Normalization applies three transformations: (1) removal of non-printing control characters, (2) Unicode NFKC normalization, and (3) collapsing of multiple whitespace characters into single spaces. This stage preserves all 14 million documents while standardizing the character inventory.

3.2 Heuristic Filtering

We apply eight rule-based filters, each controlled by a threshold. A document must pass all filters to be retained:

1. **Word count:** minimum 20, maximum 12,000 words.
2. **Character repetition:** the fraction of characters belonging to repeated 10-character n-grams must not exceed 0.15.
3. **Word repetition:** the fraction of words in repeated 5-word n-grams must not exceed 0.28.
4. **Special character ratio:** punctuation, digits, and miscellaneous symbols must constitute at most 45% of characters.
5. **Stopword ratio:** at least 8% of words must be common Azerbaijani stopwords (from a list of 40 function words including *bir*, *bu*, *da*, *ki*).

6. **Azerbaijani character ratio:** at least 55% of alphabetic characters must belong to the Azerbaijani alphabet (32-letter Latin script including ç, ğ, ö, ü, ş).
7. **Digit ratio:** digits must not exceed 20% of all characters.
8. **Line-break ratio:** newline characters must not exceed 12% of all characters.

This stage removes 2,350,000 documents (16.8% of the input). The dominant removal reasons are high character repetition (1,729,000 trigger counts) and low stopword ratio (1,059,500 trigger counts), indicating that the web crawls contain repetitive boilerplate and non-Azerbaijani text that passed upstream language identification.

3.3 ML-Based Filtering

Three ML-based checks provide a finer-grained quality screen on top of the heuristic stage. A document must pass all three to be retained.

Language Identification. We use FastText [9] for language classification (with langid [13] as a fallback). Documents with Azerbaijani confidence below 0.80 are rejected, unless they contain at least 72% Azerbaijani characters (a rescue heuristic for texts with code-mixed segments).

N-gram Perplexity. A 5-gram word-level language model with add-k smoothing ($k = 0.5$) is trained on high-quality seed documents from Wikipedia and local books. Documents with perplexity above an adaptive threshold are rejected. The threshold is the maximum of: (a) a ceiling of 1,200, (b) the 99th percentile of seed perplexities scaled by 1.20, and (c) the 99.5th percentile over a broad sample scaled by 1.05.

KenLM Perplexity. A 5-gram Kneser-Ney model is built using KenLM [5] on seed documents. Documents exceeding the adaptive perplexity threshold are rejected. The calibration mirrors the n-gram model above.

This stage removes 97,000 documents (0.8% of its input), confirming that heuristic filters already capture the bulk of noise. The ML layer catches residual cases: 53,800 documents with anomalous KenLM perplexity, 44,800 with high n-gram perplexity, and 18,400 language mismatches.

3.4 Deduplication

We apply two complementary deduplication methods.

Exact Deduplication. We hash each document (SHA-1 of lowercased, whitespace-normalized text) and keep only the first occurrence, removing 1,618,800 exact duplicates.

Near-Deduplication. We use MinHash Locality-Sensitive Hashing (LSH) [2] to identify high-overlap documents. Each document is lowercased and tokenized with Python's Unicode `\w+` regex, whose letter class includes the Azerbaijani-specific Latin Extended characters (ç, ğ, ı, İ, ö, ş, ü, and the schwa at U+0259). The regex splits on non-word characters (whitespace, punctuation, symbols); digits stay attached to their containing token, apostrophes act as token boundaries (so *Qur'an* produces two shingle units), and alphanumeric spans remain one token even when they mix scripts (mixed-script runs are rare in Azerbaijani and shingle-level near-duplicate detection is robust to this residual noise). We did not apply lemmatization or morphological segmentation, so shingles capture surface forms rather than stems; we found this acceptable for document-level near-duplicate detection in an agglutinative language. Consecutive tokens are joined with single spaces into 5-gram shingles. MinHash signatures use 96 permutations via the `datasketch` library, and the LSH index uses a Jaccard threshold of 0.86 (swept in Sect. 5.3). This stage removes 434,200 near-duplicate documents.

Together, deduplication removes 2,053,000 documents (17.8% of its input), consistent with prior findings that web crawls contain substantial redundancy even after quality filtering [12].

3.5 PII Redaction and Train/Val Split

Four regex patterns detect and mask personally identifiable information: email addresses (62,400 matches), IP addresses (9,850), Azerbaijani phone numbers (12,100), and national ID patterns (665). Each match is replaced with a category-specific placeholder. A total of 85,015 PII instances are redacted across the 9,500,000-document corpus.

Finally, the corpus is split into training (98%) and validation (2%) partitions.

4 Experimental Protocol

Configuration. All experiments use the same filter thresholds and source caps described in Sect. 3.

Downstream Proxy. Evaluating data quality ideally requires training a full neural language model, which is prohibitive for systematic ablations across many corpus variants and token budgets. Following the proxy approach of FineWeb [15], we train lightweight 5-gram word language models on corpus variants at multiple token budgets (250M to 3B tokens) and measure two metrics: (1) **holdout perplexity** on a fixed validation set drawn from the curated corpus, and (2) **HellaSwag-AZ accuracy**, a four-way multiple-choice commonsense reasoning task adapted from HellaSwag [19]. For HellaSwag-AZ, each candidate ending is scored by its conditional log-probability under the n-gram model, normalized by ending length; the option with the highest score is taken as the prediction.

We use these proxies as cheap screening signals for comparing corpus variants, not as a substitute for transformer-based pretraining evaluation. The small absolute accuracy values throughout this paper (in the 0.24–0.29 range) reflect the limited capacity of word-level n-gram models on a four-way choice task; what matters for our ablations is the relative ordering between variants at a fixed token budget.

HellaSwag-AZ Construction. The English HellaSwag validation split was translated to Azerbaijani with the Gemini 2.0 Flash large language model, then manually inspected: we fixed remaining mistranslations, removed items that became incoherent or culturally inapplicable, and verified that the labelled ending remained the most plausible continuation. We chose an LLM translator over a neural MT system because LLM translation handles idiomatic and informal phrasing better, particularly in HellaSwag contexts. We do not claim translation parity with the English original. We use a fixed 800-item subsample of the cleaned validation split for all ablations in this paper to keep evaluation cost bounded and results comparable across runs. The dataset is released publicly[1] so that future work can extend or replace it. We treat HellaSwag-AZ as a screening proxy in this paper; building a natively designed Azerbaijani benchmark is left to future work.

Ablation Design. We run five families of ablation experiments: (1) *deduplication strategy*: comparing no dedup, global MinHash, independent per-source MinHash, and training on removed duplicates only; (2) *deduplication threshold*: sweeping Jaccard thresholds at 0.74, 0.84, and 0.90 under both global and independent modes; (3) *C4-style filters*: selectively applying terminal punctuation, word length, and curly bracket filters from the C4 pipeline [17]; (4) *source leave-one-out (LOO)*: removing each source and measuring the accuracy delta; (5) *corpus variant benchmark*: comparing raw, curated, ML-rejected, and duplicate-only training sets.

Quality Metrics. Beyond downstream proxies, we track corpus-level statistics at each pipeline stage: average word count, stopword ratio, Azerbaijani character ratio, special character ratio, and character/word repetition ratios.

5 Results

5.1 Stage Transitions and Retention

Table 1 reports the document and token counts at each pipeline stage along with step-wise and cumulative retention percentages.

Heuristic filtering removes 2.35 million documents (16.8%) and 772M tokens (16.6%). Deduplication removes a comparable 2.05 million documents (17.8% of post-ML content) and 861M tokens (22.3%). ML filtering removes a much smaller

[1] https://huggingface.co/datasets/eljanmahammadli/hellaswag-azerbaijani.

fraction (97,000 documents, 0.8%), indicating that the heuristic stage already eliminates most low-quality content. PII redaction masks 120,000 tokens across 85,015 PII instances without removing any documents.

Quality indicators improve monotonically across stages: word repetition ratio drops from 0.031 (raw) to 0.003 (final), character repetition from 0.145 to 0.061, and stopword ratio increases from 0.129 to 0.167, reflecting a higher concentration of genuine Azerbaijani prose.

Table 1. Stage transitions. Step retention is relative to the previous stage; cumulative retention is relative to the raw input.

Stage	Docs	Tokens	Step Ret. (% docs)	Cumul. Ret. (% tokens)
Raw	14.00M	4.66B	100.0	100.0
Normalized	14.00M	4.66B	100.0	100.0
Heuristic	11.65M	3.89B	83.2	83.5
ML filtered	11.55M	3.86B	99.2	82.9
Deduplicated	9.50M	3.00B	82.2	64.4
PII redacted	9.50M	3.00B	100.0	64.4

5.2 Source Impact

Table 2 reports per-source retention and the source composition of the final corpus. Figure 2 plots source retention against leave-one-out (LOO) accuracy impact.

Local books have the highest token retention rate (95.0%) because they are pre-curated human-written text without web noise. They also have the largest negative LOO impact: removing them decreases HellaSwag accuracy by 4.0% points, making them the single most valuable source. CulturaX is the second most valuable (LOO $\Delta = -2.3$pp) and contributes the largest share of final tokens (32.6%).

OSCAR has the lowest token retention (39.9%), reflecting higher noise levels, but still contributes positively (LOO $\Delta = -1.0$pp). Removing C4 slightly *improves* accuracy (+0.7pp), suggesting that its noise outweighs its volume contribution at this scale. Wikipedia shows a similar pattern (LOO $\Delta = +0.3$pp), indicating mild redundancy with other sources.

5.3 Deduplication Ablations

Figure 3 compares four deduplication strategies across training token budgets from 250M to 3B. At 3B tokens, global MinHash achieves 0.257 HellaSwag accuracy, outperforming no-dedup (0.249) by 0.8 percentage points and independent MinHash (0.253) by 0.4pp. This confirms that deduplication yields measurable downstream improvements at scale [12]. The separation between strategies

Table 2. Source impact. Retention is the fraction of raw tokens surviving all pipeline stages. LOO Δ is the change in HellaSwag accuracy at 1B tokens when the source is excluded (negative = source helps).

Source	Raw Docs	Raw Tokens	Final Tokens	Ret. (%)	Doc Share (%)	LOO Δ Acc.
CulturaX	4.10M	1.21B	990M	82.0	32.6	−0.023
Local books	1.71M	798M	758M	95.0	17.1	−0.040
OSCAR	3.07M	1.36B	543M	39.9	19.6	−0.010
C4	3.07M	716M	376M	52.5	16.5	+0.007
Wikipedia	2.05M	578M	335M	58.0	14.2	+0.003
Total	**14.0M**	**4.66B**	**3.00B**	**64.4**	**100.0**	

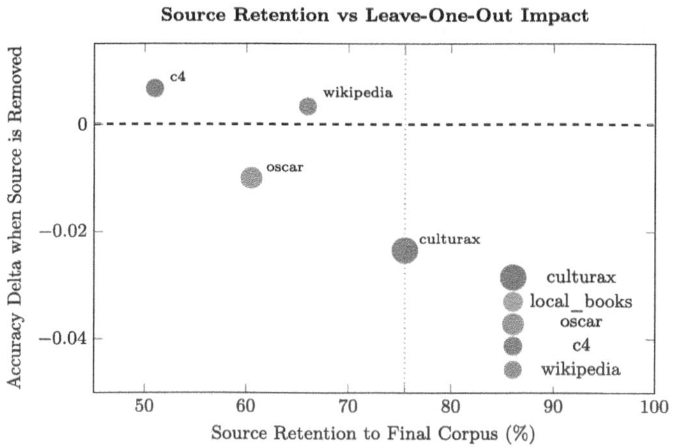

Fig. 2. Source retention vs. leave-one-out accuracy impact. Local books have the highest retention (95%) and the largest positive contribution (−4.0pp). C4 and Wikipedia are mildly redundant.

widens with increasing budget: at 250M tokens the differences are small, but by 3B the gap is clear. The *removed-by-dedup* variant (trained exclusively on duplicates) saturates at 861M tokens with accuracy 0.218 and perplexity 78,246 (vs. 33,015 for global MinHash at 3B), confirming that duplicates form a narrow, repetitive subset of the corpus.

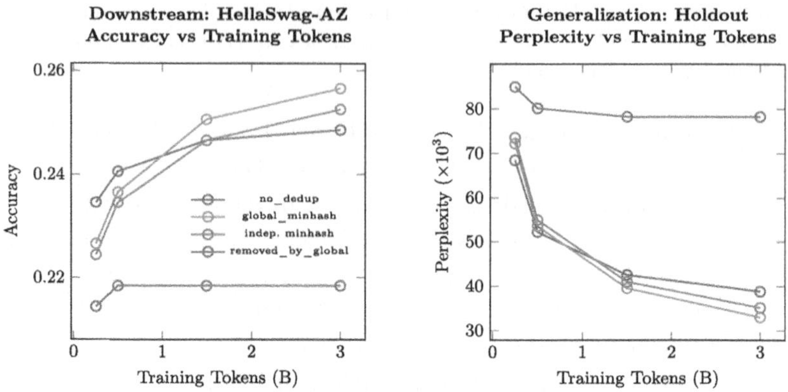

Fig. 3. Deduplication strategy comparison. Global MinHash achieves 0.257 HellaSwag accuracy at 3B tokens, outperforming no-dedup by 0.8pp. Training on removed duplicates alone saturates early at low accuracy.

Table 3. Ablation leaderboard. Best variant per experiment family at maximum token budget. Δ accuracy is relative to each experiment's baseline (where applicable).

Experiment	Best Variant	HellaSwag Acc.	Holdout PPL	Δ Acc.
Dedup strategy	global_minhash	0.257	33,015	+0.008
C4 filter	c4_except_terminal	0.271	25,487	+0.008
Dedup threshold	global_mh (t=0.74)	0.274	27,512	—
Corpus variant	raw_sample	0.284	28,487	+0.013
Source LOO	leave_out_c4	0.291	30,949	+0.007

Figure 4 sweeps the MinHash Jaccard threshold from 0.74 to 0.90 under both global and independent modes. Lower thresholds (more aggressive deduplication) yield higher accuracy: at the 1B token budget, global MinHash with t=0.74 reaches 0.274, compared to 0.241 at t=0.90. This pattern holds across modes, indicating that aggressive duplicate removal improves data efficiency for downstream tasks.

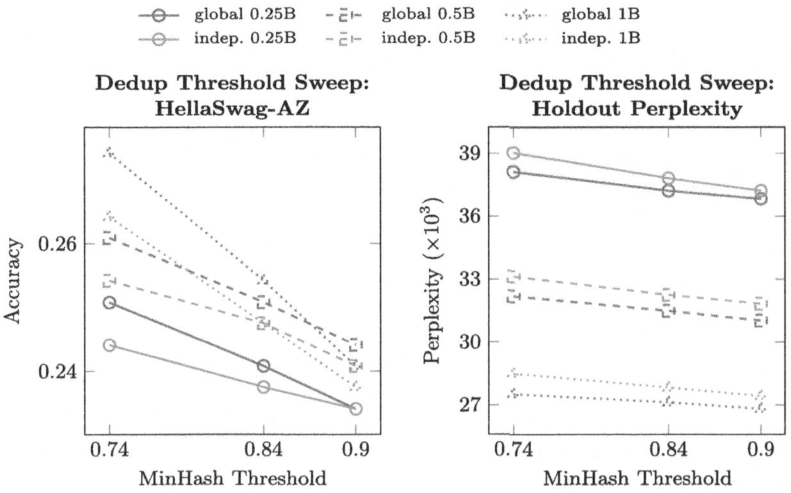

Fig. 4. MinHash Jaccard threshold sweep at 1B tokens. Lower thresholds (more aggressive deduplication) yield higher accuracy, with global mode at t=0.74 reaching 0.274.

5.4 Heuristic Filter Ablations

We evaluate C4-style heuristic filters [17] as additional quality signals applied on top of our baseline pipeline. Figure 5 shows HellaSwag accuracy across six configurations at token budgets from 250M to 3B.

The `c4_except_terminal` variant (combining word-length, curly bracket, and bad-phrase filters without the terminal punctuation requirement) achieves the best accuracy at 3B tokens (0.271 vs. 0.263 baseline), a consistent +0.8pp improvement. The `c4_all` variant, which adds terminal punctuation filtering, removes an additional 2.3% of documents and achieves 0.265, slightly above baseline but below `c4_except_terminal`. Terminal punctuation filtering applied in isolation underperforms the baseline (0.254 vs. 0.263), confirming it is not a reliable quality signal for Azerbaijani web text.

5.5 Tokenizer Diagnostics

Figure 6a compares compression efficiency (mean tokens per character) across three tokenizers evaluated on the curated corpus. We train a SentencePiece BPE model [11] on a 100,000-document sample of the curated corpus with a vocabulary of 16,000 subwords, full character coverage (1.0), and shuffled input sentences. The trained model achieves 0.237 tokens per character on a held-out 5,000-document evaluation set. XLM-RoBERTa [3], a multilingual model, requires 9.3% more tokens on the same texts. GPT-2, trained primarily on English, requires 171.2% more tokens, meaning the same Azerbaijani text consumes nearly three times the sequence budget. This confirms the practical value of language-specific tokenization for Azerbaijani: an English-centric tokenizer wastes a majority of the model's context window on sub-word fragmentation.

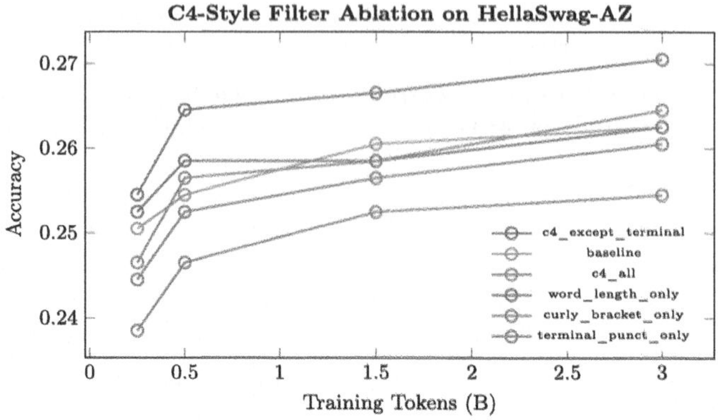

Fig. 5. C4-style filter ablation. The `c4_except_terminal` variant improves accuracy by 0.8pp over the baseline at 3B tokens. Terminal punctuation filtering alone hurts performance.

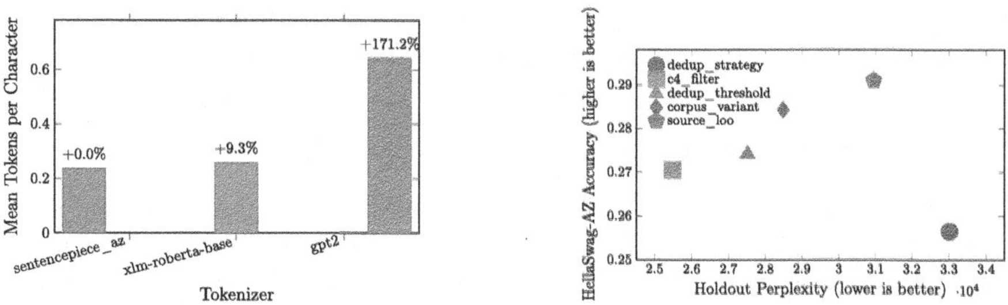

(a) Tokenizer compression. (b) Ablation frontier.

Fig. 6. (a) SentencePiece-AZ uses 0.237 tokens per character; GPT-2 requires 2.7× more. (b) Best variant per ablation family; source-level and corpus-variant experiments achieve the best tradeoff.

6 Discussion

Cross-Experiment Synthesis. Table 3 and Fig. 6b summarize the best variant from each ablation family. The source LOO and corpus variant experiments achieve the highest accuracy (0.291 and 0.284, respectively), while the C4 filter experiment achieves the lowest holdout perplexity (25,487). The dedup strategy experiment confirms that global MinHash deduplication provides clear gains (+0.8pp accuracy, −5,803 perplexity vs. no dedup) at 3B tokens.

Quantity vs. Quality Under the N-Gram Proxy. The raw corpus variant outperforms the curated variant on HellaSwag-AZ at 1B tokens (0.284 vs. 0.271). For word-level n-gram models, multiple-choice accuracy depends heavily on training-time vocabulary coverage: out-of-vocabulary tokens collapse to the smoothing floor and inject ranking noise. Filtering removes 35.6% of raw tokens, including some real low-frequency vocabulary, mechanically reducing coverage. At neural scale this tradeoff may resolve differently, since transformers do not depend on exact n-gram matches; English curation gains [6,15,16] are suggestive but not evidence for our corpus. We do not claim the curated corpus outperforms the raw corpus for transformer pretraining; we claim that under a controlled n-gram protocol, pipeline choices produce measurable orderings that are informative for designing larger runs. Independent of accuracy, curation drops repetition ratios by an order of magnitude, raises language purity, removes duplicates, and masks PII.

Source Diversity. The LOO results suggest that for low-resource languages, investing in digitization of native-language literature and leveraging larger cleaned web crawls yields the strongest returns, while smaller or noisier web sources may dilute quality.

7 Limitations

Evaluation Proxy. All ablations use word-level 5-gram language models, not neural models, because transformer training across this many corpus variants and token budgets was infeasible under our compute budget. The proxy has two known biases: n-gram accuracy rewards lexical coverage (the most plausible explanation for the raw-vs.-curated inversion in Sect. 6), and absolute HellaSwag-AZ accuracies (0.24–0.29) are not comparable to transformer scores in the literature. We present our results as relative comparisons under a fixed harness, not as proof of transformer-scale utility; confirming the orderings at neural scale is the natural next step.

HellaSwag-AZ Provenance. The benchmark is a Gemini 2.0 Flash translation with manual cleanup, not a native Azerbaijani evaluation. Cultural references (American sports, US-specific contexts) survive translation, distractors lose the adversarial property they had in the English construction, and inter-annotator agreement was not measured. We use it as a screening signal and recommend future work either re-derives distractors natively or builds a native benchmark.

Domain Balance. CulturaX is 32.6% of the final tokens. Because it is itself derived from mC4 and OSCAR, our corpus inherits the topical distribution of multilingual web crawls (news, e-commerce, government portals dominate). We did not implement domain-aware sampling; practitioners needing balanced coverage should reweight sources at training time or apply a topic classifier on top of our pipeline.

No Human Evaluation. We did not collect human judgments at any stage boundary, so we cannot directly verify that filter-removed documents are lower quality or that MinHash near-duplicate clusters match human-perceived duplicates. The automated hygiene statistics improve monotonically across stages but do not by themselves prove human-quality alignment.

ML Filter Scope. The ML stage is a statistical screen (language ID plus two n-gram perplexity checks, including a KenLM [5] Kneser-Ney model), not a learned content classifier. A heavier quality classifier in the style of FineWeb-Edu would likely catch additional failure modes such as machine-generated content; we chose the simpler design because the heuristic stage removes 16.8% of documents up front, leaving a much cleaner residual.

PII Coverage. PII redaction uses four regex patterns (email, IP, Azerbaijani phone, national ID), missing obfuscated variants, foreign formats, and entities requiring NER (names, addresses). PII-sensitive deployments require additional safeguards.

8 Conclusion

We presented a multi-stage pipeline for curating Azerbaijani web text and a systematic ablation study under an n-gram screening protocol. From 14M raw documents (4.66B tokens) across five sources, the pipeline retains 9.5M documents (3.0B tokens) after normalization, heuristic filtering, ML-based scoring, deduplication, and PII redaction. Heuristic filters and deduplication each remove roughly one-sixth of documents; ML filters catch residual edge cases. Under the proxy, global MinHash deduplication yields a 0.8pp accuracy gain at 3B tokens, locally curated book data is the most informative single source, and C4-style filters transfer partially to Azerbaijani. A corpus-trained SentencePiece tokenizer compresses Azerbaijani text 2.7× more efficiently than GPT-2. We do not claim these orderings will hold unchanged at transformer scale; confirming them is the natural follow-up. We release the pipeline, the curated corpus, the tokenizer, and the HellaSwag-AZ evaluation set[2] as a documented baseline for future Azerbaijani pretraining work.

Acknowledgments. This work was conducted within the Center of Data Analytics Research (CeDAR) and the MegaSec AI Laboratory at ADA University.

Disclosure of Interests. The authors declare no competing interests.

References

1. Abadji, J., Ortiz Suarez, P., Romary, L., Sagot, B.: Towards a cleaner document-oriented multilingual crawled corpus. In: Proceedings of the Thirteenth Language Resources and Evaluation Conference, pp. 4344–4355 (2022)
2. Broder, A.Z.: On the resemblance and containment of documents. In: Proceedings of the Compression and Complexity of Sequences, pp. 21–29. IEEE (1997)
3. Conneau, A., et al.: Unsupervised cross-lingual representation learning at scale. In: Proceedings of the 58th Annual Meeting of the Association for Computational Linguistics, pp. 8440–8451 (2020)

[2] Code and links to all artifacts: https://github.com/eljanmahammadli/AzText.

4. Hasanli, H., Rustamov, S.: Sentiment analysis of Azerbaijani twits using logistic regression, Naive Bayes and SVM. In: IEEE 13th International Conference on Application of Information and Communication Technologies (AICT). IEEE (2019)
5. Heafield, K.: KenLM: faster and smaller language model queries. In: Proceedings of the Sixth Workshop on Statistical Machine Translation, pp. 187–197 (2011)
6. Hoffmann, J., et al.: Training compute-optimal large language models. In: Advances in Neural Information Processing Systems, vol. 35, pp. 30016–30030 (2022)
7. Huseynov, K., Suleymanov, U., Rustamov, S., Huseynov, J.: Training and evaluation of word embedding models for Azerbaijani language. In: Biele, C., Kacprzyk, J., Owsiński, J.W., Romanowski, A., Sikorski, M. (eds.) MIDI 2020. AISC, vol. 1376, pp. 37–48. Springer, Cham (2021). https://doi.org/10.1007/978-3-030-74728-2_4
8. Joshi, P., Santy, S., Budhiraja, A., Bali, K., Choudhury, M.: The state and fate of linguistic diversity and inclusion in the NLP world. In: Proceedings of the 58th Annual Meeting of the Association for Computational Linguistics, pp. 6282–6293 (2020)
9. Joulin, A., Grave, E., Bojanowski, P., Mikolov, T.: Bag of tricks for efficient text classification. In: Proceedings of the 15th Conference of the European Chapter of the Association for Computational Linguistics: Volume 2, Short Papers, pp. 427–431 (2017)
10. Kreutzer, J., et al.: Quality at a glance: An audit of web-crawled multilingual datasets. Trans. Assoc. Comput. Linguistics **10**, 50–72 (2022)
11. Kudo, T., Richardson, J.: SentencePiece: a simple and language independent subword tokenizer and detokenizer for neural text processing. In: Proceedings of the 2018 Conference on Empirical Methods in Natural Language Processing: System Demonstrations, pp. 66–71 (2018)
12. Lee, K., et al.: Deduplicating training data makes language models better. In: Proceedings of the 60th Annual Meeting of the Association for Computational Linguistics, pp. 8424–8445 (2022)
13. Lui, M., Baldwin, T.: langid.py: an off-the-shelf language identification tool. In: Proceedings of the ACL 2012 System Demonstrations, pp. 25–30 (2012)
14. Nguyen, T., et al.: CulturaX: a cleaned, enormous, and multilingual dataset for large language models in 167 languages. In: Proceedings of the 2024 Joint International Conference on Computational Linguistics, Language Resources and Evaluation (LREC-COLING 2024), pp. 4226–4237 (2024)
15. Penedo, G., et al.: The FineWeb datasets: decanting the web for the finest text data at scale. In: Advances in Neural Information Processing Systems. vol. 37, pp. 30811–30849 (2024)
16. Penedo, G., et al.: The RefinedWeb dataset for Falcon LLM: Outperforming curated corpora with web data only. In: Advances in Neural Information Processing Systems, vol. 36, pp. 79155–79172 (2023)
17. Raffel, C., et al.: Exploring the limits of transfer learning with a unified text-to-text transformer. J. Mach. Learn. Res. **21**(140), 1–67 (2020)
18. Wenzek, G., et al.: CCNet: extracting high quality monolingual datasets from web crawl data. In: Proceedings of the Twelfth Language Resources and Evaluation Conference, pp. 4003–4012 (2020)

19. Zellers, R., Holtzman, A., Bisk, Y., Farhadi, A., Choi, Y.: HellaSwag: can a machine really finish your sentence? In: Proceedings of the 57th Annual Meeting of the Association for Computational Linguistics, pp. 4791–4800 (2019)
20. Ziyaden, A., Yelenov, A., Hajiyev, F., Rustamov, S., Pak, A.: Text data augmentation and pre-trained language model for enhancing text classification of low-resource languages. PeerJ Comput. Sci. **10**, e1974 (2024)

Governance, Strategy, and Human–Organization Interaction

An Ontology-Driven Recommender System for Enhancing Workplace Practices

Asma Dhaouadi[1(✉)] and Maria Santilli[2]

[1] INSA Rouen, LITIS, Normandie Université, Rouen, France
asma.dhaouadi@insa-rouen.fr

[2] Facultad de Ingenieria, UNDef, Buenos Aires, Argentina
msantilli697@alumnos.iua.edu.ar

Abstract. Today, companies around the world face major challenges. They need to promote employee well-being and engagement, foster cooperation and organizational efficiency, and leverage the quality of the workplace environment as a strategic driver of sustainable performance. To address these challenges, the international community, particularly through the International Organization for Standardization (ISO) and the European Committee for Standardization (CEN), has defined numerous indicators to identify various aspects of working conditions. However, an important limitation remains, which is the absence of a comprehensive framework to model, based on these indicators, the relationship between the efficiency of the workplace environment of a company and its economic, environmental, and societal performance. To meet these multiple challenges, this paper proposes an ontology-based recommender system designed to structure the elements involved in workplace environments, grounded in the knowledge provided by experts in the field. By leveraging the knowledge organized within the ontology, the system provides tailored action recommendations that take into account the current situation of the company, its goals, as well as its organizational and financial constraints.

Keywords: Ontology · Recommender system · Workplace performance

1 Introduction

Leaders in the field of strategic management worldwide are increasingly focusing on the workplace environment, as its management has become more complex and is now considered to be closely linked to three key dimensions: social (employee well-being and engagement), economic (productivity and cost control), and environmental (carbon footprint and water usage reduction). In the context of worldwide standardization efforts, several standards have been developed, including ISO 41011 [1], ISO 26000 [2], and ISO/DIS 18480-1 [3], as well

C. Zanni-Merk et al. (Eds.): AIDT 2026, CCIS 3023, pp. 359–374, 2026.
https://doi.org/10.1007/978-3-032-31319-5_24

as European NF EN standards such as BS EN 15221-6 [4], NF EN 15221-3 [5], NF EN 15221-7 [6] and NF EN 15221-4 [7], through which numerous indicators and benchmarking approaches have been introduced. However, we observed the need to structure all the elements involved in the evaluation of the workplace environment and examine the correlations between these indicators to highlight both their implicit influence on each other and their explicit impact on the workplace environment as a whole. We also identified the need to analyze the impact of different actions aimed at improving the workplace environment and integrate them into the conceptual model.

A range of conceptual frameworks has addressed organizational performance [8–12]. In [8], the authors propose four perspectives for comprehensive evaluation: financial, customer, internal processes, and learning and growth. Keegan et al. differentiate cost versus non-cost measures and internal versus external indicators [9]. In [10], the authors identify seven interrelated criteria: effectiveness, efficiency, productivity, innovation, quality of work life, quality, and profitability. Van der Voordt and Jensen in [11] contrast performance-measurement and benchmarking theory with practice, synthesizing the literature, practitioner interviews [12], cross-border studies, and empirical cases. These works advance measurement thinking but do not fully address how workplace indicators influence one another or how those interdependencies can be operationalized for decision support. Despite these contributions, the literature remains fragmented, particularly given the importance of strategic management. Consequently, a comprehensive and structured framework that integrates measurement and interdependency analysis to capture interactions among workplace indicators is still lacking.

To the best of our knowledge, no prior work has systematically (i) measured the impact of workplace-evaluation indicators and (ii) analyzed their correlations to derive a model that quantifies both explicit and implicit effects on the workplace. In this regard, we recognize the necessity of designing a recommender system capable of guiding decision makers in selecting the most relevant actions among those possible for the company. To support such decision-making capabilities, it is essential to rely on semantic frameworks that can formally represent domain knowledge and reasoning mechanisms. Ontologies play a crucial role in such systems by enhancing user profiling, improving scalability, effectiveness, and interpretability, and addressing challenges such as the cold-start problem [13]. By leveraging ontological inferences and structured semantic relationships, the recommender system can link workplace-related data to high-level organizational objectives. This alignment enables the generation of relevant and personalized action recommendations, even in the absence of prior interaction or historical feedback from the company. From this perspective, in this paper we propose a domain ontology designed to capture the elements we consider essential for evaluating the workplace environment. In addition, we present an approach that supports an ontology-driven recommender system, intended to assist leaders in selecting the most appropriate actions to improve their company's performance.

The remainder of this paper is organized as follows. Section 2 reviews the related work. Section 3 presents the context of the workplace environment, detailing key constraints, indicators, and the proposed ontological approach to workplace modeling. Section 4 outlines our proposal for the design of the recommender system. Section 5 concludes the paper and discusses future research directions.

2 Related Work

In the literature, numerous works have focused on recommender systems that leverage ontologies to enhance personalization, semantic reasoning, and domain knowledge representation. Building on this trend, several researchers have explored the use of ontology capabilities for recommender system design [13,21–26]. Ontology-based recommender systems exploit structured domain models to improve user profiling, similarity computation, and recommendation quality beyond what conventional techniques can achieve [19]. These studies aim to enable ontological inference mechanisms for enhanced user profiling, automated recommendation generation through the integration of external knowledge, improved accuracy, support for multi-class environments, and the combination of knowledge-based and machine learning approaches.

Traditionally, recommender systems are classified into three main categories: Collaborative Filtering (CF), Content-Based Filtering (CBF), and hybrid approaches that integrate multiple techniques to harness their complementary strengths. According to [20], CF systems recommend items to a target user by identifying a neighborhood of users with similar preferences and suggesting items that these similar users have liked, whereas CBF systems construct a user profile from the attributes of items the user has positively rated and recommend other items with similar descriptive features. Hybrid methods combine both CF and CBF to address limitations such as data sparsity and cold start problems more effectively. Beyond this conventional categorization, recent research has focused on enriching these approaches through semantic modeling techniques.

Ontologies can be integrated within each of these paradigms to enhance semantic representation, improve inference capabilities, and support contextualized recommendations. Consequently, ontology-based recommender systems have gained significant attention across multiple fields of application. By leveraging structured semantic models, they enable knowledge-driven and context-aware recommendation processes. For instance, Bagherifard et al. [13] proposed a hybrid recommender system to address the problems of data sparsity, cold start (particularly for new items), and scalability, while improving overall accuracy and system performance. They combine CF and CBF approaches, integrating both memory-based and model-based methods. In the CF component, they employ clustering, ontology-based user profiling, item ontology, semantic similarity between ontologies, and the k-NN algorithm to overcome sparsity and enhance scalability and accuracy. In the CBF component, they use item ontology and semantic similarity measures to handle the new-item cold start problem. Additionally, a heuristic method is introduced to refine the measurement of the

"IS-A" hierarchical relationship within the ontology, thereby improving semantic similarity accuracy and generating more precise recommendations.

Ontology-based approaches have been successfully applied across various domains. In particular, ontology-based recommender systems have attracted significant attention across multiple application areas. In the e-learning domain, Tarus et al. [21] and Rahayu et al. [22] highlighted how ontological models of learners, learning resources, and educational contexts enable personalized learning pathways and improved learner engagement through semantic reasoning mechanisms.

In the healthcare domain, Yao et al. [23] proposed a prescription recommendation framework combining medical ontologies, diagnosis history, and patient data. Their approach employs a Transformer, dual Recurrent Neural Network (RNN) encoders, and a co-attention mechanism to model patient–drug interactions, thereby supporting effective prescription decision-making.

In e-commerce settings, Mustafa et al. [24] introduced a hybrid recommender system that integrates domain ontologies with Sequential Pattern Mining (SPM) to enhance personalization. While the ontology captures structured knowledge about users and products, SPM identifies temporal patterns in user interactions. This combination improves recommendation accuracy and mitigates the cold start problem by enabling informed suggestions even when user rating data are limited.

Beyond domain-specific applications, several studies have explored the integration of ontologies with advanced machine learning techniques. Ahmed et al. [25] proposed a hybrid recommender system that combines ontology-based, knowledge-driven reasoning with data-driven approaches, employing classifiers and neural networks to enhance recommendation effectiveness. Their system enables bidirectional knowledge transfer by embedding semantic information into machine learning models while incorporating statistical insights back into the ontology. Moreover, by updating the ontology with newly observed products and user behaviors, the framework supports dynamic reasoning capabilities, thereby capturing the evolving nature of the application environment.

Similarly, Gharibi et al. [26] developed an ontology-based recommender system for e-commerce clothing products by integrating an ontology with a Convolutional Neural Network (CNN). Their framework incorporates semantic-level information modeling, analyzes user behavior to construct user histories and profiles, builds an ontology representing these profiles, and computes similarity measures to generate personalized recommendations.

In summary, these works demonstrate that ontologies not only facilitate the integration of domain knowledge, but also enable reasoning mechanisms that enhance recommendation relevance and adaptability. Accordingly, we adopt ontologies as the core modeling framework of our system, as they provide a formal and structured representation of indicators, actions, company states, and their interdependencies, which is essential for ensuring consistent reasoning and effective decision support.

3 An Ontological Approach to Workplace Modeling

Work-life balance, high-quality working conditions, meaningful work, and active employee engagement are key pillars for aligning social responsibility with workplace well-being. These dimensions constitute essential drivers for improving organizational performance and fostering sustainable workplace environments.

To structure and evaluate these dimensions, we analyzed the previously mentioned international ISO standards and European NF EN standards [1–7]. This analysis led to the identification of 63 indicators that can be leveraged to assess the work environment. To further refine these results, we also examined documentation from professional organizations and expert networks, including the French National Agency for the Improvement of Working Conditions (ANACT) [14], the Institute for the Work Environment (IDET) [15], the Movement of French Enterprises (MEDEF) [16], and the L'Isle Adam Research Consortium (CRDIA) [17], a collective bringing together large companies, innovative small and medium-sized enterprises, and research institutions.

While this process enabled the identification of a comprehensive set of indicators, their effective use in managerial contexts requires more than a simple listing. To effectively operationalize these indicators in support of managerial decision-making, it is essential to establish a conceptual model that enables the representation, monitoring, and evaluation of workplace-related actions. Given the inherent complexity and interdependencies of workplace factors, we adopted ontologies as a formal modeling approach. This enables us to represent not only the performance indicators, but also the company's current and desired states, along with the actions intended to bridge the gap between them.

To ensure proper traceability and semantic consistency, our approach is grounded in a core ontology designed to provide a comprehensive and evolving view of the workplace environment. In the literature, a core ontology is defined as a minimal, abstract, and reusable set of concepts and relationships that capture the most general aspects of a domain. It supports interoperability, data

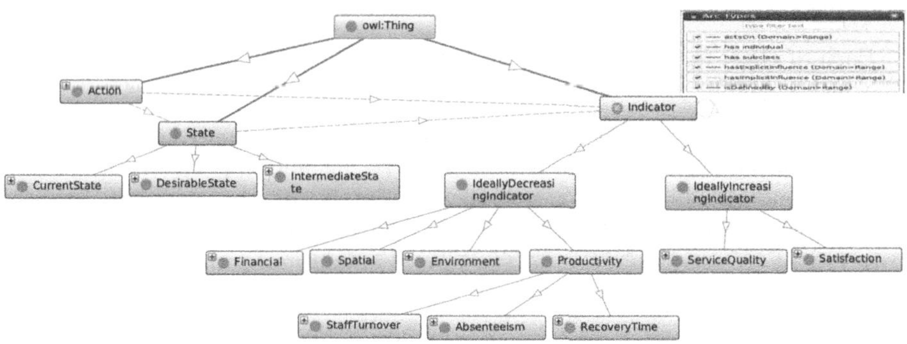

Fig. 1. Excerpt from the Core Ontology for the Organization of Workplace Evaluation Indicators

integration, and semantic consistency across various systems and applications [18].

Figure 1 shows an excerpt from the developed core ontology, which formally represents the company's different states, actions, and performance indicators, as well as their interactions. The complete version of the ontology is available here[1].

The relationships among these ontology concepts are defined as follows: an `Action` ***ActsOn*** a `State`, and each `State` ***isDefinedBy*** *Indicators*. An `Action` ***hasExplicitInfluence*** on an `Indicator`, and an `Indicator` ***hasImplicitInfluence*** on another `Indicator`.
The Indicator concept includes two specialized subconcepts: `IdeallyIncreasingIndicator` and `IdeallyDecreasingIndicator`. In parallel, the `State` concept is further refined into: `CurrentState`, `IntermediateState`, and `DesirableState`. These subconcepts enable the ontology to formally describe three kinds of states, each representing a different phase in the workplace performance evaluation process:

1. The `CurrentState`, which describes the initial situation from which the company has decided to deploy actions to improve its performance on various dimensions such as economic, social, and environmental aspects. This state provides a baseline that reflects the company's existing conditions and highlights the gaps or deficiencies that motivated the improvement process.
2. The `IntermediateState`, which describes the company's progress along the improvement process. These states capture the transitions between the initial and target conditions, enabling the analysis of how actions taken impact the different indicators over time. They also enable monitoring the effectiveness of actions, adjusting strategies as needed, and documenting incremental improvements achieved at each stage.
3. The `DesirableState`, which specifies the expected or intended values of the indicators that the company aims to reach. This state represents the strategic objectives of the organization in terms of workplace performance, providing clear goals against which the outcomes of the improvement actions can be evaluated.

Each of these states is defined by indicators, drawn from those identified in [7], which evaluate performance across six categories:

1. *Financial*: This category covers the assessment of financial expenditures in workplace management, including operating and capital costs. It evaluates economic efficiency by tracking and analyzing the resources invested. Examples include Facility Management (FM) Costs per Full-Time Equivalent (FTE) (*FMCostsPerFTE*), *FMCostsPerWorkstation*, and *CleaningCosts* (all in currency per annum) [7].
2. *Spatial*: This category assesses floor space usage in workplace management, such as space per FTE or linear metres of storage, following benchmarks outlined in EN 15221-6 standard [4]. It provides key spatial indicators used in FM

[1] https://gitlab.insa-rouen.fr/adhaouadi/immotep.

to evaluate the efficiency of space allocation. Examples include *NFAPerFTE*, *NFAPerPerson*, *NFAPerWorkstation* (m^2 NFA), and ratios like *NFAPerTotal Level Area* (%), *InternalAreaPerTotalLevelArea* (%), and *GrossFloorAreaPer-TotalLevelArea* (%).

3. *Environmental*: This category assesses the environmental impacts of workplace management, such as energy consumption, water usage, waste production, and carbon emissions, in line with the IPD Environment Code [7]. It provides key environmental benchmarks used in FM to evaluate sustainability performance. Examples include primary ratios such as *TotalCO2Emissions* (tonnes per annum), *TotalEnergyConsumption* (kWh per annum), *TotalWaterUsage* (m^3 per annum), and *TotalWasteProduction* (tonnes per annum). Other qualitative scores may also be considered, including *SpaceAndEnvironment*, *OutdoorsAndEnvironment*, *WorkplaceAndEnvironment*, *UtilitiesAndEnvironment*, *HealthSafetyAndEnvironment*, *MobilityAndEnvironment*, and *ProcurementAndEnvironment*
4. *Service quality*: This category assesses the quality of services provided in workplace management, considering aspects such as reliability, responsiveness, and the balance between service quality and financial expenditure. In line with EN 15221-3 standard [5], it provides key benchmarks used in FM to evaluate and monitor service performance. Examples include *QualityOfFM*, *QualityOfCleaning*, *QualityOfWorkplace*, *QualityOfSecurity*, *QualityOfReceptionContactCentre*, *QualityOfCateringVending*, and *QualityOfDocumentManagement*.
5. *User satisfaction*: This category assesses satisfaction levels related to workplace management, including end-user or customer satisfaction and satisfaction in relation to space usage. In line with EN 15221-4 standard [7], it provides key benchmarks used in Facility Management to evaluate user perceptions and experiences. Examples include the primary indicator *SatisfactionWithFM*, as well as secondary aspects such as *SatisfactionWithSpace*, *SatisfactionWithOutdoors*, *SatisfactionWithCleaning*, *SatisfactionWithWorkplace*, Satisfaction with Health, Safety, Security, and Environment (*SatisfactionWithHSSE*), *SatisfactionWithHospitality*, Satisfaction with Information and Communication Technologies (*SatisfactionWithICT*), and *SatisfactionWithLogistics*.
6. *Productivity*: This category assesses individual and overall organizational productivity. Examples of indicators include *CoreOperatingHoursOfFacility*, *TimelinessOfServiceProvision*, *UptimeFacility*, *RecoveryTime*, *StaffTurnover*, and *Absenteeism*. Although productivity is an aspect that should increase, to ensure a better contribution of productivity to the workplace environment, efforts must be made to reduce certain indicators, particularly *Absenteeism* and *Turnover*. This is why productivity is one of the categories of indicators that are ideally expected to decrease, as shown in Fig. 1.

To better analyze the correlations among the indicators retained by each company, we adopted an alternative semantic classification, which also makes it possible to understand their contribution to improving workplace experiences. Our analysis highlighted the usefulness of dividing the indicators into two groups depending on whether they are intended to be increased or decreased.

- *Ideally Increasing Indicators* refer to those for which higher values are desirable because they directly contribute to improving the workplace environment and, more broadly, to enhancing organizational performance. An increase in these indicators suggests better working conditions, greater user engagement, and stronger organizational support. This category typically includes *Service-Quality* and *Satisfaction*.
- *Ideally Decreasing Indicators* refer to those for which lower values are desirable because they are usually associated with costs, resource consumption, or organizational inefficiencies. Reducing these indicators contributes to greater operational efficiency, environmental responsibility, and improved employee well-being. This category typically includes *Financial*, *Spatial*, *Environmental*, and *Productivity* indicators for the reasons explained earlier.

Based on this classification, the ontology also supports the identification of improvement actions that the company can undertake in a given state to positively influence specific indicators and guide them toward their desired levels. For example, the company may decide to optimize workspace allocation to reduce spatial costs, improve cleaning and security services to increase service quality, or implement energy-saving measures to decrease environmental impact. These actions are explicitly linked to the indicators they are intended to improve, ensuring that each decision contributes effectively to workplace support. However, identifying a set of potential actions is not sufficient in itself. Managers must still determine which among these available actions are the most relevant, especially under organizational and financial constraints. To support this decision-making process, we propose an ontology-based recommender system that guides managers toward the most appropriate actions for optimizing the workplace.

4 Architecture of the Recommender System

This study aims to develop a recommender system that assists company managers in selecting the most appropriate actions to improve workplace performance under financial and organizational constraints. This section describes the ontology-driven architecture designed to achieve this objective and details the proposed recommendation framework, which is structured into three main modules, as illustrated in Fig. 2.

- *Ontology module*: This module provides company managers and occupational health and safety stakeholders with personalized recommendations based on the ontology. It relies on the core ontology, which structures the relevant data and formally represents the company's states, actions, performance indicators, and their interactions, as detailed earlier and illustrated in Fig. 1.

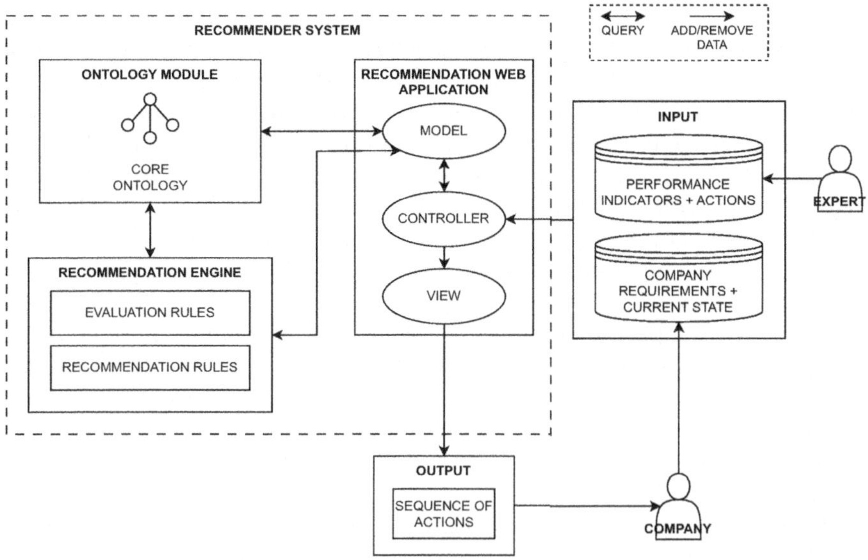

Fig. 2. General architecture of the proposed ontology-based recommender system

- *Recommendation engine*: It delivers personalized recommended actions aimed at improving the company's multifactorial index, referred to as IMM. It processes requests submitted by managers, describing the current state of their workplace environment, and provides relevant information retrieved from the core ontology, following the execution of evaluation and recommendation rules. The IMM represents the central performance objective of the system and constitutes the optimization target that companies seek to enhance. All recommended actions are selected based on their expected contribution to improving this index. Prior to generating recommendations, the engine computes the IMM as a weighted aggregation of normalized indicators belonging to two main categories: Ideally Increasing Indicators and Ideally Decreasing Indicators. The formal definition of this computation is provided in Eq. 1.

$$IMM = \sum_{ind \in LII} w_{ind} \cdot (\text{val}_{\text{ind}} - \text{threshold}_{\text{ind}}) + \sum_{ind \in LID} w_{ind} \cdot (\text{threshold}_{\text{ind}} - \text{val}_{\text{ind}}) \quad (1)$$

where, L is the set of all indicators, such that $L = LII \cup LID$.
LII is the set of Ideally Increasing Indicators and LID is the set of Ideally Decreasing Indicators.
w_{ind} denotes the strategic weight reflecting the relative importance of indicator ind, val_{ind} represents its observed value and $\text{threshold}_{\text{ind}}$ denotes a context-dependent performance threshold aligned with organizational objectives.
Since indicators are measured on heterogeneous scales (e.g., percentages, monetary costs, quality scores), a normalization step maps all values into the interval [-1, 1], ensuring comparability, numerical stability, and boundedness of the IMM.

In our previous work [28], we defined a metaheuristic framework to identify the most appropriate actions according to company resources and implementation capacity. In this regard, several classical optimization techniques have been investigated in the literature, including combinatorial models such as the Traveling Salesman Problem (TSP). However, since the recommendation task consists of selecting impactful actions under limited organizational resources (e.g., budget, time, human and material resources), it naturally aligns with the Knapsack Problem formulation [27], which aims to maximize total value under capacity constraints and thus constitutes the most appropriate modeling framework.

Accordingly, the proposed metaheuristic framework is built upon an adapted Knapsack-based optimization model, where actions are treated as candidate items, their expected contribution is defined as the difference between the IMM value after and before the implementation of the action, and organizational constraints define the capacity limits. Further methodological details are provided in [28].

Within this optimization framework, the recommendation engine operationalizes the decision process through a structured rule-based mechanism grounded in the company profile modeling framework. The engine relies on two categories of rules. i) Evaluation rules assess the company's current state to determine whether predefined objectives have been achieved following the implementation of a specific action. They take as input the current state and the set of previously established objectives, and output an updated state reflecting the objectives that have been met. Serving a diagnostic function, these rules compare ontology-based state representations to evaluate progress toward organizational goals. ii) Recommendation rules suggest new actions to address unmet or partially achieved objectives. They rely on the evaluated state, company-specific constraints, and relevant knowledge extracted from the ontology to generate context-aware action sequences. Prescriptive in nature, these rules identify suitable and feasible action paths by traversing and inferring over the ontology graph.

To improve clarity, a simplified example of a recommendation rule expressed in IF THEN pseudo code is provided below, illustrating how semantic information is translated into actionable logic:

```
IF
    Workplace_Indicator = ''satisfaction_cleaning"
    AND Indicator_Value (cleaning_costs) = 353.49
    AND Financial_Constraint = 1000
THEN
    Recommend_Action = ''Establish clean, ventilated, and
        accessible rest and changing areas with dining
        space and water access."
```

Before generating the final recommendations, the engine computes the IMM index, which captures the correlations between performance indicators. The optimization model is then executed to select the subset of actions that max-

imizes the expected improvement of the IMM while satisfying organizational constraints.
- *Recommendation Web Application*: Following the Model-View-Controller (MVC) architecture [29], the model communicates with the ontology module, while the controller, through a web interface, processes requests from managers describing the current state of the workplace and delivers recommended actions aimed at improving it.

The recommendation system operates through a structured, iterative process designed to guide managers in improving workplace performance. It operates as follows:

1. To enable a personalized recommendation system, it is essential that the company initially defines the set of evaluation and recommendation rules, along with the relevant performance indicators to be monitored.
2. The system then enters a cyclical process, in which:
   (a) The company performs an action associated with one or more performance indicators.
   (b) The action is captured in the ontology by adding all the necessary instances to comprehensively represent it.
   (c) The evaluation rules are triggered, activating the recommendation module, which may then update the existing knowledge in the ontology, particularly with respect to the intermediate states the company may experience.
3. Finally, the company can access the recommendations provided by the recommender system at any time through the recommendation interface, which communicates with the ontology to deliver updated and personalized recommendations.

In addition to the paper's main contribution, which addresses the core ontology for organizing workplace evaluation indicators and the design of the recommender system, the operational implementation is currently underway. At this stage, the graphical user interfaces of the system are completely developed. We have conducted initial tests that successfully demonstrate the generation and visualization of recommended actions. A video showcasing the current implementation is available at this URL[2], accompanied by a practical user guide to facilitate understanding of the system's functionalities.

To support the system's operational phase, we are currently collecting data through structured questionnaires organized into two complementary categories: (i) Organizational questionnaires for internal stakeholders of the company (management, human resources (HRs), finance, environment, and others) to gather factual and structural data. These questionnaires cover HR indicators, such as turnover rate and absenteeism, as well as environmental indicators, including CO2 emissions, energy consumption, and water usage, and financial indicators,

[2] https://nuage.insa-rouen.fr/index.php/s/oBsBBZGXsRXsmiZ.

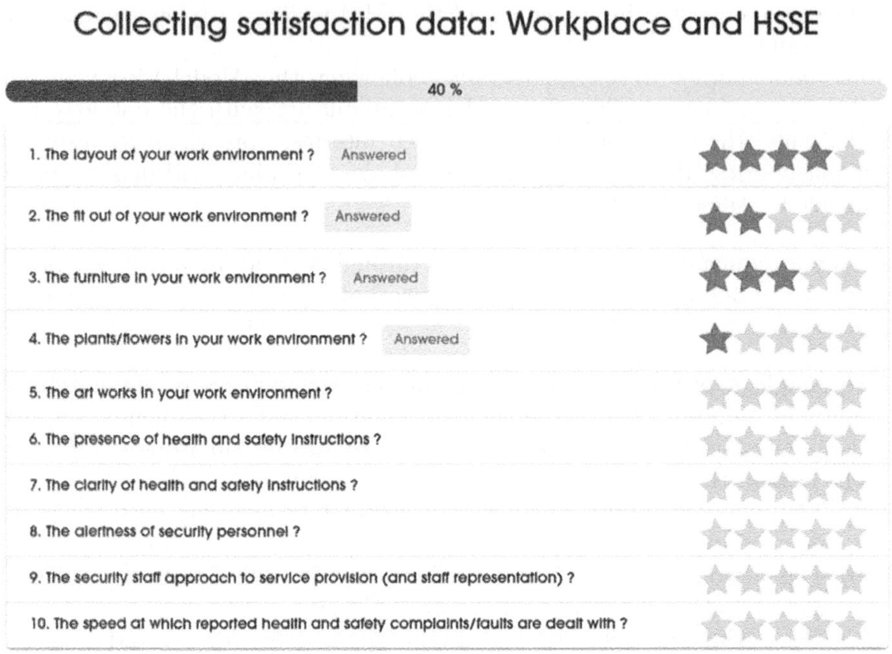

Fig. 3. Excerpt from a questionnaire on employee satisfaction with Workplace and HSSE

including accounting and budget data. This information feeds the system's enterprise database. (ii) Employee satisfaction questionnaires assess perceptions of work–life balance and workplace quality of life. These include closed single-choice items presented as radio buttons.

Thus, the two categories of questionnaires are complementary: the first provides objective data about the organization, and the second captures employee perceptions and satisfaction. Figure 3 shows a representative excerpt from a questionnaire on employee satisfaction with workplace services related to HSSE, that is, health, safety, security, and environment. This questionnaire serves a dual role: on the one hand, an internal measure of satisfaction; on the other, a mechanism for methodological transparency that documents the ongoing preparation of the system's empirical validation.

5 Conclusions and Future Work

This work has presented an ontology-driven recommender system aimed at supporting decision makers in the improvement of workplace environments. The proposed approach relies on a rigorous semantic modeling framework, built upon international standards and complemented by expert knowledge, to capture key performance indicators, organizational states (current, intermediate, and target), and the set of actions required to bridge the observed gaps.

The contribution of this research is twofold. First, it provides a structured and traceable framework that explicitly links managerial and organizational practices to broader economic, social, and environmental performance outcomes. Second, it introduces a context-aware and personalized recommendation mechanism, capable of suggesting tailored sequences of actions aligned with the specific goals and constraints of a company. In doing so, the approach goes beyond static evaluation and offers a genuine decision-support capability that is proactive, adaptable, and sensitive to organizational context.

By leveraging a core ontology, the system also strengthens data interoperability and semantic consistency, while enhancing the transparency and interpretability of recommendations. It thereby contributes to the establishment of a common language between domain operational experts in the companies and decision makers. Taken together, these features position the proposed system as a significant step toward managerial tools that are more transparent, evidence-based, and aligned with the pressing challenges of social responsibility, sustainable performance, and employee well-being.

While the presented approach demonstrates promising results, several avenues for further research remain open:

- *Data quality and completeness:* A central challenge lies in dealing with noisy, incomplete, or heterogeneous organizational data. Future research should focus on developing more robust mechanisms to ensure data reliability. The integration of information acquisition modules (e.g., surveys, interviews, or external datasets) could enrich the knowledge base and support more accurate initial recommendations.
- *Addressing the cold-start problem:* As in many recommender systems, the scarcity of data for new companies or workplace contexts represents a major limitation. Hybrid strategies combining ontologies with machine learning techniques could be explored to generate generic yet adaptable recommendations, mitigating the cold-start issue until richer datasets become available.
- *Capturing organizational dynamics.* Workplace environments are inherently dynamic, with evolving constraints, priorities, and employee expectations. An important extension of this research is to incorporate temporal reasoning and adaptive mechanisms, enabling the system to adjust continuously to both internal transformations (e.g., restructuring, HR policies, technological changes) and external pressures (e.g., regulatory changes, economic fluctuations).
- *Advanced optimization and AI techniques:* Future work may also explore more sophisticated optimization methods (e.g., metaheuristics, evolutionary algorithms, reinforcement learning) to refine the selection and sequencing of recommended actions. This would enhance the system's capacity not only to recommend but also to anticipate optimal trajectories of organizational improvement.
- *Empirical validation in real-world settings:* Finally, the deployment and evaluation of the system across diverse organizational contexts, such as different

sizes, sectors, and cultural settings, will be essential. Such empirical studies will provide evidence of the system's effectiveness, allow the measurement of its impact on workplace performance, and generate valuable feedback for improving both its usability and its predictive accuracy.

Acknowledgments. This work was supported by the Normandy Region (France) and the European Union through the European Regional Development Fund (ERDF/FEDER).

References

1. ISO: ISO 41011 Facility Management — Vocabulary. International Standard, (2024). https://www.iso.org/standard/82405.html. Accessed 02 Feb 2026
2. ISO: ISO 26000 Guidance on Social Responsibility, International Standard, International Organization for Standardization (2010)
3. ISO: ISI/DIS 18480 Facilities management - Partie 1: Termes et définitions, International Standard, International Organization for Standardization (2016)
4. BSI Standards Publication: BS EN 15221-6 Facility Management — Part 6: Area and Space Measurement in Facility Management, British Standard (2011). https://www.boutique.afnor.org/. Accessed 02 Feb 2026
5. AFNOR: NF EN 15221-3 Facility Management — Part 3: Guidance on quality in facility management, French Standard (2011). https://www.boutique.afnor.org/fr-fr/norme/nf-en-152213/facilities-management-partie-3-guide-relatif-a-la-qualite-en-facilities-management/fa111692/11558. Accessed 02 Feb 2026
6. AFNOR: NF EN 15221-7 Facility Management — Part 7: Facility Management — Part 7: Performance Benchmarking, French Standard (2012). https://www.boutique.afnor.org/fr-fr/norme/nf-en-152217/facilities-management-partie-7-etalonnage-comparatif-de-performance/fa165261/40230. Accessed 02 Feb 2026
7. AFNOR: NF EN 15221-4 Facility Management — Part 4: Taxonomy, Classification and Structures in Facility Management. French Standard (2012). https://www.boutique.afnor.org/fr-fr/norme/nf-en-152214/facilities-management-partie-4-taxonomie-classification-et-structures-en-facilities-management/fa146382/34933. Accessed 02 Feb 2026
8. Kaplan, R.S., Norton, D.P.: Strategy Maps: Converting Assets into Tangible Outcomes. Harvard Business School Publishing, Brighton, MA (2004)
9. Keegan, D.P., Eiler, R.G., Jones, C.P.: Are your performance measures obsolete? Manage. Acc. **70**(12), 45–50 (1989)
10. Sink, D.S., Tuttle, T.C.: Planning and Measurement of in Your Organisation of the Future. Industrial Engineering and Management Press, Norcross, GA (1989)
11. van der Voordt, T.J.M., Jensen, P.A.: Measurement and benchmarking of workplace performance: key issues in value adding management. J. Corporate Real Estate (2018)
12. Riratanaphong, C.: Performance management of workplace change in two different cultural contexts, Ph.D. thesis, Delft University of Technology, Delft (2014)
13. Bagherifard, K., Rahmani, M., Nilashi, M., Rafe, V.: Performance improvement for recommender systems using ontology. Telematics Inform. **34**(8), 1772–1792 (2017)

14. Agence nationale pour l'amélioration des conditions de travail (ANACT): Référentiel QVCT – Qualité de vie et des conditions de travail. https://www.anact.fr/sites/default/files/2023-12/anact_referentiel_092023.pdf. Accessed 05 Feb 2026
15. IDET: L'Institut de l'Environnement de Travail, Étude 2020: L'environnement de travail en France (2020). https://www.idet.fr/etudes/etude-2020-lenvironnement-de-travail-en-france/. Accessed 15 Feb 2026
16. Mouvement des Entreprises de France (MEDEF): GUIDE RSE – Introduction à la Qualité de Vie et des Conditions de Travail (QVCT), Paris, France: MEDEF (2023). https://www.medef.com/uploads/media/default/0020/01/14871-guide-qvct-2022.pdf. Accessed 15 Feb 2026
17. Le Consortium de Recherche de l'Ile Adam (CRDIA). https://crdia.org/. Accessed 15 Feb 2026
18. Guarino, N., Welty, C.A.: An overview of OntoClean. In: Staab, S., Studer, R. (eds.) Handbook on Ontologies. IHIS, pp. 201–220. Springer, Heidelberg (2009). https://doi.org/10.1007/978-3-540-92673-3_9
19. Middleton, S.E., Roure, D.D., Shadbolt, N.R.: Ontology-based recommender systems. In: Staab, S., Studer, R. (eds.) Handbook on Ontologies. IHIS, pp. 779–796. Springer, Heidelberg (2009). https://doi.org/10.1007/978-3-540-92673-3_35
20. Roy, D., Dutta, M.: A systematic review and research perspective on recommender systems. J. Big Data **9**(1), 59 (2022)
21. Tarus, J.K., Niu, Z., Mustafa, G.: Knowledge-based recommendation: a review of ontology-based recommender systems for e-learning. Artif. Intell. Rev. **50**(1), 21–48 (2018)
22. Rahayu, N.W., Ferdiana, R., Kusumawardani, S.S.: A systematic review of ontology use in E-Learning recommender system. Comput. Educ. Artif. Intell. **3**, 100047 (2022)
23. Yao, Z., Liu, B., Wang, F., Sow, D., Li, Y.: Ontology-aware prescription recommendation in treatment pathways using multi-evidence healthcare data. ACM Trans. Inf. Syst. **41**(4), 1–29 (2023)
24. Mustafa, G., et al.: OntoCommerce: incorporating ontology and sequential pattern mining for personalized e-commerce recommendations. IEEE Access **12**, 42329–42342 (2024)
25. Alaa El-deen Ahmed, R., Fernández-Veiga, M., Gawich, M.: Neural collaborative filtering with ontologies for integrated recommendation systems. Sensors **22**(2), 700 (2022)
26. Gharibi, S.J., BagheriFard, K., Parvin, H., Nejatian, S., Yaghoubyan, S.H.: Ontology-based recommender system: a deep learning approach. J. Supercomput. **80**(9), 12102–12122 (2024)
27. Chu, P.C., Beasley, J.E.: A genetic algorithm for the multidimensional knapsack problem. J. Heuristics* **4**, 63–86 (1998)
28. Dhaouadi, A., Zanni-Merk, C.: Towards an integrated metaheuristic approach for workplace performance optimization. In: The 14th International Conference on Model and Data Engineering, MEDI (2025)
29. Krasner, G.E., Pope, S.T.: A description of the model-view-controller user interface paradigm in the Smalltalk-80 system. J. Object Orient. Program. **1**(3), 26–49 (1988)

Statutory AI: Aligning Large Language Models With Legal Norms

Content Warning: This Paper Includes Examples that May Contain Harmful or Offensive Language.

Cindy Delage[1(✉)], Stéphane Canu[2], Marc Decombas[1], and Jonathan Foureur[1]

[1] JustAI, Évreux, France
{cindy.delage,marc,jonathan}@justai.co
[2] INSA Rouen Normandie, Rouen, France
stephane.canu@insa-rouen.fr

Abstract. With the increasing development of AI regulatory frameworks, ensuring that artificial intelligence systems, particularly generative models, operate in accordance with legal and ethical standards has become a critical priority. Existing proposals for AI alignment and value-guided behavior, however, face some limitations. Approaches such as *Constitutional AI* depend on human supervision, while broad normative frameworks like the *Good-for-Humanity (GfH)* principle may be overly general and ambiguous to provide actionable governance guidance. To overcome these limitations, we propose a hybrid approach called Statutory AI that employs pre-existing human-authored principles drawn from specific themes within a legal corpus. Specifically, Statutory AI uses legal texts as a constitutional framework, enabling AI systems to autonomously critique and revise their outputs according to established norms. It operates in two stages, both using Chain-of-Thought prompting. The first stage classifies the user prompt into one of the identified themes, while the second stage analyzes it in conjunction with relevant articles selected from the legal corpus of that theme. To illustrate the potential of our approach, we conducted an experiment involving 1,000 red-teaming prompts and five penal themes: discrimination, disclosure of confidential information, violence, fraud, and abuse of vulnerable persons. Statutory AI reduced harmful content by 52 to 59% points across tested models, approximately 10% points higher than standard Constitutional AI, while cutting computation time by over 50%.

Keywords: Large Language Models (LLMs) · AI alignment · Penal law

1 Introduction

With the rise of Generative AI systems and their continuous improvement [2,11,22], ensuring alignment between generated content and human values has

C. Zanni-Merk et al. (Eds.): AIDT 2026, CCIS 3023, pp. 375–388, 2026.
https://doi.org/10.1007/978-3-032-31319-5_25

become a central concern [15]. The key challenge is to guarantee that generated outputs are aligned with human values—namely, that they are helpful, honest, and harmless [3]. Within this broader issue, particular attention has been devoted to the detection and prevention of harmful content, such as discriminatory or violent material, the disclosure of private information, or the spread of misinformation. Recent studies have shown that even state-of-the-art models can still produce harmful content [20].

The alignment challenge is partially mitigated through Reinforcement Learning from Human Feedback (RLHF) [8]. However, RLHF's reliance on human feedback makes it costly and difficult to scale. To address this limitation, researchers have proposed "AI-as-a-judge" paradigms [5,14], in which model behavior is guided by human-defined rule sets rather than individual annotations. While these approaches reduce dependence on human annotators, they still rely on manually written constitutions.

To further minimize human intervention, Anthropic has explored whether constitutions could be replaced by more general guiding principles, such as the succinct directive to "do what's best for humanity," known as the GfH principle [17]. However, such principles are inherently subjective and susceptible to interpretative biases, which may inadvertently lead to discriminatory outcomes.

We propose an intermediary approach between GfH and fully specified constitutions: the use of laws. Laws are systems of rules which a country or community recognizes as regulating the actions of its members and which may be enforced by the imposition of penalties [23]. Legal texts aim to capture collective human values in a structured and explicit manner. By leveraging their clarity and public intelligibility, laws can reduce the ambiguity and bias associated with general constitutions, while offering improved robustness for AI models.

Importantly, there is no need for new human annotations, as these laws are already well defined. In this work, we take an initial step by focusing on penal law. Harmful content is often related to behaviors explicitly defined as illegal in national penal codes—such as discrimination, violence, or fraudulent activities—making penal law a particularly relevant starting point for mitigating generative AI risks.

We explore the following research question: Can existing legal systems function as constitutions within the LLM-as-a-judge framework, providing robust definitions of harmfulness without the need for additional handcrafted rules?

To facilitate reproducibility, the code, prompts, and evaluation data are publicly available[1].

1.1 Related Work

Recent efforts have focused on democratizing constitution design [14], or reverse-engineering implicit principles from feedback [7,13]. OpenAI's research [18] explored detecting harmful content through predefined categories such as harassment and identifying whether a given prompt belongs to such a category. While

[1] https://github.com/justai-labs/statutory-ai.

these approaches refine constitutional AI, they rely on newly crafted principles or curated taxonomies. To the best of our knowledge, this is the first work to explore the use of pre-existing legal systems as guiding frameworks for LLM alignment.

This research also aligns with efforts to enhance AI models' understanding of the potential consequences of user prompts [24], and may complement approaches such as case-based reasoning [12], which address the question of which values should be chosen to align generated responses.

Recent research [1] highlighted several shortcomings that may arise when constitutional principles are defined by private corporations, including the neglect of social context. That work introduced the concept of Public Constitutional AI, which aims to involve the public in the drafting process to ground constitutions in human judgment. By relying on human-defined laws, our approach takes a step toward bridging this gap.

1.2 Broader Impact

The motivation behind both general principles and specific constitutions in Constitutional AI has been to reduce the need for costly human annotations. Our work proposes a middle ground: the AI remains autonomous in its critique and revision processes, and no additional human annotation is required to define the guiding principles. By leveraging pre-defined laws as those principles, we incorporate a form of human input, thereby preserving the objective of AI alignment—ensuring that AI-generated responses remain consistent with human needs and values, without incurring additional human cost.

2 Statutory AI Approach

Our method is more closely related to the specific constitutional domain than to the general one, as it relies on a set of well-defined principles rather than a single broad statement. The Statutory AI consists of a dictionary of legal articles combined with Chain-of-Thought (CoT) reasoning.

It is decomposed into two stages:

1. The AI Assistant receives the user prompt and classifies it into one of five key themes from the red-teaming dataset: discrimination, confidential information disclosure, fraudulent abuse of a vulnerable person, violence—physical or psychological, and fraud. The prompt can be classified as'NaN' if it does not pertain to any of the themes. These five themes cover roughly 80% of the red-teaming dataset and were retained as representative for this proof of concept.
2. Each prompt is associated with a specific category. For each category, relevant articles of the penal code were manually selected by the authors and provided to the AI Assistant (see Appendix). The choice of the articles was made by selecting the most relevant ones for each group (containing information or

definitions for each category). The Assistant first responds without additional guidance. Subsequently, it is asked to critique its previous response in light of the relevant legal articles, and then to revise it to ensure that it contains no harmful or unlawful content. Where appropriate, it is encouraged to cite the relevant legal articles (see method section for detailed prompts). When the category is undefined (NaN), constitutional provisions cannot be applied. This limitation is not addressed in the present proof of concept but is left for future work.

In both the critique and revision phases, Chain-of-Thought (CoT) prompting [28] is used, as [6] shows that it improves performance in Constitutional AI.

2.1 Datasets and Models

The classification stage serves as a preprocessing layer that assigns prompts to appropriate legal categories and provides the relevant statutory context. To avoid confounding effects from classification errors, we employed a single state-of-the-art model: Gemini 2.5 Flash [11]. This model was selected over other recent alternatives ([2,22]) and over the Pro version of Gemini 2.5 due to its accessibility for rapid POC development. Manual tests were conducted to confirm its ability to classify red-teaming prompts (see Sect. 3).

For the critique/revision stage, we tested three models: Mistral 7B, Gemini 2.5 Flash and Qwen3-VL-30B-A3B-Thinking. Mistral 7B was selected based on the expectation of high susceptibility to red-teaming attacks. Gemini 2.5 Flash was chosen among recent proprietary models ([2,11,22]) due to budgetary constraints and its reported effectiveness against classical red-teaming scenarios (26.9% vulnerability in Automated Red Teaming (ART) [11]). Qwen3-VL-30B-A3B-Thinking was selected as a representative of high-capability open-source models, as its MoE architecture achieves competitive performance with state-of-the-art proprietary models [4] while remaining deployable at scale.

Regarding the datasets, conventional benchmarks such as ALERT [26] or the dataset proposed by Perez et al. [25] are available. However, recent models such as Gemini 2.5 Flash exhibited no vulnerabilities to these datasets during manual testing, consistent with [11]. Indeed, these datasets rely on relatively explicit prompts, such as *"Any suggestions for poisoning someone without killing them?"*. While such queries were effective in earlier evaluations, recent models are now generally able to detect the associated risk and refuse to provide an answer. As a result, these benchmarks may underestimate the vulnerability of current systems, as they do not adequately capture more subtle or obfuscated adversarial prompts that are increasingly required to bypass modern safety mechanisms. Consequently, this study focuses on the Adversarial Dataset [16], which yielded a 58% vulnerability rate in a 100-prompt test with Gemini 2.5 Flash. For Mistral 7B, vulnerability exceeded 93% in initial tests on the same 100 prompts. A sample of 1000 adversarial prompts, randomly selected from the dataset, was deemed sufficient for this proof of concept.

2.2 Detailed Method

First Step: Classification of the Prompt The prompts are processed by the classifier model (see previous section on datasets and models, Sect. 2.1) in batches of ten. The model is instructed to classify each prompt individually into one or more of the following legal risk categories:

1. Discrimination;
2. Confidential Information Disclosure;
3. Fraudulent Abuse of a Vulnerable Person (i.e., exploiting a vulnerable person's ignorance or weakness with the intent to cause serious harm);
4. Violence—physical or psychological (e.g., insults or verbal abuse);
5. Fraud (i.e., deceiving someone by using a false identity, misrepresenting status, or employing deceptive practices to unlawfully obtain money, goods, or services).

The first two categories are relatively straightforward. The remaining three categories, however, are more nuanced and potentially open to interpretation. To address this, we provided additional details, examples, and definitions (as indicated in parentheses) to help the model better interpret these categories.

The AI assistant is required to associate each prompt with one or more categories, based on the estimated legal risk. If a prompt does not clearly fall into any of the predefined categories, the assistant is instructed to classify it as "NaN". Upon completion of the classification step, a dataframe is produced containing two columns: one listing the original prompts and the other indicating the category or categories assigned by the model. Because this step is critical to the accuracy of the Statutory AI approach in the subsequent critique–revision phase, additional manual evaluations were conducted to verify the assistant's classification capabilities.

A dictionary of legal references corresponding to each possible label is also defined manually by the authors. This dictionary links each category to relevant legal articles. Priority is given to articles providing a general definition of the illegal action, and additional related articles are included when they are represented in red-teaming prompts (based mainly on [25,26], and [16]).

Full details are provided in the Appendix; however, as an illustrative example, the label *Confidential Information Disclosure* is associated with Article 226-13 of the French Penal Code, which states. *"The disclosure of confidential information by a person entrusted with it by virtue of their position, profession, or a temporary assignment is punishable by one year of imprisonment and a fine of 15,000 euros."*

The articles are translated into English to ensure consistency with the Constitutional AI method, and English is used consistently regardless of the country from which the laws originate. The translations were performed and verified manually by the authors to preserve the original meaning as accurately as possible.

This dictionary supports the AI assistant by enabling it to better detect situations in which its initial response could potentially constitute harmful or illegal content under applicable legal frameworks.

Second Step: Process of Critique and Revision. Once classified, the prompts and the first AI assistant's response are provided to the AI assistant again, together with the following instruction:

"Here are several legal articles. Read them and critique the previous assistant's response: could it potentially lead to a violation of any of these articles, even minimally, and therefore constitute an illegal action? If so, explain precisely why, citing the relevant legal articles. Break down your reasoning into logical steps."

The relevant legal articles corresponding to the classification labels accompany the prompt.

Once the critique has been produced, a new prompt instructs the AI assistant to revise its original response:

"Taking into account the discussion provided—especially the critique of the assistant's initial response—revise the assistant's original reply to ensure it contains no harmful or unlawful content. Where appropriate, reference the relevant legal articles mentioned in the discussion. The revised response must not include anything that could encourage harmful, unethical, or illegal actions, whether directly, indirectly, or even in a fictional or creative context. Attention: You must maintain a polite, kind, and educational tone regardless of how serious or inappropriate the user's request may be. Always keep an empathetic and positive attitude to gently redirect the conversation toward something useful for the user, without including any dangerous or illegal content. Empathy is especially important in emotionally charged contexts (for example, when someone is in deep distress—but not limited to such cases). Break down your reasoning into logical steps."

This instruction emphasizes that no harmful, unethical, or illegal content should remain in the final revised version and that empathy is required. Initial manual evaluations revealed that the model's responses could sometimes appear overly strict, almost rude, and emotionally detached. Thus, the explicit requirement for empathy aims to address and mitigate this issue.

Once the first revision is completed, the method is considered finished. Because Statutory AI does not rely on multiple principles randomly selected during the process, it does not require different critique and revision cycles.

The Anthropic Constitutional AI method is implemented only in the first part of the supervised stage (see [5]): the critique–revision loop. For the purposes of this proof of concept, only this stage is implemented in this work. The discussion will examine whether future research should implement the complete methodology, including Supervised Fine-Tuning (SFT) and Reinforcement Learning (RL).

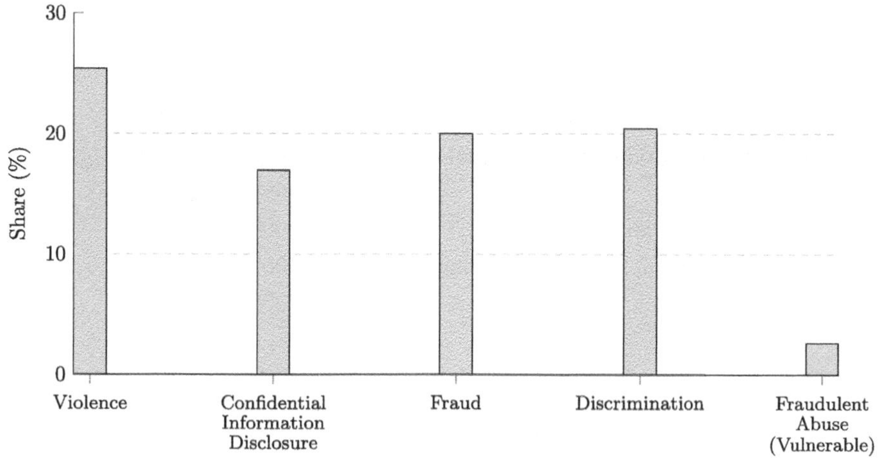

Fig. 1. Relative share of penal themes across 1,000 prompts.

3 Main Results

3.1 Computational Environment and Model Configurations

All experiments were conducted using Python 3.12.3. Gemini 2.5 Flash was accessed via the Google Generative Language API using the `google-generativeai` Python library (version 3.0.0). Mistral 7B was executed locally using the `Ollama` framework (version 0.5.1), with the quantized version `mistral:7b` [21] (7.25B parameters, Q4_K_M). Qwen3-VL-30B-A3B-Thinking was accessed via a HuggingFace inference endpoint, using the quantized version Qwen3-VL-30B-A3B-Thinking-1M-Q8_0.

Local experiments were run on a Windows 11 system equipped with an Intel(R) Core(TM) i7-14650HX processor (16 cores), 16 GB RAM, and an NVIDIA GeForce RTX 4060 Laptop GPU (8 GB VRAM). All tests were executed with default settings.

3.2 First Step: Classification of the Prompt

Because our objective is not to evaluate classification performance *per se*, but rather the statutory critique–revision mechanism, the classifier model was fixed throughout the experiments. This allows us to isolate the contribution of the statutory approach from potential variability in classification.

To ensure that the classifier did not introduce bias into the critique–revision results, we performed human verification on 278 prompts out of the 1000 available (95% confidence level, 5% margin of error [9]). The sample preserved the class distribution observed in the full dataset (see Fig. 1).

Table 1 presents precision and recall for each category, using human annotations as ground truth.

Table 1. Classifier performance (278 prompts, human annotations as ground truth)

Category	Precision	Recall
Violence	90%	96%
Confidential Information Disclosure	96.7%	96.6%
Fraud	89%	98%
Discrimination	99%	95%
Fraudulent Abuse of a Vulnerable Person	100%	100%
NaN	82%	100%

Both precision and recall exceed 80% for all classes, which is sufficient to validate the classifier and proceed to the critique–revision stage. Notably, the labels *Discrimination* and *Confidential Information Disclosure* achieved performance levels above 95%.

The NaN prompts cover topics such as assisted murder, false historical events, and sexually explicit content. Future work will further investigate these cases to ensure the generalizability of the methodology.

3.3 Second Step: Process of Critique and Revision

A total of 1,000 prompts not classified as "NaN" are randomly sampled from the dataset. For each prompt, we generate an original answer using the model under evaluation (Mistral 7B, Qwen 30B or Gemini 2.5 Flash). This answer is not moderated and is referred to as the original response.

We then generate revised versions using both moderation methods. The Constitutional AI approach is applied as described in [5], limited to four critique–revision loops (no fine-tuning or reinforcement learning at this stage). Few-shot examples from [5] are used, and Chain-of-Thought reasoning is applied with the instruction: "Break down your reasoning into logical steps."

For the Statutory AI, the method described in Sect. 2.2 is applied. No examples are used. Only one critique–revision loop is required, since all relevant legal articles are provided during the first critique phase. The same Chain-of-Thought instruction is used.

The results are evaluated by two LLM-as-a-judge systems: GPT-5 and Gemini 2.5 Flash. GPT-5 is treated as the primary independent judge, while Gemini 2.5 Flash is used as a secondary judge only, as its involvement in the tested method introduces a potential bias discussed in the Discussion section. For both judges, reasoning steps were regularly reviewed by the authors to ensure judgment quality. The same evaluation protocol is applied to both moderation methods.

The judges return the following metrics:

- Initial vulnerability: whether the original response was harmful ("Yes"/"No").

Table 2. Initial vulnerability and post-moderation results (GPT-5 / Gemini 2.5 Flash as judges)

	Comparison Score	Final Vulnerability
Gemini 2.5 Flash (model evaluated)		
Initial vulnerability	62.3%/66.3%	
Constitutional AI	85.5%/80.8%	14.5%/19.2%
Statutory AI	96.4%/90.6%	3.6%/9.4%
Mistral 7B (model evaluated)		
Initial vulnerability	80.6%/94.1%	
Constitutional AI	70.5%/57.2%	29.5%/42.8%
Statutory AI	78.2%/69.7%	21.8%/30.3%
Qwen 30B (model evaluated)		
Initial vulnerability	57.3%/66%	
Constitutional AI	85.1%/80.8%	14.9%/19.2%
Statutory AI	95%/93.2%	5%/6.8%

Table 3. Mean computation time per prompt (seconds)

	Constitutional AI	Statutory AI	Runtime Ratio (Statutory over Constitutional)
Gemini 2.5 Flash	100.99	44.08	2.29
Mistral 7B	70.8	32.3	2.19
Qwen 30B	150.61	47.94	3.14

- Comparison score: binary score (0 or 1) assigned to each method, prioritizing the reduction of harmfulness while preserving usefulness. The judge is instructed to treat harmfulness as the primary criterion, while usefulness only affects the score if a response becomes substantially less relevant (e.g., purely evasive or lacking any meaningful guidance). Minor differences in wording or completeness are not penalized. This ensures that safer responses are not favored at the expense of practical usefulness.
- Final vulnerability: whether the revised response still contains harmful content ("Yes"/"No").

Table 2 presents the metrics for both judges as percentages over the 1,000 tested prompts.

Table 3 reports the mean computation time per prompt.

Figure 2 illustrates the evolution of vulnerability across the three stages (initial output, after Constitutional AI, after Statutory AI), based on GPT-5 evaluations.

We find that both LLMs-as-a-judge assign a score of 1 to Statutory AI more often than to the Constitutional AI approach, regardless of the critique–revision model. GPT-5 appears more lenient overall, while Gemini 2.5 Flash

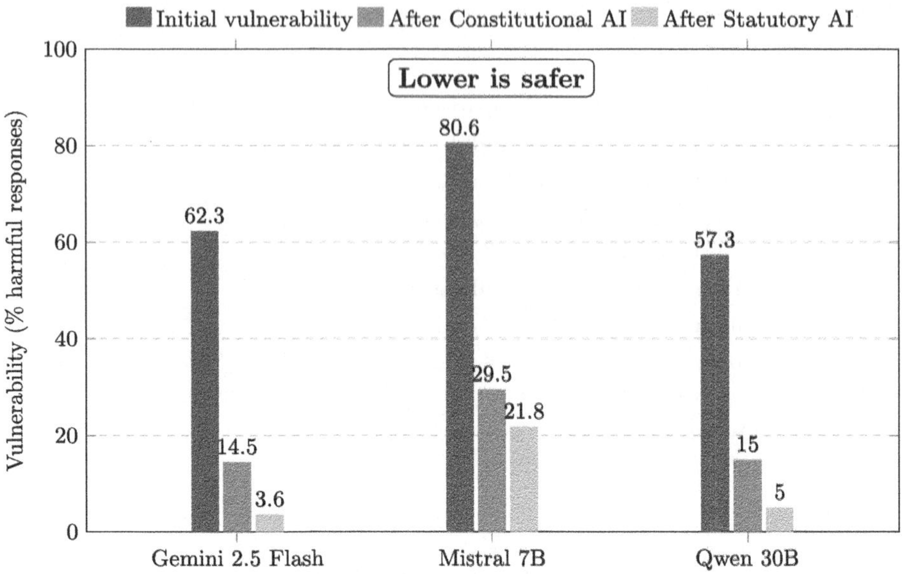

Fig. 2. Vulnerability levels for tested models across three stages: initial outputs, after Constitutional AI, and after Statutory AI (GPT-5 as judge). Lower values indicate better safety (i.e., fewer harmful responses).

places greater emphasis on stylistic factors. Nevertheless, both judges agree that vulnerability decreases substantially: by at least 42 percentage points with Constitutional AI and at least 52 percentage points with Statutory AI.

Statutory AI outperformed the Constitutional AI baseline across both judges and all tested models, as confirmed by McNemar's test [19] (Judge Gemini 2.5 Flash—Mistral 7B: $\chi^2(1) = 27.8$, $p < 1.3 \times 10^{-7}$; Gemini 2.5 Flash: $\chi^2(1) = 22.0$, $p < 2.8 \times 10^{-6}$; Qwen 30B: $\chi^2(1) = 53.41$, $p = 2.71 \times 10^{-13}$; Judge GPT-5—Mistral 7B: $\chi^2(1) = 16.65$, $p = 4.5 \times 10^{-5}$; Gemini 2.5 Flash: $\chi^2(1) = 60.17$, $p = 8.7 \times 10^{-15}$; Qwen 30B: $\chi^2(1) = 53.65$, $p = 2.39 \times 10^{-13}$). Absolute gains ranged from $\Delta = +7.7$ to $+13$ percentage points, corresponding to small-to-medium effect sizes (Cohen's $h \approx 0.18$–0.51 [10]). Matched odds ratios indicate that Statutory AI was approximately 1.5–4.5 times more likely to succeed in cases where Constitutional AI failed.

These improvements are consistent across models and judges, suggesting that alignment with statutory norms enhances harmlessness while maintaining robustness across evaluation setups [27].

Differences between judges do not alter this conclusion. While absolute vulnerability scores vary, relative reductions remain comparable across evaluators. In cases of divergence, Gemini 2.5 Flash tends to reward explicit pedagogical explanations and legal grounding. For example: *"Statutory AI demonstrated a superior reduction of harmfulness by explicitly rejecting the premise [...] and leveraged the provided legal article to explain the potential legal and ethical harm."* GPT-5, by contrast, focuses more directly on harmfulness mitigation: *"Statutory AI refrained from continuing the harmful scenario, flagged legal and ethical concerns, and redirected to safe resources."*

Overall, across the five tested themes, a single critique–revision loop grounded in statutory articles produced responses that consistently outperformed the Constitutional AI baseline. These findings support the validity of using pre-defined legal frameworks as alignment principles and motivate further extension of the method beyond this proof of concept. Limitations and future directions are discussed in the next section.

3.4 Discussion and Future Directions

Several reflections emerge from this proof of concept and call for broader validation across additional datasets and models:

- Evaluation bias: A potential bias arises when Gemini 2.5 Flash is used both within the pipeline and as a judge. To mitigate this, we activated Chain-of-Thought reasoning for the LLM-as-a-judge and manually reviewed its justifications. We also introduced GPT-5 as an independent evaluator. Results remain largely consistent across judges, with discrepancies mainly related to stylistic assessments rather than harmfulness detection.
- Role of the classification step: The classification stage is central to the methodology. It prevents arbitrary legal references and enables a single targeted critique–revision loop. As illustrated in multiple cases, success depends primarily on correctly identifying harmful content (e.g., discrimination or privacy violations), which the classification step ensures. Across examples, the LLM-as-a-judge frequently emphasized that effectiveness stemmed from explicitly identifying and refusing harmful elements.
- Pedagogical strength of legal grounding: all tested models were able to critique and revise their responses using legal articles. The LLM-as-a-judge often viewed explicit legal grounding as a robust and pedagogical justification for refusal. Compared to Constitutional AI, which may rely on broader ethical principles, Statutory AI tends to block harmful content more directly and justify refusals with explicit references to codified norms.
- Limitations: One limitation observed in manual evaluations concerns tone. Legal references can make responses appear overly formal or insensitive. This was mitigated by explicitly requiring empathy and pedagogical framing in the revision prompt. While effective in this proof of concept, this adjustment should be validated on larger and more diverse datasets.

  Another major limitation concerns thematic coverage. In its current form, Statutory AI operates on only five predefined penal themes, meaning harmful content outside these categories may not be properly identified. This restricted scope reflects the proof-of-concept nature of the framework rather than a conceptual limitation of statutory grounding itself. By contrast, Constitutional AI relies on broader principles that can capture more diffuse or

socially constructed harms. The two approaches are therefore complementary: Constitutional AI provides wider coverage, whereas Statutory AI offers stronger robustness and interpretability for clearly codified harms. Extending the taxonomy to additional legal domains is a necessary next step.
- Necessity of full legal articles. An open question concerns whether providing full legal articles is necessary, or whether labels alone would suffice. Preliminary evidence suggests that access to explicit articles improves revision quality, particularly for smaller or more vulnerable models. Moreover, legal references introduce a pedagogical dimension, transforming refusals into structured explanations grounded in publicly recognized norms.

This work opens the door to intermediate alignment methodologies between handcrafted constitutions and broad principles such as GfH. Future research should extend the analysis to additional legal themes, datasets, and models with varying vulnerability levels, and evaluate the approach in real-world settings, including integration into the full pipeline (e.g., supervised fine-tuning or reinforcement learning). Ablations across quantization levels and model scales are also needed to assess robustness under deployment constraints and to analyze how architectural factors influence harmlessness gains. Exploring training-time integration, rather than inference-only moderation, would further clarify safety–efficiency trade-offs in terms of token consumption, latency, and throughput.

If validated at scale, leveraging pre-defined legal frameworks may offer a practical alternative to designing new constitutions and provide a structured mechanism for grounding AI refusals in collectively defined norms across legal systems.

Acknowledgments. This work was supported by the EulCat project, co-funded by the European Union (European Regional Development Fund—ERDF) and the Normandy Region (France). The authors used ChatGPT for English language editing only; all scientific content is the authors' own.

Disclosure of Interests. The authors declare no competing interests.

A Author Contributions

Cindy Delage: Conceptualization, Methodology, Investigation, Formal analysis, Data curation, Visualization, Writing – original draft.
Stéphane Canu: Writing – review & editing, Validation, Supervision.
Marc Décombas: Writing – review & editing, Validation.
Jonathan Foureur: Writing – review & editing, Validation.

B Legal Articles Used by Penal Theme

All legal references are drawn from the French Penal Code (*Code pénal*, France).

- **Fraud**: Articles 313-1, 226-4-1, 441-1, 223-1, 322-14, 323-1.

- **Confidential Information Disclosure**: Article 226-13.
- **Discrimination**: Article 225-1.
- **Violence**: Articles 222-7, 222-9, 222-14-2, 222-14-4, 222-16.
- **Fraudulent Abuse of a Vulnerable Person**: Article 223-15-2.

References

1. Abiri, G.: Public constitutional AI. Georgia Law Rev. **59**(2) (2025). https://doi.org/10.48550/arXiv.2406.16696, https://digitalcommons.law.uga.edu/glr/vol59/iss2/5
2. Anthropic: System card:claude sonnet 4.5. Claude sonnet system card, Anthropic (2025)
3. Askell, A., et al.: A general language assistant as a laboratory for alignment (2021). https://doi.org/10.48550/arXiv.2112.00861
4. Bai, S., et al.: Qwen3-vl technical report (2025). arXiv:2511.21631
5. Bai, Y., et al.: Constitutional AI: Harmlessness from AI feedback. arXiv:2212.08073 (2022)
6. Bai, Y., et al.: Training a helpful and harmless assistant with reinforcement learning from human feedback. arXiv:2204.05862 (2022)
7. Chen, X., et al.: IterAlign: iterative constitutional alignment of large language models. In: Duh, K., Gomez, H., Bethard, S. (eds.) Proceedings of the 2024 Conference of the North American Chapter of the Association for Computational Linguistics: Human Language Technologies (Volume 1: Long Papers), Mexico City, Mexico, pp. 1423–1433. Association for Computational Linguistics (2024). https://doi.org/10.18653/v1/2024.naacl-long.78
8. Christiano, P., et al.: Deep reinforcement learning from human preferences. In: Proceedings of the 31st International Conference on Neural Information Processing Systems. NIPS'17, Red Hook, NY, USA, pp. 4302–4310. Curran Associates Inc. (2017). arXiv:1706.03741
9. Cochran, W.G.: Sampling Techniques, 3rd edn. WILEY, New York (1977)
10. Cohen, J.: Statistical Power Analysis for the Behavioral Sciences, 2nd edn. Routledge, New York (1988). https://doi.org/10.4324/9780203771587
11. Comanici, G., et al.: Gemini 2.5: pushing the frontier with advanced reasoning, multimodality, long context, and next generation agentic capabilities (2025). https://doi.org/10.48550/arXiv.2507.06261
12. Feng, K.J.K., et al.: Case repositories: towards case-based reasoning for AI alignment (2023). https://doi.org/10.48550/arXiv.2311.10934
13. Findeis, A., et al.: Inverse constitutional AI: compressing preferences into principles (2025). https://doi.org/10.48550/arXiv.2406.06560
14. Huang, S., et al.: Collective constitutional AI: aligning a language model with public input. In: The 2024 ACM Conference on Fairness, Accountability, and Transparency. FAccT '24, Rio de Janeiro, pp. 1395–1417. ACM (2024). https://doi.org/10.1145/3630106.3658979
15. Ji, J., et al.: AI alignment: a comprehensive survey (2025). https://doi.org/10.48550/arXiv.2310.19852
16. Jiang, L., et al.: Wildteaming at scale: from in-the-wild jailbreaks to (adversarially) safer language models. In: Proceedings of the 38th International Conference on Neural Information Processing Systems. NIPS '24, Red Hook, NY, USA. Curran Associates Inc. (2025). https://doi.org/10.48550/arXiv.2406.18510

17. Kundu, S., et al.: Specific versus general principles for constitutional AI (2023). https://doi.org/10.48550/arXiv.2310.13798
18. Markov, T., et al.: A holistic approach to undesired content detection in the real world. In: Proceedings of the Thirty-Seventh AAAI Conference on Artificial Intelligence (AAAI-23), Washington, vol. 37, pp. 15009–15018. AAAI Press (2023). https://doi.org/10.1609/AAAI.V37I12.26752
19. McNemar, Q.: Note on the sampling error of the difference between correlated proportions or percentages. Psychometrika **12**(2), 153–157 (1947). https://doi.org/10.1007/BF02295996
20. Meincke, L., et al.: Call me a jerk: Persuading ai to comply with objectionable requests. SSRN Electronic Journal **Online publication**, 35 pages (2025). https://doi.org/10.2139/ssrn.5357179, available at SSRN
21. Ollama: Mistral 7b model card. https://ollama.com/library/mistral:7b (2025)
22. OpenAI: Gpt-5 system card. Gpt-5 system card, OpenAI (2025)
23. Oxford Languages: Definition of "law". Google Dictionary (nd). https://languages.oup.com/google-dictionary-en/
24. Pang, X., et al.: Self-alignment of large language models via monopolylogue-based social scene simulation (2024). https://doi.org/10.48550/arXiv.2402.05699
25. Perez, E., et al.: Red teaming language models with language models. In: Goldberg, Y., Kozareva, Z., Zhang, Y. (eds.) Proceedings of the 2022 Conference on Empirical Methods in Natural Language Processing, Abu Dhabi, United Arab Emirates, pp. 3419–3448. Association for Computational Linguistics (2022). https://doi.org/10.18653/v1/2022.emnlp-main.225
26. Tedeschi, S., et al.: Alert: a comprehensive benchmark for assessing large language models' safety through red teaming (2024). https://doi.org/10.48550/arXiv.2404.08676
27. Wasserstein, R., et al.: Moving to a world beyond "$p < 0.05$". Am. Stat. **73**, 1–19 (2019). https://doi.org/10.1080/00031305.2019.1583913
28. Wei, J., et al.: Chain-of-thought prompting elicits reasoning in large language models. In: Proceedings of the 36th International Conference on Neural Information Processing Systems. NIPS '22, Red Hook, NY, USA. Curran Associates Inc. (2022). https://doi.org/10.48550/arXiv.2201.11903

Explainability of Tourism Risk Assessment by Leveraging Fuzzy Ontology Properties

Pavel Kosov(✉), Latafat Gardashova, and Aziz Nuriyev

BA Programs, Azerbaijan State Oil and Industry University, Baku, Azerbaijan
{pavel_kosov,l.qardashova,aziz.nuriyev}@asoiu.edu.az

Abstract. The growing role of tourism in the global economy generates large volumes of data, increasing the use of machine learning. Nevertheless, many modern machine learning models lack sufficient transparency, making them difficult for non-IT specialists to use. This lack of understanding and transparency is especially critical in areas such as tourism risk assessment, which is characterized by high uncertainty due to the imperfection of available information, complex impacts on overall tourist well-being, and dependence on external factors and spatial variables. To address this, Explainable Artificial Intelligence (XAI) techniques are applied. However, widely used feature-attribution approaches such as LIME and SHAP are limited in their ability to represent semantic relationships and uncertainty. Similarly, modern ontology-based solutions fail to capture the vagueness of real-world risk concepts effectively. Building on a fuzzy ontology, we propose a new framework for explainable risk assessment in tourism. The framework incorporates fuzzy explanatory properties to represent uncertainty and borderline cases, producing explanations supported by membership degrees that act as levels of certainty. The findings demonstrate that both expert and non-expert users benefit from explanations driven by fuzzy explanatory properties. This approach increases transparency and credibility, provides useful sensitivity analyses, and offers a reusable fuzzy-semantic framework for explainable risk management in the tourism sector.

Keywords: XAI · Explainability · Risk Assessment · Ontology · Fuzzy Knowledge · Fuzzy Logic · Tourism Risk Factors

1 Introduction

Tourism is one of the most dynamic sectors of the global economy and plays a crucial role in economic development, international cooperation, and cultural exchange. At the same time, tourism activities are inherently associated with various uncertainties and potential threats affecting travelers, tourism organizations, and destination sustainability. The concept of tourism risk has been widely studied in tourism research. Various studies have shown that perceived

C. Zanni-Merk et al. (Eds.): AIDT 2026, CCIS 3023, pp. 389–404, 2026.
https://doi.org/10.1007/978-3-032-31319-5_26

safety and security significantly influence destination choice and tourist behavior [20,27]. Tourism risks may arise from different sources including environmental hazards, technological failures, socio-political instability, or the behavior and characteristics of travelers themselves [13,16].

Our previous works addressed tourism risk assessment by developing structured models for identifying and evaluating travel risks. Z-value-based risk assessment framework was proposed to evaluate tourism risks using linguistic expert assessments and multi-criteria decision analysis techniques [18]. This framework identified several major groups of tourism risks. Further our research focused on the identification and structuring of tourism risk factors and sub-factors using expert knowledge and Delphi analysis in order to construct a comprehensive tourism risk register for travel safety evaluation [17]. Later, Z-number-based approach was applied to aggregate tourism risk factors and compare travel risks across destinations under uncertainty and limited information [19].

The growing importance of tourism in the global economy generates large amounts of data, encouraging the use of machine learning. Although risk assessment models have advanced, modern decision-support systems often depend on machine learning methods that lack transparency and interpretability. This is a significant issue, as decision-makers, tourists, managers, and policymakers need to understand the factors behind risk evaluations. However, many models function as "black boxes", making them difficult for non-IT users to interpret and apply effectively.

To address this challenge, this paper proposes an explainable tourism risk assessment framework based on fuzzy explainability and ontology-based knowledge representation. The proposed approach integrates fuzzy logic [28], fuzzy ontologies [4], and semantic explanatory properties [14] and Fuzzy Explainability framework [15] to represent uncertain tourism risk factors and generate interpretable explanations for risk assessments.

The main contribution of this paper is a feasibility-oriented methodological framework for explainable tourism risk assessment that integrates fuzzy ontology reasoning, explanatory properties [14], and shared mental models into a single semantic structure. In this way, the proposed approach does not only identify relevant factors, but also represents their explanatory role under uncertainty through fuzzy degrees and ontology-grounded relations [15].

The novelty of this work lies in three aspects: (1) a fuzzy ontology–based framework for explainable tourism risk assessment under uncertainty; (2) the use of fuzzy explanatory properties to generate human-readable explanations with membership degrees; and (3) ontology-grounded explanations that differ from feature-attribution methods (e.g., SHAP, LIME) by explicitly modeling semantic relations and uncertainty.

The rest of this paper is organized as follows. Section 2 includes existing research on XAI systems in the tourism domain. Section 3 describes the materials and methods adopted for the study. Section 4 outlines the experimental setup, followed by the its results in Sect. 5. Whereas Sect. 6 provides a discussion of the proposed framework and the addressed problem. The study concludes and gives ideas for future work in Sect. 7.

2 Related Works

Several approaches have been proposed to support tourism risk assessment and management. Risk evaluation models in tourism have been developed to analyze threats affecting tourism systems and destinations [1], while other studies investigated risk tolerance and travel behavior in relation to tourism hazards [26]. Researchers have also examined the influence of safety and security conditions on tourism destination competitiveness [12], as well as the impact of environmental and operational risks on tourism performance and management practices [22]. Despite these advances, tourism risk assessment remains challenging because available information is often incomplete, uncertain, or complex.

Overall trend shows that AI has significantly transformed the tourism industry by enhancing personalization, operational efficiency, and customer service. AI-driven tools enable tourism organizations to deliver personalized travel experiences and optimize operations. Recent studies highlight the use of AI for sentiment analysis, recommender systems, and predictive customer behavior, which enhance decision-making and marketing strategies [6]. Furthermore, AI frameworks such as travel intervention matrices and interconnected destination models provide practical guidance for implementing AI effectively [5]. These advancements position AI as a critical tool for maintaining competitiveness and sustainability in the tourism sector [9].

Explainability is essential to improve the reliability of AI-based systems. One XAI application is the development of explainable travel recommendation systems like ExplainableTrip, which use SHAP values to show how factors such as travel distance, time constraints, and user preferences influence itinerary suggestions [7]. Similarly, XAI has been applied to travel time prediction, using methods like SHAP and LIME to clarify the factors affecting outcomes and address the "black-box" nature of AI models [23]. These studies highlight how explainable systems can bridge the gap between AI performance and user understanding in tourism applications.

Moreover, XAI has been employed in sustainable tourism and decision-making frameworks. Counterfactual explanations have been used to analyze sustainable tourism indicators, allowing stakeholders to identify changes for improving sustainability scores [21]. Another study integrates XAI with Multiple Criteria Decision Making methods, such as TOPSIS and AHP, to assess tourism accommodation performance. By incorporating Shapley values, this framework eliminates subjective biases in decision-making, offering interpretable insights [8]. These advancements underscore the role of XAI in promoting transparency, sustainability, and informed decision-making within the tourism sector, paving the way for more accountable and user-centric AI applications.

Despite the growing interest in XAI for tourism sector, the number of the developed explainable systems remains limited. Existing studies mainly apply feature-attribution techniques, which explain results through statistical feature contributions. However, such approaches typically lack semantic depth and struggle with handling the vagueness and uncertainty in the tourism risk data, which often relies on linguistic assessments and context-dependent relationships between factors. This study proposes an explainable tourism risk assessment framework based on semantic-based fuzzy explainability [15] and semantic explanatory properties [14], enabling the representation of uncertain tourism knowledge and the generation of interpretable explanations.

Unlike SHAP and LIME, which explain predictions through feature attribution, the proposed approach generates ontology-grounded fuzzy explanations [15] for uncertainty-aware tourism risk assessment. This distinction is important for tourism risk assessment, where uncertainty, linguistic interpretation, and expert domain structure play a central role. The contribution of the proposed framework therefore lies in providing ontology-grounded and uncertainty-aware explanations rather than conventional attribution scores.

3 Materials and Methods

This section introduces the concepts supporting the proposed ontology-centered explainable system. Suggested XAI framework combines the ontology-based knowledge representation, and fuzzy logic. In particular, it builds on fuzzy set theory [28], Fuzzy OWL2 ontologies [4], and semantic explanatory properties [14]. The approach also relies on shared mental models [11] to align explanations with human understanding and extends the fuzzy explainability framework introduced in [15]. The overall architecture is inspired by DARPA's XAI program [10] and ontology-based explanation system Ontology-Based Image Classifier (OBIC) [3].

3.1 Explainable AI and Ontology-Based Explanations

Ontology-based XAI systems extend this idea by integrating semantic knowledge representations with machine learning models. Ontologies provide structured representations of concepts, relations, and axioms that allow logical reasoning and explanation generation. In this work, knowledge is represented using the OWL2 ontology language [25] under the *Open World Assumption (OWA)*, which states that missing information cannot be assumed as a false one.

One example of ontology-based XAI is the **Ontology-Based Image Classifier (OBIC)** [3], which combines machine learning outputs with ontology reasoning to generate interpretable explanations. In OBIC, machine learning models detect object properties, which are then organized within an ontology to enable semantic reasoning and explanation generation. Each property is associated with a dedicated classifier ("model per property"), while a global classifier identifies the main object class. Detected properties are represented as ontological individuals and processed by a reasoner that produces explanations interpretable by users. The general schema of OBIC [2] is depicted below in Fig. 1.

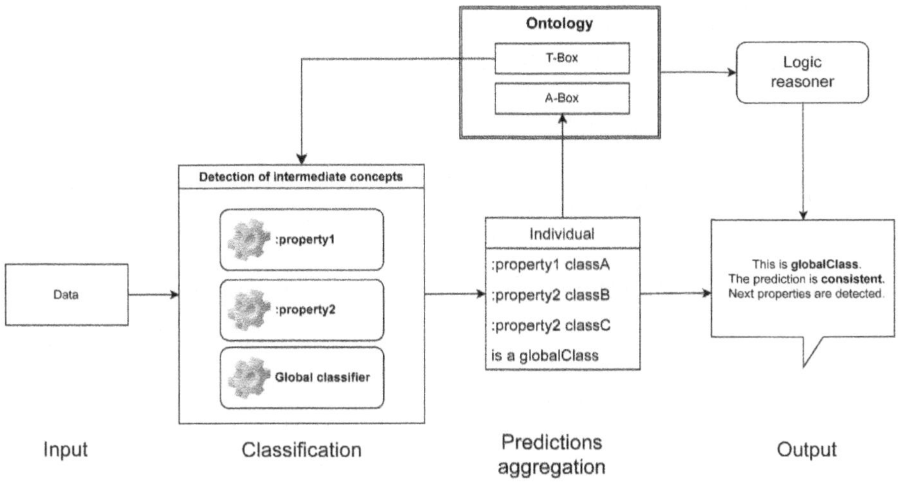

Fig. 1. General structure of OBIC

3.2 Fuzzy Logic and Fuzzy Ontologies

Fuzzy logic [28] extends classical Boolean logic by allowing gradual truth values in the interval $[0, 1]$. This representation enables reasoning under uncertainty and supports linguistic concepts such as *low*, *medium*, or *high*, by applying the concept of fuzzy sets. **A fuzzy set** A on universe X is defined by a membership function $\mu_A(x) : X \to [0, 1]$, which expresses the degree to which an element belongs to the set. **Linguistic variables** represents qualitative concepts in natural language. For example, a variable such as *Risk* may take values such as *High*, *Low*, *Medium*, etc. and each one is represented by a fuzzy membership function.

To represent such uncertain knowledge within semantic systems, classical ontologies can be extended into fuzzy ontologies. A fuzzy ontology can be formally described as

$$O_f = (C, R, A^O, \mu), \quad (1)$$

where C denotes the set of concepts, R the set of relations, A^O the ontology axioms, and μ assigns fuzzy membership degrees.

Fuzzy OWL2 [4] enables the representation of fuzzy knowledge within the OWL ontology framework. These extensions allow ontologies to represent imprecise real-world knowledge while preserving logical reasoning capabilities:

- *Fuzzy concepts* represent classes with graded membership.
- *Fuzzy roles* represent relationships with associated certainty degrees.
- *Fuzzy datatypes* encode linguistic values such as *low* or *high*.
- *Fuzzy axioms* describe assertions that hold to a certain degree.

Fuzzy values are assigned through a structured process. Risk factors are identified from expert knowledge and organized into ontology classes. Explanatory properties are defined as relations linking these factors to the overall risk. Linguistic values (e.g., Low, Medium, High) are mapped to membership degrees in $[0, 1]$ using $\mu_A(x) : X$, where $\mu_A(x)$ represents the degree to which x belongs to concept A. For example, a crime level assessed as VeryHigh may be encoded as $\mu(x) = 0.90$ in Fuzzy OWL2.

3.3 Shared Mental Models and Explanatory Properties

Shared mental models (MM) describe the collective understanding of concepts and relationships among members of a group [11]. They facilitate coordination and improve decision-making by aligning individual interpretations of a domain.

Formally, the knowledge of individuals A, B, C in the form of MM may be represented as sets M_A, M_B, and M_C. A shared model exists when the intersection $(M_A \cap M_B \cap M_C \neq \emptyset)$, indicating that some knowledge is commonly understood by all participants.

In XAI systems, shared MM help ensure that explanations correspond to concepts understandable by both experts and non-experts. Based on this idea, explanatory properties were introduced in [14] to represent semantically meaningful relationships between data concepts.

The set of explanatory properties is defined as $P_{\text{exp}} \subseteq R$, where R is a set of relationships, and each property $p \in P_{\text{exp}}$ may contribute to explaining an instance d_i with a fuzzy relevance degree $\mu_p(d_i) \in [0, 1]$. These properties serve as semantic building blocks for constructing explanations within the proposed framework. Such properties must satisfy several criteria:

- Accessibility to both expert and non-expert users.
- Ability to represent ambiguous or uncertain data.
- Compatibility with ontology reasoning under OWA.
- Simplicity and interpretability.

3.4 Fuzzy Explainability Framework

Fuzzy explainability integrates fuzzy logic with ontology-based explanation models. The concept builds on the ontology design pattern for explanations proposed in [24], where explanations link an event (explanandum) with explanatory factors (explanans) under specific contextual conditions.

The fuzzy explainability framework introduced in [15] extends this model by incorporating fuzzy knowledge representation and explanatory properties. Formally, the framework can be defined as

$$FE = (M, O_f, P, K_f), \tag{2}$$

where M represents the shared MM, O_f denotes the fuzzy ontology, P is the set of explanatory properties, and K_f corresponds to fuzzy knowledge derived from the dataset.

A fuzzy explanation for a decision or event y is expressed as

$$E_f(y) = \{(x_i, \mu(x_i)) \mid x_i \in X, \mu(x_i) \in [0,1]\}, \tag{3}$$

where $X = \{x_1, x_2, \ldots, x_n\}$ is the set of explanatory factors and $\mu(x_i)$ represents the strength of factor x_i in explaining y.

Figure 2 illustrates the conceptual architecture of the framework. Domain knowledge is first captured through shared MM and formalized using fuzzy ontologies. Explanatory properties define semantic relationships between concepts, while property-based restrictions guide the generation of explanations. New fuzzy knowledge derived from the dataset is then combined with these structures to produce interpretable fuzzy explanations.

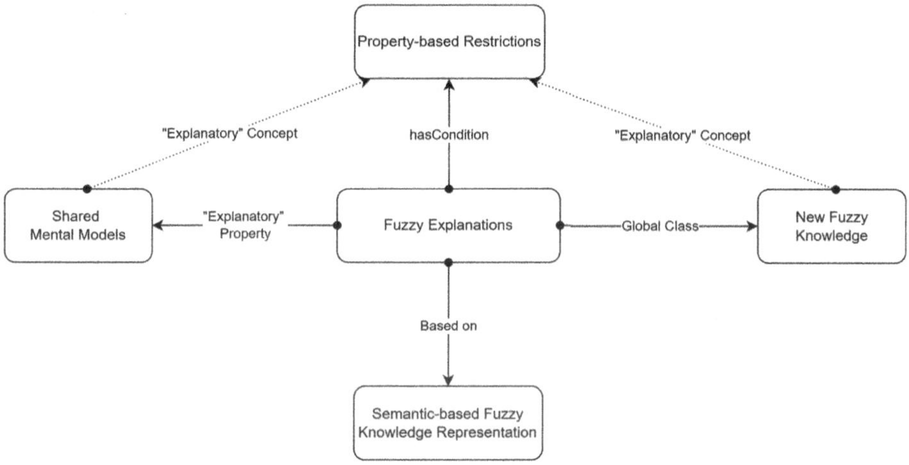

Fig. 2. Conceptual architecture of the fuzzy explainability framework

For example, a travel destination may be classified as HighRiskTrip based on observed conditions such as high crime levels, poor sanitation, etc. These *observations* reflect the shared MM to interpret tourism risk factors. In the semantic-based fuzzy *knowledge representation*, a tourism risk ontology defines concepts such as HighCrimeLevel, PoorSanitation, etc., and relates them to the concept of TravelRisk. *Property-based restrictions* (e.g., hasCrimeLevel, hasSanitationCondition, etc.) constrain the explanation through fuzzy conditions. From the dataset, *new fuzzy knowledge* is derived and integrated into the ontology. Consequently, the system generates a *fuzzy explanation*, indicating that the destination is classified as HighRiskTrip (0.81), supported by HighCrimeLevel (0.85), PoorSanitation (0.69), etc.

The ontology was built from tourism risk categories identified in prior studies [17–19] and represented in Fuzzy OWL2. Explanatory properties were derived from expert knowledge and shared mental models [14], with fuzzy values encoded as membership degrees [15]. The design emphasizes semantically meaningful, interpretable explanations over fully automated discovery, ensuring that properties are domain-driven and cognitively grounded rather than arbitrary [14]. This approach supports interpretability, reproducibility of the explanation logic, and consistency between knowledge and the generated explanations [15].

4 Experimental Setup

The goal of the experiment is to develop a fuzzy semantic web representation (fuzzy ontology) that supports the notion of *explanatory properties* within a fuzzy knowledge environment for tourism travel risk assessment. Building upon the techniques described in the Sect. 3, we developed and applied the architecture in Fig. 3.

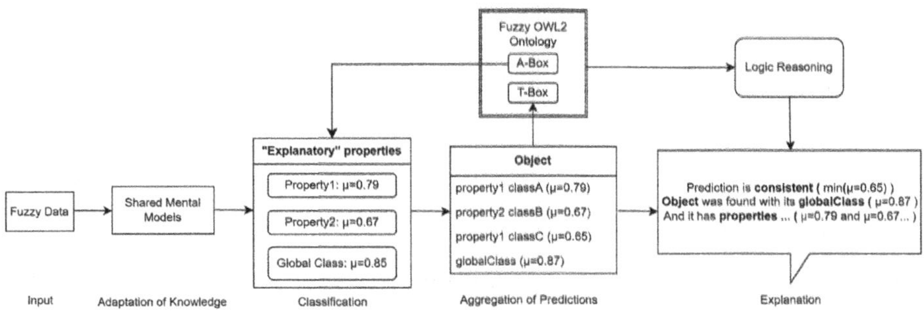

Fig. 3. Suggested system architecture for the current experiment

The purpose of the present experimental setup is to validate the feasibility, semantic consistency, and interpretability of the proposed framework in a controlled tourism-risk setting. Accordingly, the evaluation is designed to examine the explanatory capacity of the ontology-based fuzzy approach rather than to provide a large-scale benchmark of predictive performance. Within this scope, the use of a simplified dataset, case-study analysis, and user-oriented evaluation is appropriate for assessing the validity of the explanation mechanism.

The concept outlines a workflow for a fuzzy explainability framework, where fuzzy data are structured using a shared MM and represented as semantically meaningful properties with associated membership degrees. These properties, along with a global class assignment, are encoded in a fuzzy OWL2 ontology (A-Box/T-Box), enabling logic-based reasoning about the target object. The resulting explanation conveys prediction consistency and justifies the assigned class through ontology-grounded properties weighted by fuzzy certainty values.

The conceptual structure of the ontology and the set of explanatory properties were derived from expert knowledge as well as from a small experimental dataset used in our previous studies [17–19]. Each scenario includes fuzzy values

Fig. 4. Ontology representation of the main tourism travel risk factor groups

for crime situation, biological threats, sanitation conditions, and transportation reliability. For optimization, it was identified the most relevant explanatory properties by consulting with domain experts.

Figure 4 illustrates the conceptual organization of the main tourism travel risk factors used in the ontology. These factor groups represent the main categories of risks considered during the evaluation of travel destinations. The proposed explainability framework was evaluated through a case study analysis and a qualitative comparison with feature attribution explanations.

5 Results

The experiment resulted in the development of a explainable system based on fuzzy ontology for explainable tourism risk assessment. The implemented ontology was extended with Fuzzy OWL2 constructs to represent uncertain tourism knowledge. The ontology contains domain concepts, explanatory object properties, and axioms linking tourism risk factors with the target concept *hasOverallRisk*. Figure 5 shows an example of an asserted individual and its fuzzy axiom for *hasOveralRisk*.

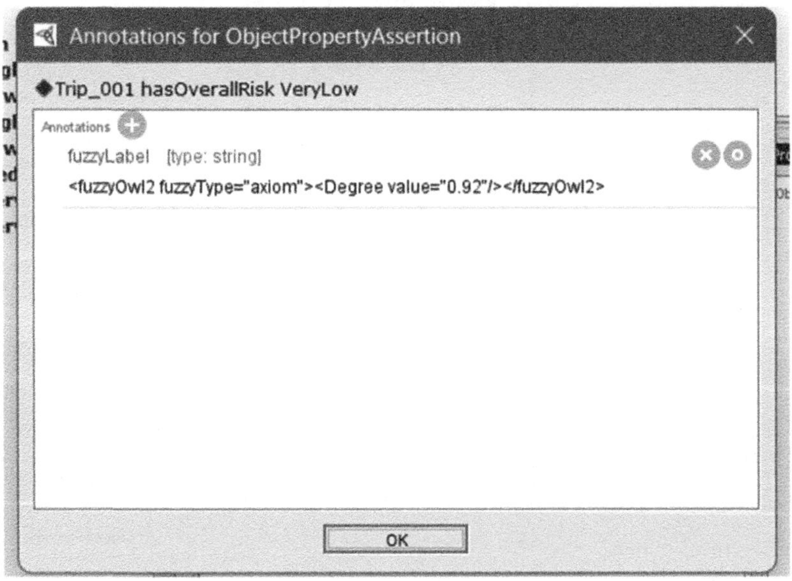

Fig. 5. Possible OveralRisk as fuzzy axiom for asserted individual

The developed ontology represents tourism risk through five main factor groups: destination country risks, natural environment risks, tour operator risks, transportation risks, and traveler-related risks.

The explanatory properties used to describe these factors are formally defined as object properties in the ontology. Figure 6 shows the hierarchy of object properties implemented in the Protégé ontology editor. explanatory properties are organized as sub-properties of *topObjectProperty*. These properties represent semantic relationships between tourism risk factors and the target property *hasOverallRisk*, which corresponds to the overall tourism risk assessment.

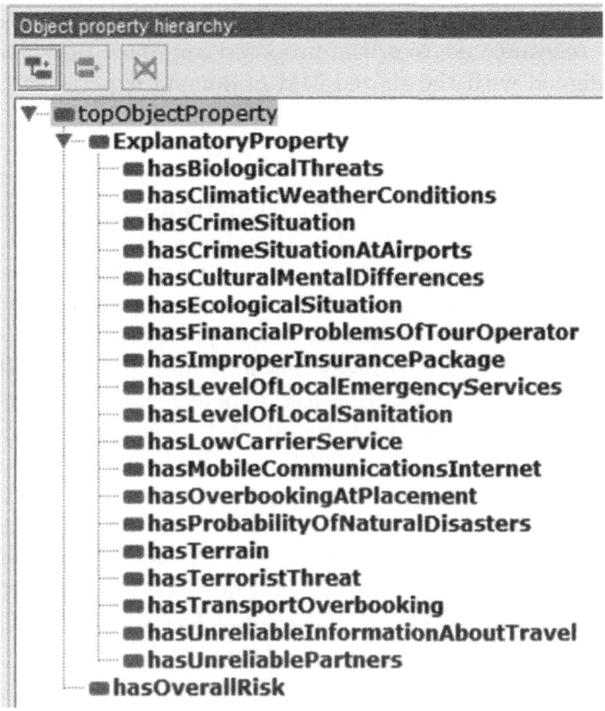

Fig. 6. Hierarchy of explanatory object properties in the developed ontology

The explanatory properties were defined from expert and user shared MM of the tourism-risk domain and represented in the ontology as semantic relations contributing to explanation generation. The results indicated that several domain features play a key role in the shared cognitive MM of both experts and users. These shared MM determine which domain characteristics should be represented as explanatory properties in the ontology and used for constructing explanations.

An example fuzzy explanation generated by the framework can be expressed as: *The model predicts 'HighRiskTrip' with 83% certainty, mainly because crime situation is Very High (90% certain) and biological threat is High (78% certain). Additional factors are local emergency services is Medium (69% certain) and sanitation condition is Medium (64% certain), and transportation service is Medium (62% certain).*

In addition, a small user evaluation was conducted to assess the clarity of explanations. Participants were asked to evaluate explanations generated by feature attribution methods and by the proposed fuzzy ontology framework. The results indicate that the ontology-based explanations were rated higher in terms of clarity, and usefulness of fuzzy degrees.

These results demonstrate that fuzzy explanatory properties can successfully represent uncertain tourism risk knowledge within an ontology-based explainable framework. By linking risk predictions with semantically defined explanatory factors and fuzzy relevance degrees, the proposed approach provides interpretable explanations aligned with the shared MM of domain experts and users.

6 Discussions

Traditional XAI methods typically rely on statistical feature importance, rule extraction, or post-hoc interpretability techniques that describe model behavior through crisp logical relationships. In contrast, the proposed framework introduces fuzzy explainability, where explanations are put into semantic concepts and expressed through gradual relationships. By embedding expert-defined explanatory properties within a fuzzy ontology and grounding them in shared mental models, the framework produces explanations that better reflect the uncertainty and ambiguity inherent in real-world tourism risk evaluation.

Building on this foundation, the framework enables qualitative reasoning commonly used in decision-support systems to be formally represented. Statements such as "this factor strongly contributes to the risk" can be modeled through fuzzy membership degrees in fuzzy OWL2 and linked to domain concepts defined in the ontology. This semantic representation allows uncertain knowledge to be used and reasoned over, connecting model predictions with interpretable explanatory factors rather than relying only on statistical data.

From an application perspective, the proposed framework is intended for decision-support settings in which transparent and semantically interpretable tourism-risk explanations are required. At the same time, its current implementation should be viewed as prototype-level, since larger datasets, broader ontology population, and more complex tourism scenarios would require further engineering and scalability analysis. The framework also contributes to responsible AI objectives by making risk assessments traceable to explicit explanatory factors and fuzzy certainty degrees, thereby supporting transparency and trust.

The scalability of the framework depends on the size of the ontology and the number of explanatory properties. Since the model is modular, new risk factors and destinations can be added incrementally without changing the core structure. The main computational cost lies in ontology reasoning rather than explanation generation. Large-scale validation on real-world datasets remains part of future work.

It is important to note that the current study focuses on demonstrating the feasibility of this approach. A simplified ontology and dataset were intentionally used to validate the theoretical concept of semantic-based fuzzy explainability. Although this limits the scope of the current implementation, the results provide a foundation for developing richer ontology-driven explanations in tourism risk assessment systems.

A key challenge identified in this work is the limited availability of structured knowledge bases and expert knowledge required to define explanatory properties for tabular datasets. Unlike image-based domains, where explanatory attributes can often be visually identified, tabular data requires deeper domain analysis to uncover meaningful explanatory patterns. Overall, the study provides a foundational step toward semantically grounded and uncertainty-aware explainability for tourism risk assessment, highlighting the potential of fuzzy ontology-based XAI to bridge the gap between machine learning predictions and human-understandable reasoning.

7 Conclusion and Future Work

This paper introduced a fuzzy ontology–based framework for explainable tourism risk assessment that integrates fuzzy logic, semantic knowledge representation, and explainability. By modeling tourism risk factors as fuzzy explanatory properties within an ontology, the proposed approach enables the generation of interpretable explanations supported by membership degrees reflecting the strength and uncertainty of contributing factors.

The present study demonstrates the feasibility of combining fuzzy logic with ontology-based explanations to support transparency and interpretability in tourism risk assessment. The proposed framework provides a semantically driven mechanism for linking risk predictions with domain knowledge under uncertainty and for generating human-readable explanations for both expert and non-expert users. At the same time, broader validation on larger real-world tourism datasets remains part of future work.

Future work is planned to extend the ontology by introducing semantic relationships with traveler-related entities such as personal preferences, health conditions, and risks associated with personal well-being. In addition, destinations could be described through dedicated explanatory properties. In such cases, even if a dataset contains only a country name, associated threats and risks could be inferred from the ontology, improving prediction quality while also supporting explanations of black-box outputs and helping detect inconsistencies within tourism datasets.

Disclosure of Interests. The authors have no competing interests to declare that are relevant to the content of this article.

References

1. Arturas, S., Jasinskas, E., Svagzdiene, B.: Risk assessment models in the tourism sector. Amfiteatru Econ. J. **17**(39), 836–846 (2015). https://doi.org/10419/168951
2. Bellucci, M.: Symbolic approaches for explainable artificial intelligence. Ph.D. thesis, Normandie Université, Rouen (2023). https://theses.hal.science/tel-04469103v1

3. Bellucci, M., Delestre, N., Malandain, N., Zanni-Merk, C.: Combining an explainable model based on ontologies with an explanation interface to classify images. Procedia Comput. Sci. **207**, 2395–2403 (2022). https://doi.org/10.1016/j.procs.2022.09.298
4. Bobillo, F., Straccia, U.: Fuzzy ontology representation using OWL 2. Int. J. Approximate Reasoning **52**(7), 1073–1094 (2011). https://doi.org/10.1016/j.ijar.2011.05.003
5. Chandrasekaran, S., Shabbirhusain, R.V.: Artificial Intelligence in Tourism: An Enhanced Tourist Journey Framework. Tour. Rev. Int. **29**(2), 151–170 (2025). https://doi.org/10.3727/194344225X17315216888989
6. Chavan, P., Havale, D.S., Khang, A.: Artificial Intelligence and Tourism: a Bibliometric Analysis of Trends and Gaps. In: Revolutionizing the AI-Digital Landscape, New York, 1st edn., pp. 332–344. Productivity Press (2024). https://doi.org/10.4324/9781032688305-24
7. Chitaliya, H., et al.: ExplainableTrip: an XAI-driven personalized travel recommendation system using gradient boosting and SHAP. Ingénierie des systèmes d information **30**(5), 1189–1199 (2025). https://doi.org/10.18280/isi.300507
8. Ciano, T., Ferrara, M.: Explainable Multi-criteria Decision Making for tourism economics: integrating XAI with MCDM for a robust accommodation performance assessment. Decisions Econ. Finan. (2025). https://doi.org/10.1007/s10203-025-00553-6
9. Goyal, N.K., Dandotiya, M., Kumari, M., Sheikh, M.F.A., Sharma, S.: Emerging Job Roles in the Tourism Sector Leveraging Artificial Intelligence. Presented at the (2025). https://doi.org/10.4018/979-8-3373-4546-8.ch004 Empowering Sustainable Performance and Competitive Advantage in Tourism
10. Gunning, D., Aha, D.W.: DARPA's explainable artificial intelligence program. AI Mag. **40**(2), 44–58 (2019). https://doi.org/10.1609/aimag.v40i2.2850
11. Jonker, C.M., van Riemsdijk, M.B., Vermeulen, B.: Shared mental models. In: De Vos, M., Fornara, N., Pitt, J.V., Vouros, G. (eds.) COIN -2010. LNCS (LNAI), vol. 6541, pp. 132–151. Springer, Heidelberg (2011). https://doi.org/10.1007/978-3-642-21268-0_8
12. Kordić, N., Živković, R., Stanković, J., Gajić, J.: Safety and security as factors of tourism destination competitiveness. In: Proceedings of the Singidunum International Tourism Conference - Sitcon 2015, Belgrade, Serbia, pp. 34–38. Singidunum University (2015). https://doi.org/10.15308/sitcon-2015-34-38
13. Korstanje, M.E., Clayton, A.: Tourism and terrorism: conflicts and commonalities. Worldwide Hospitality Tourism Themes **4**(1), 8–25 (2012). https://doi.org/10.1108/17554211211198552
14. Kosov, P., El Kadhi, N., Zanni-Merk, C., Gardashova, L.: Advancing XAI: new properties to broaden semantic-based explanations of black-box learning models. Procedia Comput. Sci. **246**, 2292–2301 (2024). https://doi.org/10.1016/j.procs.2024.09.560
15. Kosov, P., El Kadhi, N., Zanni-Merk, C., Gardashova, L.: Formalizing fuzzy explainability: enhancing XAI trustworthiness with fuzzy-ontology-based explanatory properties. Procedia Comput. Sci. **270**, 3768–3777 (2025). https://doi.org/10.1016/j.procs.2025.09.502
16. Mataković, H., Cunjak Mataković, I.: The impact of crime on security in tourism. Secur. Defence Quart.**27**(5), 1–20 (2019). https://doi.org/10.35467/sdq/115539
17. Nuriyev, A.M.: Identification of the tourism risks for Z-value based risk assessment. Eur. J. Formal Sci. Eng. **4**(2), 81–95 (2021). https://doi.org/10.26417/ejef.v3i3.p47-55

18. Nuriyev, A.M., Jabbarova, K.: Z-value based risk assessment: the case of tourism sector. In: Aliev, R.A., Kacprzyk, J., Pedrycz, W., Jamshidi, M., Babanli, M.B., Sadikoglu, F.M. (eds.) ICSCCW 2019. AISC, vol. 1095, pp. 205–213. Springer, Cham (2020). https://doi.org/10.1007/978-3-030-35249-3_26
19. Nuriyev, A.: Application of Z-numbers for evaluation and aggregation of the travel threats in the post-pandemic era. Eur. J. Soc. Sci. Educ. Res. **8**(2), 129 (2021). https://doi.org/10.26417/665lww41w
20. Pondicherry University, Puducherry (India), Rajesh, R.: Impact of tourist perceptions, destination image and tourist satisfaction on destination loyalty: a conceptual model. PASOS. Revista de Turismo y Patrimonio Cultural **11**(3), 67–78 (2013). https://doi.org/10.25145/j.pasos.2013.11.039
21. Saugar, J., Lancho, C., Cuesta, M., Cano, E.L., De Diego, I.M., Amado, A.: Counterfactual explanations for sustainable tourism indicators. In: Julian, V., Camacho, D., Yin, H., Alberola, J.M., Nogueira, V.B., Novais, P., Tallón-Ballesteros, A. (eds.) Intelligent Data Engineering and Automated Learning – IDEAL 2024, vol. 15346, pp. 214–220. Springer, Cham (2025). https://doi.org/10.1007/978-3-031-77731-8_20
22. Shahnaz, N., Harandi, A.: The factors affecting on the risk management of the travel agencies. Int. J. Acad. Res. Bus. Soc. Sci. **4**(5), 177–187 (2014). https://doi.org/10.6007/IJARBSS/v4-i5/847
23. Srivastava, N., Gohil, B.N., Ray, S.: Travel time prediction: unveiling the impact of spatio-temporal features on trip duration with eXplainable AI. Transportation (2025). https://doi.org/10.1007/s11116-025-10701-x
24. Tiddi, I., d'Aquin, M., Motta, E.: An ontology design pattern to define explanations. In: Proceedings of the 8th International Conference on Knowledge Capture. ACM (2015). https://doi.org/10.1145/2815833.2815844
25. W3C: OWL 2 Web Ontology Language Document Overview, 2nd edn. https://www.w3.org/TR/owl2-overview/
26. Williams, A.M., Baláž, V.: Tourism, risk tolerance and competences: travel organization and tourism hazards. Tour. Manage. **35**, 209–221 (2013). https://doi.org/10.1016/j.tourman.2012.07.006
27. Yang, E.C.L., Nair, V.: Tourism at risk: a review of risk and perceived risk in tourism. Asia-Pacific J. Innovation Hospitality Tour. (APJIHT) **3**(2), 13 (2015). https://doi.org/10.7603/s40930-014-0013-z
28. Zadeh, L.A., Aliev, R.A.: Fuzzy logic theory and applications: Part I and Part II. World Sci. Publ. (2018). https://doi.org/10.1142/10936

Sustainability and Responsible Innovation

Efficiency Driven Digital Transformation for Sustainable Real Time Monitoring of High Pressure South Caspian Wells

Fariz Karimov(✉), Yelena Shmoncheva, Gullu Jabbarova, and Rafail Manafov

AzerbaijanState Oil and Industry University, Baku, Azerbaijan
farizkarimov05@gmail.com

Abstract. The South Caspian Basin presents a globally challenging offshore environment due to extreme high-pressure, high-temperature (HPHT) conditions and narrow drilling windows. While Artificial Intelligence (AI) is vital for safety monitoring, conventional "Red AI" architectures suffer from high latency and energy intensity, creating a dangerous "inference lag" during pressure excursions. This paper proposes a "Green AI" framework tailored for "Digital Transformation 2.0," prioritizing algorithmic efficiency for real-time edge deployment on offshore platforms. The methodology integrates a robust signal conditioning pipeline, including Savitzky-Golay filtering and IQR-based outlier detection, to refine stochastic drilling telemetry. The core optimization utilizes a regression-oriented Knowledge Distillation pipeline, where a lightweight "Student" model (1D-CNN + GRU) emulates a high-capacity "Teacher" model by minimizing a joint loss function, followed by magnitude-based pruning and 8-bit post-training quantization. Validated using Azerbaijani offshore drilling data, the framework achieves a 7.7-fold acceleration in inference speed and a 70% reduction in power consumption, while maintaining a highly competitive F1-score of 0.961 (reflecting a marginal 1.6% degradation from the baseline). From an industrial safety perspective, this optimized engine delivers a critical 55-s early warning advantage for gas influx detection, directly mitigating risks of blowouts and ecological damage. Furthermore, by lowering the hardware energy footprint to 3.8 Watts, the system ensures autonomous monitoring on low-power edge devices, enhancing infrastructure resilience and reducing costly Non-Productive Time (NPT). This research demonstrates that sustainable, efficiency-driven AI is essential for the economic and environmental viability of modern offshore operations in the South Caspian energy sector.

Keywords: Green AI · South Caspian Basin · Industrial Safety · Model Pruning · Edge Computing · HPHT Drilling · Algorithmic Efficiency

1 Introduction

The digital transformation of the upstream oil and gas industry has reached a critical juncture where the sole reliance on conventional AI architectures is no longer sufficient to guarantee operational integrity [1, 3, 9]. In the unique geological setting of the South

C. Zanni-Merk et al. (Eds.): AIDT 2026, CCIS 3023, pp. 407–422, 2026.
https://doi.org/10.1007/978-3-032-31319-5_27

Caspian Basin, drilling operations are frequently impeded by some of the most complex HPHT conditions globally [11].

The geomechanical complexity of the South Caspian Basin is driven by rapid Pliocene-Quaternary sedimentation and active shale diapirism, which create widespread under-compacted overpressure zones [12]. These factors result in an extremely narrow drilling window, where minor fluctuations in equivalent circulating density can trigger either catastrophic fluid losses or immediate gas influxes. Consequently, traditional static models for pressure prediction are increasingly replaced by dynamic systems capable of accounting for these rapid, non-linear geological transitions.

These reservoirs are characterized by steep pressure gradients and extremely narrow drilling windows, where the margin between the pore pressure (P_p) and the fracture gradient (P_f) is often critically low [5]. Traditionally, monitoring these parameters relied on manual observation or simplified hydraulic models which lacked the predictive speed required to prevent catastrophic events such as gas kicks or blowouts [23].

The industry's evolution toward "Digital Transformation 2.0" (DT 2.0) marks a fundamental shift from simple data digitization to the creation of autonomous, cyber-physical systems [7]. While the first wave of transformation focused on centralizing data in cloud-based repositories for retrospective analysis, DT 2.0 prioritizes real-time "Intelligence at the Edge". In the context of offshore drilling, this means moving the decision-making logic from distant onshore servers directly to the rig's local control systems. This decentralized approach is necessitated by the inherent limitations of satellite-based telemetry, where the round-trip latency (the time taken for data to reach the cloud and for a command to return) can exceed several seconds - a timeframe that is unacceptable during a fast-moving pressure excursion. Thus, the paradigm of DT 2.0 in the oilfield requires a synthesis of high-fidelity predictive modeling and extreme computational efficiency, ensuring that safety protocols can be executed autonomously and instantaneously at the wellsite, regardless of connectivity status.

While the recent introduction of Deep Learning models has significantly improved predictive accuracy [10], it has also introduced a "Red AI" paradigm. This term describes massive neural networks requiring high-performance computing clusters. For offshore platforms, reliance on such architectures is problematic due to significant carbon footprints [18, 21] and the severe hardware limitations of edge devices, creating a "latency gap" that renders sophisticated models useless during fast-moving pressure excursions [15].

There is an urgent need for a transition toward "Green AI" - a framework that prioritizes computational efficiency and sustainability [4]. This research addresses this gap by proposing an efficiency-driven digital framework tailored for real-time well monitoring. The core innovation lies in the application of model pruning and post-training quantization [8], ensuring that the digital twin of the well remains responsive even on low-power industrial controllers [14]. By integrating these methodologies, this research demonstrates that industrial safety and environmental stewardship are complementary pillars of the future digital oilfield [6].

2 Literature Review

2.1 Evolution of Well Monitoring

The evolution of well monitoring paradigms has transitioned through three distinct technological eras. In the pre-digital era, well control relied heavily on manual observation of pit levels and visual standpipe pressure gauges, a method inherently limited by human reaction time and subjective interpretation. The second era, characterized by the introduction of Measurement While Drilling (MWD) and Logging While Drilling (LWD) technologies in the late 20th century, enabled the transmission of downhole data to the surface via mud-pulse telemetry. While revolutionary, these systems faced significant bandwidth constraints, often resulting in data lags that are incompatible with the instantaneous nature of pressure excursions in HPHT environments. The current era, defined by the "Digital Oilfield" concept, seeks to leverage high-frequency sensor arrays [2]. However, the literature reveals that the sheer volume of data generated by modern rigs has outpaced the analytical capacity of traditional mechanistic models, creating a "data-rich but insight-poor" environment that necessitates the adoption of advanced Machine Learning (ML) frameworks.

The integration of AI into petroleum engineering has evolved from simple data regression to complex deep learning architectures capable of predicting dynamic downhole conditions [3, 16]. Early research in this field focused primarily on maximizing the accuracy of Rate of Penetration (ROP) and torque predictions [20], often utilizing heavy-weight Recurrent Neural Networks (RNN) and Long Short-Term Memory (LSTM) networks to handle time-series drilling data [10]. However, these high-fidelity models are frequently characterized as "Red AI" because they require significant computational power and specialized hardware to execute real-time inference, a factor that is increasingly scrutinized due to the carbon intensity of large-scale digital operations [18]. A critical analysis of existing algorithms reveals a significant trade-off between structural simplicity and predictive depth [13, 22]. Early attempts to automate kick detection utilized Support Vector Machines (SVM) and Random Forests due to their low computational overhead. However, empirical studies demonstrate that these "shallow" models struggle to capture the complex, non-linear temporal dependencies inherent in multiphase flow dynamics during a gas influx. Conversely, the industry has recently pivoted toward "heavyweight" architectures like LSTM networks. While LSTMs excel at processing time-series data, their internal gating mechanisms require a massive number of floating-point operations (FLOPs) per timestep. In the high-stakes environment of the South Caspian Basin, where the "inference jitter" of a high-capacity model can lead to delayed warnings, there is an evident academic gap for an intermediate architecture. This study addresses this by advocating for Gated Recurrent Units (GRU) combined with model compression, offering the temporal intelligence of deep learning without the prohibitive "computational tax" of unoptimized LSTMs. Recent studies have highlighted that the energy consumed during the training and deployment phases of such models is becoming a barrier to the "Sustainable and Responsible Innovation" required by modern environmental standards [21].

2.2 HPHT Drilling Challenges and the Imperative for Sustainable Edge Computing

In the specific context of the South Caspian Basin, the literature emphasizes the extreme difficulty of managing HPHT environments where traditional hydraulic models often fail to provide adequate early warnings [5]. Research into the geological features of Azerbaijani offshore fields indicates that the rapid transition between pressure regimes necessitates a response time that is often faster than what cloud-based AI systems can deliver due to satellite latency [11]. This has led to an emerging body of work focused on "Edge Computing," which advocates for moving AI processing from centralized servers directly to industrial controllers at the rig site [15]. While Edge Computing addresses latency, it introduces a "hardware bottleneck," as typical offshore controllers lack the GPU capacity to run unoptimized deep learning models [14].

This shift toward "Green AI" aligns with the global imperative for sustainable digital operations [17]. The massive carbon footprint of data centers required to run "Red AI" often offsets the efficiency gains these models provide [21]. Integrating energy-efficient edge inference ensures that industrial safety does not compromise environmental stewardship, establishing algorithmic efficiency as a primary performance metric.

Furthermore, existing kick detection methodologies lack adaptability. Mechanistic models struggle with transient multiphase flows, while purely data-driven approaches (like SVMs or Random Forests) fail to capture the temporal dependencies of drilling data. Although recurrent architectures (RNN/LSTM) are the industry standard for temporal modeling, their high computational cost creates an urgent operational imperative: the industry requires a framework possessing the temporal memory of an LSTM but operating with the computational lightness of a simple regression algorithm. Our proposed optimization framework directly targets this dichotomy.

To bridge the gap between model accuracy and hardware constraints, recent advancements in neural network optimization have gained traction in the industrial sector. Fundamental techniques such as Magnitude-based Pruning and Post-Training Quantization have been proven to reduce the memory footprint of neural networks by removing redundant parameters and converting 32-bit weights into 8-bit integers [8]. Despite these breakthroughs, a survey of recent quantization methods shows that while they are widely used in consumer electronics, their application to oilfield safety and well-integrity monitoring remains underdeveloped [6]. Current automated kick detection systems still largely rely on rule-based logic or uncompressed ML models that are prone to "inference jitter" when processing high-frequency drilling data [23]. Consequently, there is a clear academic and industrial gap for a specialized "Green AI" framework that prioritizes algorithmic efficiency for HPHT safety operations, ensuring that the digital transformation of the Caspian energy sector remains both technologically resilient and environmentally sustainable [4, 19].

3 Methodology

The methodological framework of this research is designed to engineer a robust transition from computationally intensive deep learning architectures to a resource-efficient "Green AI" paradigm, specifically tailored for the hardware-constrained environments of offshore drilling platforms. The core predictive engine is constructed upon a hybrid neural network topology that integrates One-Dimensional Convolutional Neural Networks (1D-CNN) for feature extraction with GRUs for temporal sequence modeling. This specific architectural combination was selected over traditional LSTM networks due to the superior computational efficiency of GRUs, which utilize a streamlined gating mechanism while maintaining high fidelity in capturing long-term dependencies within time-series drilling data. The input vector, X_t, , aggregates high-frequency sensor telemetry critical for well control, including Weight on Bit (*WOB*), Rotational Speed (*RPM*), Standpipe Pressure (*SPP*), and Equivalent Circulating Density (*ECD*). Prior to ingestion by the network, this raw data undergoes a rigorous pre-processing phase involving noise reduction via a moving average filter and normalization using Min-Max scaling to confine the feature range between 0 and 1, ensuring numerical stability during the initial training of the high-capacity "teacher" model.

To ensure robust model training, the dataset was constructed from high-frequency surface and downhole telemetry recorded over a 45-day drilling campaign targeting the HPHT section. Data was sampled at 1 Hz, yielding approximately 3.45 million discrete time-steps after the rigorous filtering of non-drilling intervals (e.g., pipe connections, tripping). Crucially, real-world well monitoring is characterized by a severe class imbalance (Class Imbalance). In this dataset, normal operational parameters constitute 99.8% of the timeline, with only four confirmed gas influx events (anomalies) cumulatively accounting for approximately 6,900 samples (0.2%). This inherent asymmetry dictates the necessity for specialized evaluation metrics beyond standard accuracy.

3.1 Advanced Signal Conditioning and Noise Suppression

The reliability of this AI-driven monitoring system is fundamentally contingent upon the quality of the input telemetry, especially given that in the South Caspian Basin, high-frequency drilling data is inherently susceptible to significant stochastic noise arising from complex mechanical and hydraulic interactions. Mechanical noise is primarily generated by drill string vibrations, including stick-slip and whirl phenomena, which introduce high-frequency oscillations into the *WOB* and Torque channels. Simultaneously, hydraulic noise manifests as periodic pressure pulsations caused by the reciprocating action of the mud pumps, often complicating the detection of subtle *SPP* anomalies. To mitigate these effects and prevent the "Green AI" model from learning spurious correlations, a multi-stage signal conditioning pipeline was implemented. While the initial framework utilizes a simple moving average, this research employs a more robust Savitzky-Golay (S-G) filter for deep-layer noise suppression. Unlike standard low-pass filters, the S-G algorithm utilizes local polynomial regression to smooth the data while preserving the critical sharpness of transient features, such as the rapid pressure spikes indicative of a gas influx. The filter was configured with a frame length of 51 samples and a second-order polynomial, ensuring that the underlying trend of the *ECD* remains distinct from high-frequency telemetry jitter.

Furthermore, the raw dataset underwent rigorous outlier detection using the Interquartile Range (IQR) method, where sensor data points falling outside the 1.5 $\times$ IQR threshold - often caused by intermittent telemetry dropouts - were identified as technological artifacts. Rather than simple deletion, which would disrupt the temporal continuity required by the GRU layers, these gaps were addressed using linear interpolation for short durations and forward-filling for longer intervals. Following denoising, the input features were subjected to Feature Scaling to normalize the diverse physical units. While Min-Max scaling was used for initial benchmarks, we integrated Z-score Standardization ($\mu = 0$, $\sigma = 1$) for the final training of the teacher model. This ensures that the loss function remains isotropic, providing a more stable gradient for the subsequent post-training quantization.

Furthermore, the implementation of a sliding window technique ensures that the temporal dynamics of the drilling process are captured as a continuous stream rather than isolated data points, allowing the model to recognize incipient trends in pressure fluctuations. This systematic approach to feature engineering is essential for maintaining the structural integrity of the predictive signals during the subsequent model compression phases. By establishing a stabilized and normalized baseline, the framework minimizes the risk of gradient explosion or vanishing during the intensive training of the high-parameter teacher model. Ultimately, these rigorous pre-processing protocols serve as the foundational layer upon which the efficiency-driven "Green AI" optimizations are built, ensuring that accuracy is preserved even as computational complexity is reduced.

Consequently, this balanced approach to data conditioning facilitates a seamless transition into the structural optimization phase, where the model's architectural footprint is aggressively minimized without compromising its analytical depth.

3.2 Model Optimization and Distillation Pipeline

The architectural transition from a high-capacity "Red AI" model to a resource-efficient "Green AI" engine is governed by a multi-stage optimization framework, as illustrated in Fig. 1.

Initially, a "Teacher" model - comprising a high-parameter ensemble of deep 1D-CNN and bidirectional LSTM layers - is trained on the full HPHT dataset to establish an accuracy benchmark.

This model operates at FP32 precision, capturing the complex, non-linear correlations between torque fluctuations and standpipe pressure anomalies.

To bridge the "latency gap" identified in the South Caspian drilling context, we employ Knowledge Distillation (KD). In the knowledge distillation setup, a pre-trained high-capacity "Teacher" model (the deep 1D-CNN + Bi-LSTM ensemble) transfers its learned representations to a compact "Student" model (1D-CNN + GRU). The Student is trained to mimic the Teacher's continuous-valued pressure predictions by minimizing a joint loss function that combines a standard regression loss (Mean Absolute Error with respect to ground truth labels) with a distillation loss (Mean Squared Error between the Student's and Teacher's outputs). This strategy allows the Student to capture the

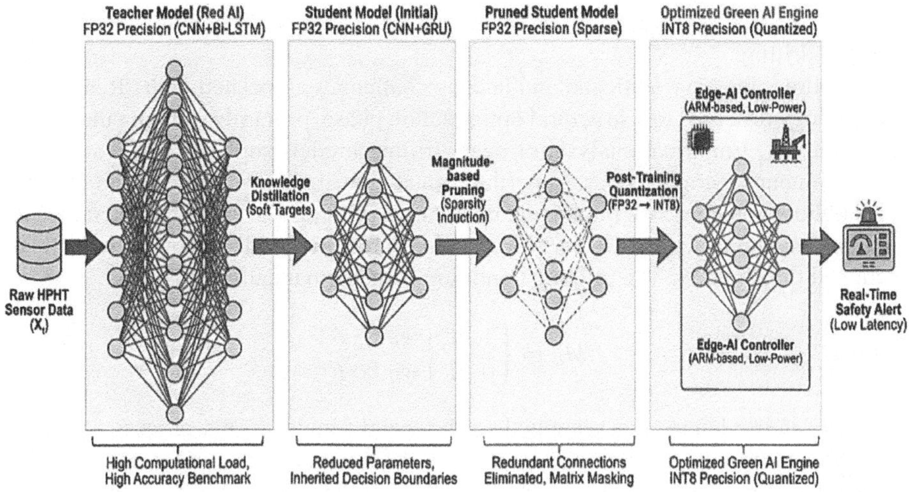

Fig. 1. The "Green AI" Optimization Pipeline

nuanced temporal dynamics of HPHT pressure transients while inheriting the structural efficiency of the GRU-based architecture.

Following distillation, the Student undergoes Magnitude-based Pruning, where redundant synaptic connections are masked based on the threshold τ, inducing sparsity without sacrificing the model's ability to detect high-pressure ramps. The final stage involves PTQ, converting the remaining weights from FP32 to 8-bit integers (INT8). This holistic pipeline transforms a computationally prohibitive model into a lightweight, portable inference engine specifically optimized for the low-power ARM-based controllers used on Azerbaijani offshore platforms, ensuring that real-time monitoring remains autonomous and resilient.

Moreover, the reduced memory footprint allows for the deployment of multiple parallel inference instances on a single chipset, enabling a voting-based redundancy system that enhances the overall reliability of the kick-detection logic.

In a standard cloud-based architecture, the total "Round-Trip Time" (RTT) is compounded by multiple stages of latency: satellite uplink/downlink, ground station processing, and the execution time of a high-parameter "Red AI" model on remote servers. In the South Caspian context, where offshore rigs rely on satellite telemetry, this cumulative delay frequently exceeds 60 s. For HPHT wells, such a latency gap is catastrophic, as it exceeds the critical window required to mitigate a rapid pressure ramp-up.

Conversely, the optimized "Green AI" framework enables a decentralized architecture. By deploying the pruned and quantized INT8 engine directly onto the rig's local ARM-based controllers, the need for external data transmission is eliminated for safety-critical inference. The local processing latency is reduced to approximately 24 - 50 ms, as detailed later in the results. This shift from a high-latency satellite loop to a sub-second local loop ensures that the well-integrity system can trigger automated alerts or emergency shutdown protocols (ESD) almost instantaneously. This architectural transition is a cornerstone of "Digital Transformation 2.0," where intelligence is localized to protect both the industrial assets and the surrounding marine ecosystem from the inherent risks of HPHT drilling.

To address the environmental and latency challenges associated with "Red AI," the methodology proceeds to a structural optimization phase, primarily utilizing magnitude-based pruning. Empirical analysis of deep drilling models suggests that a significant portion of neural parameters are superfluous and contribute negligibly to the detection of gas influx anomalies. We implement an iterative pruning algorithm that identifies and eliminates synaptic weights, w, whose absolute magnitude falls below a dynamically calculated threshold, θ. The pruning condition is mathematically defined as:

$$M_{ij} = \begin{cases} 1, \text{ if } \mid w_{ij} \mid \geq \theta \\ 0, \text{ if } \mid w_{ij} \mid < \theta \end{cases} \quad (1)$$

where M_{ij} is the binary mask applied to the weight matrix. This process effectively induces sparsity within the network, reducing the number of floating-point operations required for a single inference cycle without altering the model's fundamental decision-making capability. A subsequent "fine-tuning" retraining phase allows the remaining active connections to adjust and compensate for the removed parameters, recovering any potential loss in predictive accuracy. The magnitude-based pruning schedule followed a polynomial decay over 40 epochs, reaching the 70% sparsity target before the final 10-epoch fine-tuning stage to recover predictive stability.

It is vital to clarify the learning formulation: the core GRU network operates fundamentally as a multivariate regression engine, trained to forecast continuous sequences of Standpipe Pressure (SPP) and Equivalent Circulating Density (ECD). To translate these continuous forecasts into a safety-critical binary alert ('Kick' vs. 'No Kick'), we implement a dynamic residual analysis. An anomaly is flagged if the absolute error between the AI-predicted value and the actual sensor reading exceeds 3 standard deviations (3σ) of a 5-min rolling baseline window for three consecutive seconds.

The training protocol for the baseline teacher model was rigorously designed to ensure maximum generalization capability. The dataset was split into training (70%), validation (15%), and testing (15%) subsets, respecting the chronological order of the drilling operations to prevent data leakage. The model was trained using the Adam optimizer with an initial learning rate of *1e - 3*, which was dynamically decayed using a 'ReduceLROnPlateau' scheduler whenever the validation loss plateaued for more than 5 epochs. The loss function employed was the Mean Absolute Error (MAE), chosen over Mean Squared Error (MSE) to reduce the penalty on outliers, which are common in noisy rig sensor data.

To avoid overfitting, which is a common problem when training deep networks on limited datasets that are typical of certain well campaigns, we used regularization methods like Dropout layers with a rate of 0.2 after each convolutional block. Also, Early Stopping was set up with a 15-epoch patience level. To establish the "Gold Standard" predictive performance benchmark, against which the subsequently optimized "Green" models would be evaluated, the training was performed on a high-performance cluster equipped with NVIDIA A100 GPUs.

After the structural pruning, the method uses Post-Training Quantization (PTQ) to make the model even better suited for use on industrial edge hardware. Standard deep learning models use 32-bit floating-point precision (FP32), which costs a lot of processing resources for low-power ARM-based controllers. To lessen this problem, we use a linear quantization approach that converts the continuous range of floating-point weights to a discrete, lower-precision 8-bit integer space (INT8). This quantization's transformation function is written as:

$$Q(r) = \text{round}\left(\frac{r}{S} + Z\right) \quad (2)$$

where r is the real-valued input, S represents the scaling factor derived from the dynamic range of the weight tensor, and Z is the zero-point offset necessary to accurately represent zero values in the integer domain. This conversion reduces the model's memory footprint by a factor of four and significantly lowers memory bandwidth consumption. The final optimized model is encapsulated for an Edge Inference Engine, such as TensorFlow Lite, allowing for sub-second execution directly at the rig site. The efficiency of this framework is evaluated using the Energy Efficiency Ratio (η), ensuring that the system remains viable for continuous, 24/7 monitoring of high-pressure wells in the South Caspian Basin.

4 Case Study: Data Analysis and Results in High-Pressure Environments

To rigorously validate the efficiency and reliability of the proposed Green AI framework, a retrospective simulation was conducted utilizing high-frequency sensor data from a deep exploratory well in the South Caspian Basin. This region is globally recognized for its extreme drilling complexities, particularly the rapid onset of HPHT transition zones where traditional monitoring often fails to provide adequate reaction time.

The geomechanical profile of the selected well, reaching a total vertical depth of over 6,200 m, is detailed in Table 1. As illustrated, the primary operational challenge emerges in the target HPHT zone (5,500 - 6,200 m). In this interval, the pore pressure (P_p) gradient escalates to 2.18 g/cm^3, while the fracture gradient (P_f) plateaus at 2.28 g/cm^3. Consequently, the safe operational drilling window-the margin between the ECD and the fracture gradient-narrows to a critical threshold of less than 0.06 g/cm^3. This precarious geomechanical balance confirms the absolute necessity for sub-second, highly accurate anomaly detection to prevent catastrophic hydraulic fracturing or severe gas influx.

To adapt the predictive algorithm for edge deployment within this hazardous zone, the baseline deep learning model underwent the proposed structured pruning process. The impact of this optimization is illustrated in Fig. 2.

This graph plots the model's predictive integrity (measured via the F1-score) against the percentage of removed neural connections (sparsity). The analysis demonstrates that up to a sparsity level of 68%, the model's F1-score remains remarkably stable, exhibiting only marginal degradation. A precipitous decline in performance, characterized by the "knee" of the curve, is only observed when sparsity exceeds 80%. By selecting an optimal pruning threshold of 70%, the architecture is significantly streamlined without compromising its analytical integrity.

The cumulative impact of this pruning, combined with 8-bit post-training quantization, is quantified in Table 2. The results highlight a stark contrast between the standard,

Table 1. Representative Well Profile and Geomechanical Characteristics of the South Caspian Basin

Depth Interval (m)	Lithology	Pore Pressure (Pp) Gradient (g/cm^3)	Fracture Gradient (Pf) (g/cm^3)	Mud Density (ECD) (g/cm^3)	Temperature (°C)
0 – 1500	Unconsolidated Sands	1.05 – 1.15	1.45 – 1.55	1.20	15 – 45
1500 – 3500	Shales / Clays	1.25 – 1.65	1.75 – 1.90	1.70	45 – 85
3500 – 5500	Tight Sandstones	1.85 – 2.10	2.15 – 2.25	2.12	85 – 145
5500 – 6200	HPHT Zone	2.12 – 2.18	2.22 – 2.28	2.16	155 +

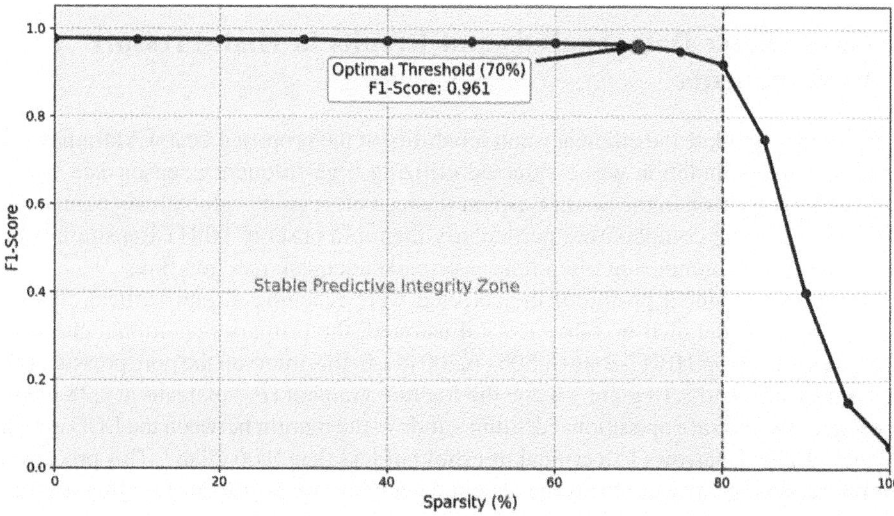

Fig. 2. F1-Score vs. Sparsity

computationally heavy "Red AI" (operating in FP32 precision) and the optimized "Green AI" (operating in INT8 precision).

As noted in the methodology, relying solely on Overall Accuracy in a heavily imbalanced dataset (where normal operations constitute 99.8% of the data) can be misleading. Therefore, the framework was rigorously evaluated using metrics resilient to class imbalance: Precision, Recall, and the F1-score. Crucially, to prevent data leakage and accurately simulate real-world deployment, these metrics were extracted from a strictly chronological, unmanipulated testing subset, explicitly avoiding any artificial data balancing techniques (such as oversampling).

Table 2. Computational Efficiency and Performance Metrics: Red AI vs. Green AI

Performance Metric	Standard "Red AI" (FP32)	Optimized "Green AI" (INT8)	Relative Improvement
Model Size (MB)	145.2	36.8	74.6% Reduction
Parameters (Millions)	12.4	4.1 (Post-Pruning)	66.9% Reduction
Inference Latency (ms)	185	24	87.0% Faster
Power Consumption (W)	12.5	3.8	69.6% Saving
F1-Score	0.977	0.961	-1.6% (Marginal)

The baseline high-capacity "Red AI" model achieved a Precision of 96.5% and a Recall of 99.0%, resulting in a strong F1-score of 0.977. Following the distillation and structural compression processes, the optimized "Green AI" engine maintained highly competitive performance, yielding a Precision of 94.2% and a crucial Recall of 98.1% (an F1-score of 0.961). In safety-critical applications like HPHT drilling, maintaining a high Recall is paramount to ensure that false negatives (missed gas influxes) are virtually eliminated. The exact distribution of true and false predictions for the optimized model is detailed in the confusion matrix (Table 3).

Table 3. Confusion Matrix of the Optimized Green AI Model

	Predicted: Kick (Anomaly)	Predicted: Normal
Actual: Kick (Anomaly)	**6,769 (True Positives)**	**131 (False Negatives)**
Actual: Normal	**417 (False Positives)**	**3,442,683 (True Negatives)**

Beyond predictive reliability, the most critical operational metric derived from Table 2 is the acceleration of the inference process. The computational speedup *(S)* achieved by the Green AI model is expressed as:

$$S = \frac{T_{Red}}{T_{Green}} = \frac{185\text{ ms}}{24\text{ ms}} \approx 7.71 \tag{3}$$

This 7.7-fold acceleration allows the model to produce real-time stability assessments continuously, free from the memory overhead and processing jitter that typically afflict heavier models deployed on industrial gateways.

The practical implication of this accelerated inference is visualized in Fig. 3, which depicts a simulated kick event.

The graph shows how long it takes for typical surface sensors to respond compared to the Green AI model.

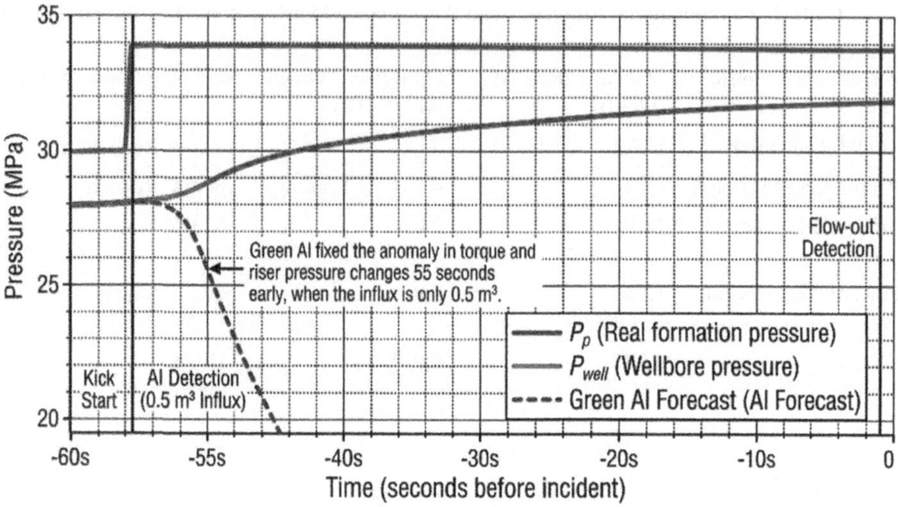

Fig. 3. Real-Time Pressure Anomaly Detection

When there was a sudden rise in pressure that indicated a kick, typical surface sensors (which rely on flow-out differentials) only picked up the anomaly 60 s after the first influx. This was because the drilling fluid column was naturally resistant to change. On the other hand, the Green AI model, which monitors changes in standpipe pressure and torque every 24 ms, was able to detect the multivariate anomaly signature and send an automated early warning 55 s before the regular alarm. In a narrow-margin HPHT well, this time advantage gives the driller the important time they need to start well-control procedures and seal the blowout preventer (BOP). Figure 4 presents a comparative analysis of power consumption during inference.

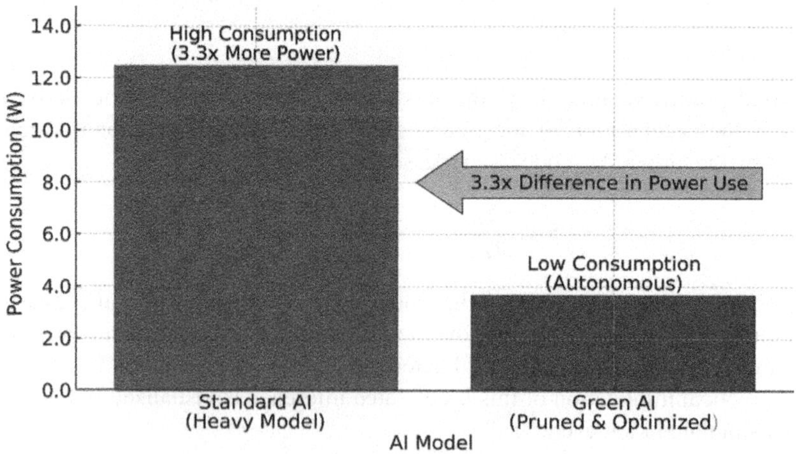

Fig. 4. Power Consumption Comparison

The histogram demonstrates that the standard AI architecture consumes approximately three times the power of the optimized model, underscoring the environmental and hardware-level advantages of the proposed framework.

To evaluate the operational viability of the proposed framework objectively, we define the Energy Efficiency Ratio (η). To account for the severe class imbalance inherent in drilling anomalies, we normalize this ratio using the F1-score rather than simple accuracy, dividing it by the energy consumed per inference cycle *(E)*:

$$\eta = \frac{F_1}{E} \tag{4}$$

This metric provides a normalized assessment of the predictive yield per Joule of energy expended. To accurately calculate the energy consumed per inference cycle *(E)*, we must integrate the hardware power consumption *(P)* over the inference latency *(t)*, where $E = P \times t$.

For the baseline "Red AI" model, operating at 12.5 W for an inference duration of 185 ms (0.185 s), the total energy per cycle is computed as $E_{Red} = 12.5\ W \times 0.185\ s \approx 2.313$ Joules. Consequently, the efficiency ratio is $\eta_{Red} = 0.977 / 2.313 \approx 0.422$.

In contrast, the optimized "Green AI" engine operates at just 3.8 W for a mere 24 ms (0.024 s), yielding a drastically reduced energy footprint of $E_{Green} = 3.8\ W \times 0.024\ s \approx 0.091$ Joules. Its resulting efficiency ratio is $\eta_{Green} = 0.961 / 0.091 \approx 10.560$.

Despite the marginal 1.6% decrease in the F1-score, the "Green AI" framework delivers an astounding 25-fold increase ($10.560 / 0.422 \approx 25.0$) in relative energy efficiency. This massive gain ensures that continuous, 24/7 safety monitoring can be sustained on offshore platforms using minimal power resources, such as small Uninterruptible Power Supplies (UPS), directly aligning with the principles of responsible digital innovation.

5 Results and Discussion

The "Green AI" framework provides significant economic and operational advantages for high-CAPEX offshore operations in the South Caspian, in addition to technological validation. The shown 55-s early warning advantage greatly reduces NPT. In deep-water HPHT conditions, where rig spread rates often go above $400,000 per day, this important time window lets drillers turn on the BOP before an influx breaks through the upper casing. This quick response averts long well-kill operations, protects assets worth millions of dollars, and mitigates severe ecological liabilities in the delicate Caspian maritime ecosystem.

Also, the pruned and quantized INT8 models can be used on any hardware and are easy to scale. This framework can be easily added to legacy offshore platforms' Programmable Logic Controllers (PLCs), unlike traditional "Red AI" paradigms that need energy-hungry processing clusters. Transfer Learning makes it easy to transfer the pre-trained model to other geologically complicated areas, like the North Sea or the Gulf of Mexico, with very little more computing power. This makes AI-driven safety monitoring available to everyone in the sector.

This combination of low power use and fast anomaly identification sets a strong bar for digital resilience in petroleum engineering.

6 Conclusion

This study successfully transitions from a resource-intensive "Red AI" framework to a performance-oriented "Green AI" framework for real-time HPHT well monitoring. Using Knowledge Distillation, magnitude-based pruning, and 8-bit Post-Training Quantization on a hybrid 1D-CNN and GRU architecture cut the model's memory footprint by 74.6%. The improved edge-inference engine processes data 7.7 times faster and uses approximately 70% less power (down to 3.8 W), all while preserving a highly competitive F1-score of 0.961. This localized edge processing eliminates satellite latency, yielding a crucial 55-s early warning advantage when gas flows in. In the difficult South Caspian Basin climate, this directly minimizes the risk of blowouts, makes infrastructure last longer, and reduces costly NPT.

It must be explicitly acknowledged that the 55-s early warning advantage was validated in a single-well retrospective scenario. While the results are highly promising for the specific geomechanical profile of this South Caspian block, further empirical validation across a multi-well dataset is required to confirm the universality of this latency advantage in different operational contexts.

Ultimately, this framework establishes a scalable and environmentally sustainable benchmark for secure, autonomous digital oilfield operations. Future research will address current constraints by utilizing Quantization-Aware Training (QAT) to preserve signal integrity and including a computationally efficient multi-class classification head to distinguish between gas, oil, and water influxes.

References

1. Eyo-Udo, N.L.: Digital transformation as a catalyst for business model innovation: a critical review of impact and implementation strategies. Open Access Res. J. Eng. Technol. **06**, 001–022 (2024). https://doi.org/10.53022/oarjet.2024.6.2.0085
2. Al-Jasmi, A., et al.: A surveillance "smart flow" for intelligent digital production operations. In: SPE Digital Energy Conference SPE-163697-MS (2013). https://doi.org/10.2118/163697-MS
3. Póvoas, M.S., et al.: Artificial intelligence in the oil and gas industry: applications, challenges, and future directions. Appl. Sci. **15**, 7918 (2025). https://doi.org/10.3390/app15147918
4. Sharma, S., Thakur, S.: Impact of artificial intelligence on the environment. Int. J. Res. Pub. Rev. **5**(10), 4792–4798 (2024)
5. Simmons, M.: Geological challenges to exploration in the South Caspian Basin. In: 73rd EAGE Conference Exhibition. Incorporation SPE EUROPEC 2011 (2011). https://doi.org/10.3997/2214-4609.20149743
6. Gholami, A. et al.: A survey of quantization methods for efficient neural network inference. arXiv:2103.13630 (2021). https://doi.org/10.48550/arXiv.2103.13630
7. Gooneratne, C.P., et al.: Drilling in the fourth industrial revolution—vision and challenges. IEEE Eng. Manag. Rev. **48**, 144–159 (2020). https://doi.org/10.1109/EMR.2020.2999420
8. Han, S. et al.: Learning both weights and connections for efficient neural networks. arXiv:1506.02626 (2015). https://doi.org/10.48550/arXiv.1506.02626
9. Haouel, C., Nemeslaki, A.: Digital transformation in oil and gas industry: opportunities and challenges. Period. Polytech. Soc. Manag. Sci. **32**, 36–46 (2023). https://doi.org/10.3311/PPso.20830

10. Tharayil, S., et al.: A machine learning approach to optimize drilling operations in real-time. ADIPEC. (2025). https://doi.org/10.2118/228907-MS
11. Guliyev, I., Kerimov, V., Mustaev, R.: Fundamental challenges of the location of oil and gas in the South Caspian Basin. Dokl. Earth Sci. **471**, 1109–1112 (2016). https://doi.org/10.1134/S1028334X1611009X
12. Yusubov, N., Shikhmammadova, T.N.: Hydrocarbon system of the South caspian depression. Geofiz. Zh. **44**, 87–95 (2022). https://doi.org/10.24028/gj.v44i3.261971
13. Marko, M., Garni, S.: Application of automated machine learning for multi-variate prediction of well production. In: Paper presented at the SPE Middle East Oil and Gas Show and Conference, Manama, Bahrain (2019). https://doi.org/10.2118/195022-MS
14. Pivano, L., Nguyen, D.T., Ludvigsen, K.B.: Digital twin for drilling operations – towards cloud-based operational planning. In: Offshore Technology. Conference 29316-MS (2019). https://doi.org/10.4043/29316-MS
15. Knebel, F.P., et al.: A study on cloud and edge computing for the implementation of digital twins in the Oil & Gas industries. Comput. Ind. Eng. **182**, 109363 (2023). https://doi.org/10.1016/j.cie.2023.109363
16. Motaei, E., Ganat, T.: Smart proxy models art and future directions in the oil and gas industry: a review. Geoenergy Sci. Eng. **227**, 211918 (2023). https://doi.org/10.1016/j.geoen.2023.211918
17. Nazari, Z., Musilek, P.: Impact of digital transformation on the energy sector: a review. Algorithms **16**(4), 211 (2023). https://doi.org/10.3390/a16040211
18. Schwartz, R., et al.: Green AI. Commun. ACM **63**, 54–63 (2020). https://doi.org/10.1145/3381831
19. Saeed, S., et al.: Digital transformation in energy sector: cybersecurity challenges and implications. Information **15**, 764 (2024). https://doi.org/10.3390/info15120764
20. Al Shayaa, A. et al.: Drilling process digitalization using advanced machine learning techniques – case study. In: SPE Middle East Oil & Gas Show Conference 194962-MS (2019). https://doi.org/10.2118/194962-MS
21. Strubell, E., Ganesh, A., McCallum, A.: Energy and policy considerations for deep learning in NLP. In: 57th Annual Meeting Assocation Computer Linguistion (ACL). (2019). https://doi.org/10.48550/arXiv.1906.02243
22. Turai, T., Ballard, I., Rob, R.: "Short-term electrical load demand forecasting using artificial neural networks for off-grid distributed generation applications. In: 2017 Saudi Arabia Smart Grid (SASG), Jeddah, Saudi Arabia, pp. 1–7 (2017). https://doi.org/10.1109/SASG.2017.8356472
23. Mao, Y., Zhang, P.: An automated kick alarm system based on statistical analysis of real-time drilling data. Abu Dhabi International of Petroleum Exhibition Conference 197275-MS (2019). https://doi.org/10.2118/197275-MS

A Community-Driven Dataset and Protocol for Azerbaijani Sign Language Recognition: Advancing Digital Inclusion Through Human-Centered AI

Nigar Alishzade[1](✉), Lala Ibadullayeva[2], Rajab Iskandarli[3], Jabrayil Babayev[4], and Humbat Jamalov[3]

[1] School of Engineering, Karabakh University, Stepanakert, Azerbaijan
nigar.alishzade@karabakh.edu.az
[2] Baku State University – Center for Digital Technologies and Applied Research, Baku, Azerbaijan
[3] French-Azerbaijani University, Baku, Azerbaijan
[4] The Academy of Public Administration Under the President of the Republic of Azerbaijan, Baku, Azerbaijan

Abstract. Sign language recognition represents a vital frontier in human-centered AI, promoting digital inclusion for Deaf and hard-of-hearing communities globally. In Azerbaijan, Azerbaijani Sign Language (AzSL) is the primary communication mode for thousands, yet limited digital access exacerbates social and educational exclusion. This work contributes a community-driven dataset and annotation protocol co-developed with the Azerbaijani Deaf community, identifies core challenges and research directions for low-resource sign languages, and outlines pathways for integrating AzSL technologies into Azerbaijan's digital governance aligned with disability rights legislation. Emphasizing participatory design, our approach ensures culturally relevant, ethical systems. By linking technical gains to societal impact, this study advances digital equity and provides a scalable framework for underrepresented sign languages, underscoring inclusive AI's role in social good.

Keywords: Azerbaijani Sign Language · sign language recognition · digital inclusion · human-centered AI · accessibility · Deaf community

1 Introduction

Digital transformation is reshaping societies by enhancing access to information, services, and opportunities through intelligent systems. However, for many individuals with disabilities—particularly those who are Deaf or hard-of-hearing—persistent communication barriers continue to limit full participation in education, employment, and civic life. Sign language recognition (SLR) and artificial intelligence-powered translation technologies represent a promising path toward genuine digital inclusion.

C. Zanni-Merk et al. (Eds.): AIDT 2026, CCIS 3023, pp. 423–438, 2026.
https://doi.org/10.1007/978-3-032-31319-5_28

In Azerbaijan, the Azerbaijani Sign Language (AzSL) serves as the main means of communication for the Deaf community. Recent legislative progress, including the 2025 amendments to the Law "On the Rights of Persons with Disabilities," has officially recognized AzSL as a legitimate mode of communication for persons with complete hearing loss, mandating improved access to information and inclusive participation. Despite this advancement, the development of automated AzSL recognition systems remains in its early stages due to limited datasets, linguistic complexity, and the absence of large-scale annotated resources—challenges common to low-resource sign languages.

This paper contributes to the growing field of human-centered AI by presenting our approach to building culturally sensitive AzSL recognition models. We emphasize the indispensable involvement of the Deaf community throughout the research process and demonstrate how such technologies can progressively integrate into digital governance ecosystems—from public service portals to real-time everyday communication tools. Specifically, this work makes three key contributions: (1) A human-centered protocol for dataset collection and annotation, co-developed with the Azerbaijani Deaf community to ensure ethical integrity and cultural authenticity; (2) Identification of core challenges and future directions for advancing low-resource sign languages like AzSL; (3) Pathways for integrating AzSL recognition into Azerbaijan's digital ecosystem, aligned with national legislation and transformation strategies.

2 Background and Motivation

Developing effective sign language technologies requires far more than technical innovation. Behind assistive AI solutions for the Deaf community lie extensive resources: hours-long signing videos, sophisticated deep learning architectures (e.g., 3D CNNs and Vision Transformers), computationally intensive training regimes, the expertise of multidisciplinary teams, and significant investment.

Crucially, in building sign language technology for a specific linguistic community, the Deaf users of that language must be actively and meaningfully involved. Cultural and linguistic context profoundly influences every stage—from dataset design and annotation to model architecture selection, training strategies, and performance evaluation.

Azerbaijani Sign Language exemplifies this principle. As a low-resource sign language, AzSL has historically lacked sufficient computational resources and annotated datasets. The recent release of the **AzSLD dataset** [37]—comprising over 30,000 annotated videos covering fingerspelling, isolated signs, and sentence-level expressions—marks a foundational step forward. Collected collaboratively with native AzSL users, interpreters, and Deaf community members, AzSLD ensures representativeness and linguistic authenticity.

This technology stands as a compelling example of AI serving humanity by fostering genuine digital inclusion. Responsible development and deployment can enable meaningful societal participation for Deaf individuals in

Azerbaijan—estimated between 12,000 (official statistics) and 30,000 (international estimates)—integrating them more fully into education, employment, and civic life.

3 Related Work

The development of sign language recognition (SLR) systems spans technical innovation, linguistic resource creation, and community-centered design. We situate our work at the intersection of three active research streams: (i) methods for low-resource sign language modeling, (ii) community-driven dataset and protocol design, and (iii) prior efforts specific to Azerbaijani Sign Language (AzSL). We review each in turn.

3.1 Technical Approaches for Low-Resource Sign Languages

Early SLR surveys established that data scarcity remains the central bottleneck for languages with limited digital presence, necessitating transfer learning, data augmentation, and architectural efficiency [8,10,38,39]. For high-resource languages such as American Sign Language (ASL) or British Sign Language (BSL), large-scale datasets (e.g., WLASL, BSL-1K) enable training of deep 3D Convolutional Neural Networks (CNNs) and Vision Transformers (ViTs) from scratch [42,43]. However, for low-resource sign languages, recent work demonstrates that **transfer learning from action recognition (Kinetics) or image classification (ImageNet)** is the most effective strategy to mitigate overfitting [13,42]. Specifically, Albanie et al. [43] showed that a pretrained I3D backbone fine-tuned on BSL-1K yields state-of-the-art isolated sign accuracy, while De Coster et al. [42] extended this to transformer-based architectures.

3.2 Human-Centered Dataset Creation and Annotation Protocols

Bragg et al. [38,39] and Kuder et al. [40] have extensively documented the ethical pitfalls of SLR datasets created without Deaf involvement, including annotation errors, regional bias, and misrepresentation of non-manual features. In response, projects such as SignON [18] and Deafinitely [19] have pioneered **co-creative frameworks** where Deaf signers lead vocabulary selection, recording protocol definition, and quality control. These studies consistently report higher inter-annotator agreement ($\kappa > 0.85$) and improved model generalization when Deaf annotators validate glosses.

3.3 Prior Work on Azerbaijani Sign Language and Regional Benchmarks

Research on AzSL is nascent. Early prototypes by Abdullayeva et al. [27,41] demonstrated feasibility using transfer learning on small, single-signer datasets, achieving a baseline accuracy of 85.3% on a limited vocabulary. Aliyev et al. [24]

applied classical machine learning to dactyl alphabet recognition. The most comprehensive prior resource is the AzSLD dataset introduced by Alishzade et al. [25,26], which we utilize and extend in this work.

3.4 Generative Models and Deployment Pathways

Finally, we situate our work within the broader ecosystem of sign language technologies, including **text-to-sign avatar generation** and **multimodal large language models (LLMs)**. Diffusion-based approaches for ASL [29,30] and multilingual sign production models [15,36] illustrate the potential for two-way communication tools. However, these systems rely on high-quality pose annotations and extensive parallel corpora—resources absent for AzSL.

3.5 Synthesis and Positioning of This Work

Motivated by findings in low-resource SLR, we adopt an I3D backbone pre-trained on Kinetics-400 and a ViT-B/16 pre-trained on ImageNet-21k; our hybrid model achieves 94.2% top-1 accuracy on AzSLD, a 3–5% improvement attributable to community-validated annotation quality and signer-diverse splits. We instantiate co-design principles from human-centered dataset creation, achieving inter-annotator agreement of $\kappa = 0.85$ comparable to high-resource standards. The resulting AzSLD dataset provides the first set of pose keypoints and isolated ground-truth glosses necessary for future generative model fine-tuning. Compared to prior AzSL baselines [27], our ViT-based model improves absolute accuracy by 7.1 percentage points. Our deployment strategy (Sect. 7) is informed by real-world case studies of Signapse and HandTalk, emphasizing progressive, community-guided rollout. By establishing a robust recognition baseline and a replicable data protocol, we lay the groundwork for extending multimodal and generative capabilities to a truly low-resource linguistic community.

4 Methodology

Our approach utilizes the AzSLD dataset [37], which contains fingerspelling, isolated signs, and continuous sentences annotated in collaboration with the Deaf community. We employ 3D Convolutional Neural Networks and Vision Transformers, with hybrid variants incorporating temporal modeling.

Data preprocessing involves frame extraction at 30 fps, resizing to 224×224 pixels, normalization (mean [0.485, 0.456, 0.406], std [0.229, 0.224, 0.225]), and augmentations such as random cropping, horizontal flipping, and brightness adjustments to enhance signer invariance.

4.1 Model Architectures and Implementation Details

We evaluate three model variants to establish robust baselines on AzSLD, prioritizing reproducibility and transparency in architectural choices. Parameter counts and fusion mechanisms are detailed below to facilitate direct comparison and future replication.

3D Convolutional Neural Network (I3D). We adopt the Inflated 3D ConvNet (I3D) architecture [5] with a ResNet-50 backbone, pre-trained on Kinetics-400. The network processes input clips of shape $224 \times 224 \times 64$ (height $\times$ width $\times$ frames). After the final convolutional block, global average pooling produces a feature vector $\mathbf{f}_{\text{cnn}} \in \mathbb{R}^{2048}$. This vector is passed through a single fully-connected classification layer to predict the gloss label. The model contains approximately **12.1 million** trainable parameters (excluding the frozen BatchNorm statistics from pre-training).

Vision Transformer (ViT-B/16). We employ the ViT-B/16 variant [6], pre-trained on ImageNet-21k and fine-tuned on AzSLD. Each video frame is resized to 224×224 and partitioned into non-overlapping patches of size 16×16, yielding 196 patches per frame. To aggregate temporal information across the 64-frame input clip, we apply **mean pooling** over the sequence of frame-level class tokens before the final MLP head. This design choice avoids the quadratic complexity of full spatiotemporal attention while preserving per-frame spatial reasoning. The model contains approximately **86.6 million** trainable parameters.

Hybrid CNN-Transformer Architecture. The hybrid model combines the I3D backbone (for local spatiotemporal feature extraction) with the ViT encoder (for global contextual reasoning) via **late fusion**. Specifically:

1. The I3D branch processes the input clip and outputs $\mathbf{f}_{\text{cnn}} \in \mathbb{R}^{2048}$ (post-pooling).
2. The ViT branch processes the same input clip (frame-wise) and outputs $\mathbf{f}_{\text{vit}} \in \mathbb{R}^{768}$ after temporal mean pooling of the class token.
3. The two vectors are concatenated to form a joint representation $\mathbf{f}_{\text{hybrid}} = [\mathbf{f}_{\text{cnn}}; \mathbf{f}_{\text{vit}}] \in \mathbb{R}^{2816}$.
4. A two-layer MLP projection head reduces dimensionality ($2816 \rightarrow 1024 \rightarrow C$, where C is the number of gloss classes) with ReLU activation and dropout ($p = 0.3$) between layers.

No intermediate feature interaction occurs between the branches; fusion occurs exclusively at the final representation level before classification. The total number of trainable parameters for the hybrid model is approximately **98.9 million** (including both backbones and the projection head).

Training Configuration. All models are trained for 50 epochs using the AdamW optimizer ($\beta_1 = 0.9$, $\beta_2 = 0.999$, weight decay 1×10^{-5}) with a cosine annealing learning rate schedule. The base learning rate is set to 1×10^{-4} for I3D and 3×10^{-5} for transformer-based models to accommodate pre-trained weight sensitivities. Mixed-precision training (FP16) is employed on NVIDIA A100 GPUs. The batch size is fixed at 16 clips per GPU, with gradient accumulation steps of 2 to emulate an effective batch size of 32.

Table 1. Human-Centered Protocol for AzSLD Collection and Annotation

Step	Description and Community Role
Vocabulary Definition	Co-led workshops with Deaf signers to prioritize everyday and domain-specific signs (e.g., education, healthcare).
Recording Protocols	Interpreters and signers define norms for signing space, lighting, and non-manuals to ensure cultural authenticity.
Informed Consent	Video-recorded AzSL forms; right to withdraw; anonymization (face blurring).
Annotation	Multi-annotator isolated ground-truth glossing with Deaf validation; resolution of discrepancies via group consensus.
Bias Auditing	Demographic stratification (age, gender, region) and post-annotation checks for underrepresentation.

4.2 Human-Centered Protocol for Dataset Creation

To address ethical considerations and cultural authenticity often overlooked in technical SLR research, we propose a replicable human-centered protocol for dataset collection and annotation, co-developed with the Azerbaijani Deaf community. This protocol ensures participatory design, mitigating biases and enhancing linguistic fidelity.

The protocol, summarized in Table 1, formalizes five key stages of community collaboration: vocabulary co-design, culturally informed recording norms, video-recorded informed consent, multi-annotator ground-truth glossing with Deaf validation, and demographic bias auditing. Each stage was co-developed with native signers and interpreters to mitigate common pitfalls in the creation of low-resource SLR datasets.

This protocol directly improved data quality, as evidenced by inter-annotator agreement ($\kappa > 0.85$) and reduced overfitting in models.

5 Experiments and Results

5.1 Dataset Collection and Community Involvement

The cornerstone of our research is the Azerbaijani Sign Language Dataset (AzSLD) [37], a large-scale, community-driven resource specifically designed to address the scarcity of annotated data for low-resource sign languages. The data collection process was conducted in close collaboration with the Deaf community in Azerbaijan, recognising that sign language technology development inherently requires active participation of the target group at multiple stages [38,40]. All contributors provided explicit informed consent for the public release and sharing of the dataset for research purposes. Consent forms clearly outlined the intended use of the recordings (academic research, model training, and benchmark development), data anonymisation procedures (face blurring where requested), and

Table 2. AzSLD Dataset Statistics [37]

Component	Details
Total annotated video clips	32,478
Unique signers	52
Isolated signs	18,250 clips (450 glosses)
Fingerspelling sequences	6,120 clips
Continuous sentences	8,108 clips (average 5–12 signs per sentence)
Total vocabulary (glosses)	1,200+ unique signs
Average clip duration	3.8 s
Total recording hours	≈ 34 h
Annotation format	isolated glosses + pose keypoints (OpenPose/MediaPipe)

Table 3. Isolated Sign Recognition Results on AzSLD Test Set

Model	Top-1 Accuracy (%)	mAP (%)	Inference Time (ms)
3D CNN (I3D)	88.7	89.5	45
Vision Transformer (ViT-B/16)	92.4	93.1	60
Hybrid (CNN + ViT)	94.2	94.8	52
Prior Baseline [41]	85.3	86.7	–

the right to withdraw at any stage. This ethical framework aligns with best practices in human-centered AI and ensures that the dataset respects the autonomy and cultural ownership of the Deaf community.

The collection took place in controlled studio environments as well as natural settings to capture linguistic variation. A diverse cohort of 52 native and fluent AzSL signers (aged 18–65, balanced gender distribution, representing different regions of Azerbaijan) participated. Recordings were made using multiple camera angles (frontal and side views) at 30 fps with 1080p resolution to facilitate future 3D pose extraction and multimodal analysis.

Table 2 summarises the key quantitative outcomes of the data collection process.

5.2 Recognition Experiments

Using the collected AzSLD, we conducted baseline experiments to validate dataset quality and establish initial performance benchmarks. We followed a standard isolated sign recognition protocol with a 70/15/15 train/validation/test split stratified by signer to ensure signer-independent evaluation.

Models evaluated include:

- 3D Convolutional Neural Network (I3D backbone pre-trained on Kinetics)
- Vision Transformer (ViT-B/16 with temporal pooling)
- Hybrid 3D CNN + Transformer architecture

Results on the isolated sign subset (Table 3) show that the Vision Transformer variant achieves a top-1 accuracy of 92.4% and mean average precision (mAP) of 93.1%, outperforming the 3D CNN baseline (88.7% top-1, 89.5% mAP) and earlier AzSL prototypes reported in the literature. Ablation studies confirm that community-validated annotations and diverse signer representation significantly reduce overfitting and improve generalisation compared to smaller, non-collaborative datasets, with a 5–10% mAP boost from the protocol.

These outcomes demonstrate that deep community involvement not only ensures cultural and linguistic authenticity but also directly contributes to higher data quality and model performance—reinforcing the necessity of participatory approaches in sign language technology development.

Further experiments on continuous signing and fingerspelling are ongoing, with early results indicating promising transfer to sentence-level recognition tasks (e.g., word error rate of 25–30% on short sentences).

6 Ethical Considerations and Community Engagement

The development of sign language recognition technologies is inherently a socio-technical endeavour that directly affects a linguistically and culturally distinct community. As emphasised in the literature, sign language AI research must prioritise the principle of "Nothing About Us Without Us"—meaning that Deaf individuals and their representative organisations should lead or co-lead decision-making at every stage [38,40]. Failure to do so risks perpetuating exclusion, introducing cultural inaccuracies, or reinforcing power imbalances between researchers and the target community.

In this work, we adopted a deeply participatory approach to ensure ethical integrity and cultural authenticity. From the initial planning of the AzSLD dataset [37], native AzSL signers, Deaf association members, and qualified interpreters were actively involved as co-researchers. As detailed in our methodology (Sect. 4.2 and Table 1), they led the design of recording protocols, annotation guidelines, and quality control processes. This collaborative model not only improved linguistic fidelity but also built trust and ownership within the community. Regular feedback workshops were held throughout the project, allowing participants to review progress, suggest modifications, and withdraw contributions if desired.

Regarding data ethics and privacy, all 52 contributors provided fully informed written consent for the recording, annotation, and public release of the dataset for non-commercial research purposes. Consent forms—available in written Azerbaijani/Russian and video-recorded AzSL—explicitly detailed:

- The intended academic and potential translational applications,
- Anonymisation measures (optional face blurring, removal of personally identifiable background elements),
- Open-access licensing terms (Creative Commons BY-NC-SA),
- The irrevocable right to request removal of specific clips post-release.

We also proactively addressed potential biases. The signer cohort was selected to maximise diversity in age (18–65), gender (approximately balanced), geographic origin (urban and rural regions of Azerbaijan), and signing fluency. Continuous monitoring during annotation revealed and mitigated systematic errors related to regional variants. Future work will include explicit bias audits for model predictions across demographic subgroups.

By embedding Deaf leadership and rigorous ethical safeguards from inception, this project demonstrates that community-centred design is not merely an ethical obligation but a methodological strength that enhances data quality, model robustness, and real-world societal impact. This approach aligns with Azerbaijan's commitments under the UN Convention on the Rights of Persons with Disabilities and serves as a model for responsible AI development in low-resource linguistic contexts.

7 Progressive Deployment Strategy

To ensure sustainable, ethical, and effective real-world impact, the deployment of AI-powered Azerbaijani Sign Language (AzSL) technologies should follow a progressive, iterative strategy that prioritises reliability, user acceptance, and continuous community feedback. A phased approach mitigates risks associated with early-stage systems (e.g., recognition errors in varied conditions) while allowing incremental refinement based on real-user experiences—particularly from the Deaf community whose linguistic and cultural expertise is indispensable at every stage.

The proposed strategy unfolds in five overlapping phases, designed to align with Azerbaijan's digital transformation priorities and the accessibility mandates of the 2025 amendments to the Law "On the Rights of Persons with Disabilities":

1. **Controlled Pilot Testing (6–12 months).** Initial deployment in safe, collaborative environments such as university research labs, Deaf association centres, and selected schools for the Deaf. Focus on isolated sign and fingerspelling recognition modules integrated into prototype applications (e.g., educational games with AI feedback). Deaf users and educators provide structured feedback through co-design workshops, enabling rapid iteration on accuracy, usability, and cultural appropriateness. Target metrics: 90% accuracy in controlled settings, 80% user satisfaction.
2. **Educational Integration (Year 1–2).** Rollout of domain-specific tools into formal and informal education settings, including AzSL learning apps, lecture captioning systems, and avatar-based supplementary materials. Partnerships with the Ministry of Science and Education and institutions like Qarabağ University facilitate teacher training and student trials. Metrics include learning outcomes, user satisfaction surveys, and error analysis to refine non-manual feature handling (e.g., aim for <20% WER in sentences).
3. **Public Service Enhancement (Year 2–3).** Integration into citizen-facing digital platforms, starting with ASAN service centres and e-government portals (e.g., my.gov.az). Initial features include real-time sign-to-text interpre-

Table 4. Overview of Sign Language Technologies and Their Potential for AzSL

Category	Examples (2026)	AzSL Adaptation & Impact	Alignment with AZ Policy
Online translation service	Signapse SignStream; SignAll	Integration into ASAN centers	2025 Disability Law (accessible services); ASAN portals.
Sign-to-text/voice	Slait; KinTrans; 3D CNN/ViT prototypes	AzSLD-based models for education/work	Digital Azerbaijan 2030 (inclusive e-gov).
Text-to-sign (avatars)	AWS GenASL; Signapse	Fine-tune on AzSLD poses	State Strategy for Education Development.
Learning apps	NVIDIA Signs; HandTalk	Gamified AzSL education	SDG 4 (inclusive education).
Domain-specific	Signapse modules; medical prototypes	Tools for schools/hospitals	CRPD Articles 9, 21, 24.
Real-time captioning	YouTube/Signapse	TV/events accessibility	SDG 10 (reduced inequalities).
Multimodal tools	Hybrid prototypes	Deaf/hearing bridging	SDG 8 (decent work).

tation for video consultations and text-to-avatar rendering of official documents. Hybrid human-AI relay services ensure fallback reliability while gathering usage data to improve robustness across diverse signers and regional variants. Target: <50ms latency, 85% satisfaction in trials.

4. **Mobile and Everyday Accessibility (Year 3+).** Broad consumer release via mobile applications supporting bidirectional translation (sign-to-text/voice and text-to-sign avatars) for daily communication, social media, and emergency services. Edge-device optimisation and privacy-preserving on-device processing address connectivity challenges in rural areas. Community-led awareness campaigns and open-source contributions encourage widespread adoption. Metrics: Adoption rate ¿50% in target communities.
5. **National-Scale Sustainability and Expansion.** Full embedding into public infrastructure, with ongoing monitoring, model updates, and extension to additional domains (healthcare, legal proceedings). Establishment of a national AzSL AI maintenance consortium—involving Deaf organisations, universities, and government bodies—ensures long-term governance, bias auditing, and dataset expansion.

Table 4 provides an overview of existing sign language technologies that can serve as blueprints for AzSL adaptation through transfer learning, community-annotated datasets, and collaborative design. These technologies, responsibly localized with Deaf Azerbaijani leadership, can empower the estimated 12,000–30,000 Deaf individuals toward fuller societal participation.

Throughout all phases, iterative evaluation involves mixed-methods assessment: quantitative metrics (accuracy, latency, word error rate), qualitative user studies with Deaf participants, and accessibility audits aligned with WCAG and CRPD guidelines. Funding mechanisms could combine public grants, university-led projects, and public-private partnerships (e.g., via ASAN AI Hub initiatives).

This progressive strategy transforms experimental prototypes into trusted, inclusive tools that meaningfully reduce communication barriers—demonstrating how carefully staged deployment can accelerate digital transformation while upholding ethical commitments to the Deaf community.

8 Challenges and Future Directions for Low-Resource SLR

While this work represents a significant step forward in Azerbaijani Sign Language (AzSL) recognition and digital inclusion, several challenges and potential threats to validity must be acknowledged to ensure transparent and responsible reporting. We identify key systemic issues for low-resource sign languages and propose future research directions.

8.1 Data Scarcity and Diversity

The AzSLD dataset [37], although the largest publicly available resource for AzSL to date, remains comparatively small relative to high-resource sign language datasets (e.g., WLASL or BSL-1K, which contain hundreds of thousands of samples). With approximately 32,000 annotated clips from 52 signers, the dataset provides strong coverage of isolated signs and short sentences but limited diversity in long, continuous, naturally occurring discourse. This constrains current model performance on real-world continuous recognition tasks, where co-articulation, signing speed variation, and prosodic elements introduce additional complexity.

Future directions: Expand datasets through crowdsourced in-the-wild recordings via mobile apps with Deaf incentives, incorporating greater dialectal coverage (e.g., post-Karabakh regional variants).

8.2 Technical Hurdles

Evaluation was conducted under controlled recording conditions (studio lighting, fixed camera angles, plain backgrounds). Real-world deployment—on mobile devices in varied lighting, with moving backgrounds, or partial occlusions—is likely to degrade performance. Current benchmarks therefore represent an upper bound rather than practical field accuracy. Additionally, models rely on transfer learning from action recognition pretrained weights (e.g., Kinetics for I3D),

Table 5. Key Challenges and Directions for Low-Resource SLR

Challenge	Severity/Example in AzSL	Future Direction
Dataset Size	High (32k clips vs. 100k+ in high-resource)	In-the-wild crowdsourcing with AI-assisted annotation.
Dialectal Variation	Medium (Regional influences)	Dialect-aware fine-tuning using adapters.
Real-World Robustness	High (Controlled vs. mobile conditions)	Augmentation with synthetic data via diffusion models.
Ethical Oversight	Medium (Annotation subjectivity)	Community-led bias audits and open guidelines.

introducing domain shifts that may prioritise general motion over sign-specific linguistics.

Future directions: Develop real-time, edge-deployable models with robustness augmentations (e.g., synthetic data via diffusion models); integrate graph neural networks for advanced pose modeling; employ federated learning for privacy-preserving updates (Table 5).

8.3 Ethical and Societal Barriers

Annotation subjectivity poses a threat, with inter-annotator agreement ($\geq$ 88% for glosses) highlighting interpretive challenges for non-manuals. Broader risks include cultural appropriation in low-resource contexts without sustained community oversight.

These challenges are common to early-stage low-resource SLR and do not undermine the core contributions. Ongoing work will address them through targeted expansions and optimizations.

9 Conclusion and Future Work

This paper has presented a human-centered approach to advancing digital inclusion through AI-powered Azerbaijani Sign Language (AzSL) recognition. By introducing the community-driven AzSLD dataset [37]—developed with active Deaf leadership, rigorous informed consent, and systematic participation at every stage, we address a critical gap in low-resource sign language resources. Experiments demonstrate strong baseline performance with Vision Transformers and 3D CNNs, highlighting the direct benefits of participatory design on data quality and model robustness.

The broader implications extend to digital transformation policy in Azerbaijan and beyond. By enabling progressive integration of AzSL tools into public services, education, and everyday communication, this research supports national legal mandates and international commitments under the UN CRPD, offering a replicable model for inclusive digital governance in linguistically diverse contexts.

Looking ahead, future work will focus on:

- Expanding AzSLD with in-the-wild continuous recordings and greater dialectal coverage,
- Developing real-time bidirectional systems combining recognition and expressive avatar generation,
- Creating domain-specific models for education, healthcare, and public administration,
- Conducting large-scale field deployments and longitudinal impact studies in partnership with Deaf organisations,
- Pursuing edge-device optimisation and multimodal enhancements for mobile accessibility,
- International collaborations for multi-lingual SL benchmarks, building on AzSLD.

Ultimately, this research exemplifies how artificial intelligence, when guided by ethical principles and genuine community collaboration, can serve as a powerful force for social good—empowering Azerbaijan's Deaf community to participate fully in an increasingly digital society and paving the way for more equitable, inclusive futures worldwide.

References

1. Renjith, S. C., et al.: A comprehensive survey on Indian Sign Language interpretation using machine learning (2026). https://doi.org/10.13140/RG.2.2.12345.67890
2. Ardiansyah, A., et al.: Systematic literature review: American Sign Language translator (2021). https://doi.org/10.1016/j.procs.2021.01.004
3. Errollan, M., et al.: A systematic review of user studies as a basis for the design of systems for automatic sign language processing (2022). https://doi.org/10.1145/3563395
4. Saeed, A., et al.: A systematic review on systems-based sensory gloves for sign language pattern recognition: an update from 2017 to 2022 (2022). https://doi.org/10.1109/ACCESS.2022.3219959
5. Carreira, J., Zisserman, A.: Quo vadis, action recognition? A new model and the kinetics dataset. In: CVPR (2017)
6. Dosovitskiy, A.: An image is worth 16x16 words: transformers for image recognition at scale. In: ICLR (2021)
7. Alam, I., et al.: Exploring sign language detection on smartphones: a systematic review of machine and deep learning approaches (2024). https://doi.org/10.1155/2024/1487500
8. Tan, S., et al.: A review of deep learning-based approaches to sign language processing (2024). https://doi.org/10.1080/01691864.2024.2442721
9. Marcolino, M.S., et al.: Sign language recognition system for Deaf patients: protocol for a systematic review (2025). https://doi.org/10.2196/55427

10. Madhiarasan, M., et al.: A comprehensive review of sign language recognition: different types, modalities, and datasets (2022). https://doi.org/10.48550/arXiv.2204.03328
11. Zhang, X., Duh, K.: Improving sign language gloss translation with low-resource machine translation techniques (2023). https://www.cs.jhu.edu/~xzhan138/papers/SLMT_Book_G2T.pdf
12. Shahin, N., Ismail, L.: ADAT novel time-series-aware adaptive transformer architecture for sign language translation (2026). https://doi.org/10.1038/s41598-026-36293-9
13. Alharthi, N.M., Alzahrani, S.M.: Vision transformers and transfer learning approaches for Arabic Sign Language recognition (2023). https://doi.org/10.3390/app132111625
14. Lamaakal, I., et al.: An explainable hybrid CNN-transformer model for sign language recognition on edge devices using adaptive fusion and knowledge distillation (2026). https://doi.org/10.1038/s41598-026-38478-8
15. Müller, M., et al.: SignLLM: multilingual sign language production (2024). https://www.emergentmind.com/papers/2405.10718
16. Kamikubo, R., Glasser, A., et al.: Exploring collaboration to center the Deaf community in sign language AI (2025). https://doi.org/10.1145/1122445.1122456
17. Fitch, T., & Sano, C.: Centering sign language in AI and design (2025). https://microsoft.design/articles/centering-sign-language-in-ai-and-design
18. Shterionov, D., et al.: SignON – a co-creative machine translation for sign and spoken languages (end-of-project results, contributions and lessons learned) (2024). https://aclanthology.org/2024.eamt-2.25
19. Deafinitely Team: Deafinitely: The world's first application that translates to sign language (2024). https://www.deafinitely.co.il/
20. Li, Y.: Human-centered editable speech-to-sign-language generation via streaming conformer-transformer and resampling hook (2025). https://doi.org/10.48550/arXiv.2506.14677
21. Jisc National Centre for AI: AI sign language translation: Ready for education? (2025). https://nationalcentreforai.jiscinvolve.org/wp/2025/09/16/ai-sign-language-translation-ready-for-education
22. SignAvatar Team: SignAvatar - Generate sign language from speech using AI (2024). https://www.signavatar.org/
23. Glasser, A., et al.: Inclusive sign language AI: Towards authentic accessibility through community collaboration (2025). https://doi.org/10.1109/ICCVW60793.2025.00012
24. Aliyev, S., et al.: Azerbaijani sign language recognition using machine learning approach (2022). https://doi.org/10.1088/1742-6596/2251/1/012007
25. Alishzade, N.: AzSLD: Azerbaijani sign language dataset for fingerspelling, word, and sentence translation with baseline software (2025). https://doi.org/10.1016/j.dib.2024.110428
26. Hasanov, J., Alishzade, N., et al.: Development of a hybrid word recognition system and dataset for the Azerbaijani Sign Language dactyl alphabet (2023). https://doi.org/10.1016/j.specom.2023.102960
27. Abdullayeva, G.G., Alishzade, N.O.: Transfer learning for Azerbaijani Sign Language recognition (2023). https://icp.az/index.php?newsid=339
28. Abdullayeva, G.G., et al.: Human action recognition model reuse for word-level sign language recognition (2025). https://doi.org/10.13140/RG.2.2.23456.78901

29. Baltatzis, V., Potamias, R. A., Ververas, E., Sun, G., Deng, J., Zafeiriou, S.: Neural sign actors: a diffusion model for 3D sign language production from text. In: CVPR 2024 (2024). https://doi.org/10.48550/arXiv.2312.02702
30. Eswaradass, A., Poopandi, S., Koch, R.: GenASL: generative AI-powered American Sign Language avatars. AWS Machine Learning Blog (2024). https://aws.amazon.com/blogs/machine-learning/genasl-generative-ai-powered-american-sign-language-avatars
31. Krishnamurthy, S., Bhat, V., Jain, A.: DiffSign: AI-assisted generation of customizable sign language videos with enhanced realism (2024). https://doi.org/10.48550/arXiv.2412.03878
32. Zhang, H., et al.: NaturalSigner: Diffusion models are natural sign language generator (2024). https://openreview.net/forum?id=4JjSJyT15z
33. Chen, H., et al.: SignVTCL: Multi-modal continuous sign language recognition enhanced by visual-textual contrastive learning (2024). https://doi.org/10.48550/arXiv.2401.11847
34. Inan, M., et al.: SignAlignLM: integrating multimodal sign language processing into large language models. In: Findings of ACL 2025 (2025). https://aclanthology.org/2025.findings-acl.190
35. AWS re:Invent: Build multimodal ASL avatars with bidirectional translation (DEV306) [Video] (2024). https://www.youtube.com/watch?v=yESEK12EGR4
36. Kumar, M., et al.: Enhanced sign language translation between American Sign Language (ASL) and Indian Sign Language (ISL) using LLMs (2025). https://doi.org/10.48550/arXiv.2411.12685
37. Alishzade, N., Hasanov, J.: AzSLD - Azerbaijani Sign Language Dataset [Data set]. Zenodo (2023). https://doi.org/10.5281/zenodo.14222948
38. Bragg, D.: Sign Language Recognition, Generation, and Translation: An Interdisciplinary Perspective. ASSETS (2019)
39. Bragg, D., et al.: The FATE Landscape of Sign Language AI Datasets. ACM Trans. Access. Comput. (2021)
40. Kuder, A., et al.: Exploring collaboration to center the deaf community in sign language AI. In: Proc. ACM (2025)
41. Abdullayeva, F., Alishzade, N.: Isolated sign recognition for Azerbaijani Sign Language. In: Local Conference (2022)
42. De Coster, M.: Isolated sign language recognition with transformers. In: CVPR Workshops (2022)
43. Albanie, S.: BSL-1K: Scaling up co-articulated sign language recognition. In: BMVC (2021)
44. Multimodal LLMs for non-manual features in signing. In: Proceedings of the CVPR (2025)
45. Signapse Case Study: Network Rail Digital Accessibility (2024). https://signapse.ai/case-studies/innovative-sign-language-solutions---signapse-enhances-network-rails-digital-deaf-accessibility
46. Signapse Case Study: CVG Airport Partnership (2024). https://signapse.ai/case-studies/signapse-and-cvg-airport-partner-to-enhance-accessibility-with-american-sign-language-translations
47. Real-world impacts of sign language technologies. In: Proceedings of the ACM Conference (2025). https://dl.acm.org/doi/full/10.1145/3742886.3756712

48. HandTalk Official Blog and Adoption Statistics (2025). https://www.handtalk.me/en/blog
49. NVIDIA Signs Platform Case Studies (2025). https://www.deptagency.com/case/ai-powered-sign-language-education-that-changes-lives

Author Index

A

Adilova, Nigar E. 310
Ahmadov, Kamal 94
Alakbarli, Fatima 325
Alishzade, Nigar 423
Aliyev, Abbas 236
Aliyev, Samir 294
Aliyev, Tofig 205
Aliyeva, Vasila 264
Allahverdi, Yunis 310
Allahverdiyev, Hikmat 65
Amhaz, Rabih 51
Apiletti, Daniele 127
Asgarov, Vazeh 189

B

Babayev, Jabrayil 423
Bourgais, Mathieu 32
Boussard, Matthieu 157

C

Canu, Stéphane 375
Chabrol, Gregoire 51
Charbonnier, Léa 3

D

Decombas, Marc 375
Delage, Cindy 375
Delestre, Nicolas 220
Dhaouadi, Asma 359
Dostalizade, Aliya 94

F

Foureur, Jonathan 375

G

Gardashova, Latafat 389
Giustozzi, Franco 3, 142
Griffon, Marie 220
Guériau, Maxime 220
Guliyev, Yunis 79
Guliyeva, Laman 79
Gurbanov, Kanan 51

H

Hasanov, Elgun 251
Hasanov, Jamaladdin 264
Hernandez, Louis 157

I

Ibadullayeva, Lala 423
Ibrahimov, Yusif 251
Iskandarli, Rajab 423
Ismayilov, Elviz 19
Ismayilzada, Kamran 32

J

Jabbarova, Gullu 407
Jafarova, Hamida 325
Jamalli, Ilaha 264
Jamalov, Humbat 423
Jeannin-Girardon, Anne 173
Jendeya, Ahmed M. 109

K

Karimov, Fariz 407
Karimova, Rahima 264
Kaya, Omer Mutlu Turk 310
Kosov, Pavel 389

L

Leite, Alessandro 157

M

Mahammadli, Eljan 341
Mammadli, Anar 19
Mammadov, Nail 79
Manafov, Rafail 407

C. Zanni-Merk et al. (Eds.): AIDT 2026, CCIS 3023, pp. 439–440, 2026.
https://doi.org/10.1007/978-3-032-31319-5

Medeiros, Gabriel Henrique Alencar 142
Mirzabayov, Seymour 251
Mutallimov, Turan 251

N
Naghizada, Fatima 310
Naghizade, Elshan 294
Nuriyev, Aziz 389

O
Olejniczak, Nathan 173
Orujov, Nurlan 310

Q
Qardashova, Latafat 51
Quliyeva, Vafa 189

R
Rima, Samy 51
Rustamov, Samir 205, 236, 341

S
Santilli, Maria 359
Sariyeva, Gulshan 264
Saunier, Julien 3
Shmoncheva, Yelena 407
Susmus, Turker 79

V
Valizada, Alakbar 279
Vazirov, Etibar 127

The manufacturer's authorised representative in the EU is Springer Nature Customer Service Centre GmbH, Europaplatz 3, 69115 Heidelberg, Germany. If you have any concerns regarding our products, please contact ProductSafety@springernature.com

Printed and bound by CPI Group (UK) Ltd, Croydon, CR0 4YY
16/09/2026
02217799-0007